Lecture Notes in Physics

For information about Vols. 1–131, please contact your bookseller or Springer-Verlag.

Vol. 132: Systems Far from Equilibrium. Proceedings, 1980. Edited by L. Garrido. XV, 403 pages. 1980.

Vol. 133: Narrow Gap Semiconductors Physics and Applications. Proceedings, 1979. Edited by W. Zawadzki. X, 572 pages. 1980.

Vol. 134: γγ Collisions. Proceedings, 1980. Edited by G. Cochard and P. Kessler. XIII, 400 pages. 1980.

Vol. 135: Group Theoretical Methods in Physics. Proceedings, 1980. Edited by K. B. Wolf. XXVI, 629 pages. 1980.

Vol. 136: The Role of Coherent Structures in Modelling Turbulence and Mixing. Proceedings 1980. Edited by J. Jimenez. XIII, 393 pages. 1981.

Vol. 137: From Collective States to Quarks in Nuclei. Edited by H. Arenhövel and A. M. Saruis. VII, 414 pages. 1981.

Vol. 138: The Many-Body Problem. Proceedings 1980. Edited by R. Guardiola and J. Ros. V, 374 pages. 1981.

Vol. 139: H. D. Doebner, Differential Geometric Methods in Mathematical Physics. Proceedings 1981. VII, 329 pages. 1981.

Vol. 140: P. Kramer, M. Saraceno, Geometry of the Time-Dependent Variational Principle in Quantum Mechanics. IV, 98 pages. 1981.

Vol. 141: Seventh International Conference on Numerical Methods in Fluid Dynamics. Proceedings. Edited by W. C. Reynolds and R. W. MacCormack. VIII, 485 pages. 1981.

Vol. 142: Recent Progress in Many-Body Theories. Proceedings. Edited by J. G. Zabolitzky, M. de Llano, M. Fortes and J. W. Clark. VIII, 479 pages. 1981.

Vol. 143: Present Status and Aims of Quantum Electrodynamics. Proceedings, 1980. Edited by G. Gräff, E. Klempt and G. Werth. VI, 302 pages. 1981.

Vol. 144: Topics in Nuclear Physics I. A Comprehensive Review of Recent Developments. Edited by T.T.S. Kuo and S.S.M. Wong. XX, 567 pages. 1981.

Vol. 145: Topics in Nuclear Physics II. A Comprehensive Review of Recent Developments. Proceedings 1980/81. Edited by T. T. S. Kuo and S. S. M. Wong. VIII, 571-1.082 pages. 1981.

Vol. 146: B. J. West, On the Simpler Aspects of Nonlinear Fluctuating. Deep Gravity Waves. VI, 341 pages. 1981.

Vol. 147: J. Messer, Temperature Dependent Thomas-Fermi Theory. IX, 131 pages. 1981.

Vol. 148: Advances in Fluid Mechanics. Proceedings, 1980. Edited by E. Krause. VII, 361 pages. 1981.

Vol. 149: Disordered Systems and Localization. Proceedings, 1981. Edited by C. Castellani, C. Castro, and L. Peliti. XII, 308 pages. 1981.

Vol. 150: N. Straumann, Allgemeine Relativitätstheorie und relativistische Astrophysik. VII, 418 Seiten. 1981.

Vol. 151: Integrable Quantum Field Theory. Proceedings, 1981. Edited by J. Hietarinta and C. Montonen. V, 251 pages. 1982.

Vol. 152: Physics of Narrow Gap Semiconductors. Proceedings, 1981. Edited by E. Gornik, H. Heinrich and L. Palmetshofer. XIII, 485 pages. 1982.

Vol. 153: Mathematical Problems in Theoretical Physics. Proceedings, 1981. Edited by R. Schrader, R. Seiler, and D.A. Uhlenbrock. XII, 429 pages. 1982.

Vol. 154: Macroscopic Properties of Disordered Media. Proceedings, 1981. Edited by R. Burridge, S. Childress, and G. Papanicolaou. VII, 307 pages. 1982.

Vol. 155: Quantum Optics. Proceedings, 1981. Edited by C.A. Engelbrecht. VIII, 329 pages. 1982.

Vol. 156: Resonances in Heavy Ion Reactions. Proceedings, 1981. Edited by K.A. Eberhard. XII, 448 pages. 1982.

Vol. 157: P. Niyogi, Integral Equation Method in Transonic Flow. XI, 189 pages. 1982.

Vol. 158: Dynamics of Nuclear Fission and Related Collective Phenomena. Proceedings, 1981. Edited by P. David, T. Mayer-Kuckuk, and A. van der Woude. X, 462 pages. 1982.

Vol. 159. E. Seiler, Gauge Theories as a Problem of Constructive Quantum Field Theory and Statistical Mechanics. V, 192 pages. 1982.

Vol. 160: Unified Theories of Elementary Particles. Critical Assessment and Prospects. Proceedings, 1981. Edited by P. Breitenlohner and H.P. Dürr. VI, 217 pages. 1982.

Vol. 161: Interacting Bosons in Nuclei. Proceedings, 1981. Edited by J.S. Dehesa, J.M.G. Gomez, and J. Ros. V, 209 pages. 1982.

Vol. 162: Relativistic Action at a Distance: Classical and Quantum Aspects. Proceedings, 1981. Edited by J. Llosa. X, 263 pages. 1982.

Vol. 163: J. S. Darrozes, C. Francois, Mécanique des Fluides Incompressibles. XIX, 459 pages. 1982.

Vol. 164: Stability of Thermodynamic Systems. Proceedings, 1981. Edited by J. Casas-Vázquez and G. Lebon. VII, 321 pages. 1982.

Vol. 165: N. Mukunda, H. van Dam, L.C. Biedenharn, Relativistic Models of Extended Hadrons Obeying a Mass-Spin Trajectory Constraint. Edited by A. Böhm and J.D. Dollard. VI, 163 pages. 1982.

Vol. 166: Computer Simulation of Solids. Edited by C.R.A. Catlow and W.C. Mackrodt. XII, 320 pages. 1982.

Vol. 167: G. Fieck, Symmetry of Polycentric Systems. VI, 137 pages, 1982.

Vol. 168: Heavy-Ion Collisions. Proceedings, 1982. Edited by G. Madurga and M. Lozano. VI, 429 pages. 1982.

Vol. 169: K. Sundermeyer, Constrained Dynamics. IV, 318 pages. 1982.

Vol. 170: Eighth International Conference on Numerical Methods in Fluid Dynamics. Proceedings, 1982. Edited by E. Krause. X, 569 pages. 1982.

Vol. 171: Time-Dependent Hartree-Fock and Beyond. Proceedings, 1982. Edited by K. Goeke and P.-G. Reinhard. VIII, 426 pages. 1982.

Vol. 172: Ionic Liquids, Molten Salts and Polyelectrolytes. Proceedings, 1982. Edited by K.-H. Bennemann, F. Brouers, and D. Quitmann. VII, 253 pages. 1982.

Lecture Notes in Physics

Edited by H. Araki, Kyoto, J. Ehlers, München, K. Hepp, Zürich
R. Kippenhahn, München, H. A. Weidenmüller, Heidelberg
and J. Zittartz, Köln

212

Gravitation, Geometry and Relativistic Physics

Proceedings of the "Journées Relativistes"
Held at Aussois, France, May 2–5, 1984

Edited by Laboratoire "Gravitation et Cosmologie Relativistes", Université Pierre et Marie Curie et C.N.R.S., Institut Henri Poincaré, Paris

Springer-Verlag
Berlin Heidelberg GmbH 1984

Editor

Laboratoire de Physique Théorique "Gravitation et Cosmologie Relativistes"
C.N.R.S./U.A. 769- Université Pierre et Marie Curie, Institut Henri Poincaré
11, rue Pierre et Marie Curie, F-75231 Paris Cedex 05, France

ISBN 978-3-540-13881-5 ISBN 978-3-540-39081-7 (eBook)
DOI 10.1007/978-3-540-39081-7

Originally published by Springer-Verlag Berlin Heidelberg New York Tokyo in 1984

2153/3140-543210

P R E F A C E

This year our laboratory[*] has organized the "Journées Relativistes" in Aussois[#] from May the 2nd to May the 5th and edited the following proceedings.

Twenty-five years ago the theoretical tools for relativistic gravitation used to look very specialized and the orders of magnitude of the effects too small for experimentation. Then the field was often thought of as rather isolated. Nowadays this opinion is no longer valid.

Since the early days the subject has exploded in different directions and merged into several topics related to almost all the fields in physics. This results in a scientific community which has no precise name but exists nevertheless. In this community, the researchers are more or less specialized but the community itself is not : on one hand, sophisticated structures and geometrical tools are studied and used in mathematics and theoretical physics ; on the other hand, technological progress and the paucity of deep empirical knowledge provide the experimentalists with a strong motivation whatever the difficulties are.

As the different possible topics cover a broad range of preoccupations, we chose to emphasize the physical points of view : theoretical and experimental physics, astrophysics and cosmology. Within this framework, the key words of the meeting were "synthesis" and "prospect".

As a synthesis our goal was i) to present the developments of the subject from the early Riemannian geometry until nowadays with physical, epistemological and historical points of view : geometry, general relativity, experimental gravitation ...

---

[*] Laboratoire de Physique Théorique, "Gravitation et Cosmologie Relativistes", C.N.R.S./U.A. 769, Université Pierre et Marie Curie, Institut Henri Poincaré, 11 rue Pierre et Marie Curie - 75231 Paris Cedex 05.

[#] Meeting supported by the University Pierre et Marie Curie, the C.N.R.S. and the D.R.E.T.

ii) to summarize the situation of several important subjects concerning relativistic gravitation and related topics : thermal background radiation, gravitational lenses, inflationary universe ...

As prospects we chose to emphasize i) the diversity and the vitality of "geometrical physics", including relativistic gravitation and relativity : general relativity, supergravity, atomic physics, solid state physics ...

ii) the necessity of extra theoretical studies and clarifications in several fields where experiments and observations display a high accuracy : geodesy, atomic physics ...

iii) the fruitfulness of experimental gravitation (and especially of gravitational wave detection experiments) which was the starting point of recent discussions and works on quantum non-demolition[†], squeezed states, addition of laser fields, high performance interferometers ...

If conclusions were to be drawn from the meeting, on one hand I would put forward that besides the traditional problems (e.g. quantum gravity, gravitational fields from given sources, early universes ...) there exists an expanding field of preoccupations in "geometrical physics" related to very different theoretical, observational and experimental topics. On the other hand, I would especially emphasize that several precise theoretical questions, originating from the increasing accuracy of experiments and observations, have been asked during this meeting. They provide theoreticians with subjects for reflection and require answers in the near future.

We all especially acknowledge F. Allix and C. Trecul for their material organization of the meeting and C. Trecul for her help in the elaboration of the following proceedings.

September 1984

Ph. TOURRENC
Directeur du laboratoire

[†] Unfortunately we could not include in these proceedings the paper of W. Unruh because it did not arrive on time.

Table of contents

page

I. GENERAL RELATIVITY

J.N. GOLDBERG. Developments and Predictions 1
L. BLANCHET. Radiative Gravitational Fields and Radiation Reaction Forces in General Relativity 18
J. MARTIN, E. RUIZ and M.J. SENOSIAIN. Multipoles Particles in General Relativity : the Weyl and Kerr Metrics 29
J. KIJOWSKI. Unconstrained Degrees of Freedom of Gravitational Field and the Positivity of Gravitational Energy 40
J. HAJJ-BOUTROS. A Method for Generating Exact Solutions of Einstein's Field Equations 51
C. BARRABES. Causal Relativistic Thermodynamics of Transitory Processes in Electromagnetic Continuous Media 54
J. EISENSTAEDT. La relativité générale : une théorie sans problème(s) ? .. 57

II. THEORETICAL PHYSICS AND GEOMETRY

A. LICHNEROWICZ. Géométrie et Physique 77
Y. CHOQUET-BRUHAT. Supergravities 88
HU HESHENG (H.S. HU). Some Nonexistence Theorems for Massive Yang-Mills Fields and Harmonic Maps 107
D.M.L.F. SANTOS. Geometrical Approach to the Physics of Random Networks .. 117
J.B. KAMMERER. The Algebra of Multiplication Operators of Star-Product in $\mathbb{R}^{2n}$ 129
D. CANARUTTO and C.T.J. DODSON. Manifold b-Incompleteness Stability Via a Structure of Principal Connections 132
X. JAEN, A. MOLINA and J. LLOSA. Front Form Predictive Relativistic Mechanics Non Interaction Theorem 134
J. CARMINATI and R.G. MC LENAGHAN. Some New Results on the Validity of Huygens' Principle for the Scalar Wave Equation on a Curved Space-Time ... 138
N. BESSIS and G. BESSIS. Atomic Fine and Hyperfine Structure Calculations in a Space of Constant Curvature 143

page

III. EXPERIMENTAL RELATIVITY AND GRAVITATION

P. TEYSSANDIER. Theories of Gravity and Experimental Tests in the Post-Newtonian Limit 154
C. BOUCHER and J.F. LESTRADE. Survey of Relativistic Effects in Geodesy and Fundamental Astronomy 174
J.P. BRIAND. Relativistic Effects in Heavy Ions 187
A. BRILLET. The Interferometric Detection of Gravitational Waves 195
J. HOUGH, S. HOGGAN, G.A. KERR, J.B. MANGAN, B.J. NEERS, G.P. NEWTON, N.A. ROBERTSON, H. WARD and R.W.P. DREVER. The Development of Long Baseline Gravitational Radiation Detectors at Glasgow University 204
R. SCHILLING, L. SCHNUPP, D.H. SHOEMAKER, W. WINKLER, K. MAISCHBERGER and A. RÜDIGER. Improved Sensitivities in Laser Interferometers for the Detection of Gravitational Waves 213
C.N. MAN and A. BRILLET. Injection Locking and Coherent Summation of Argon Ion Lasers 222
A. HEIDMANN and S. REYNAUD. Can the Photon Noise Be Reduced ? 226
N. DERUELLE and Ph. TOURRENC. The Problem of the Optical Stability of a Pendular Fabry-Perot 232

IV. ASTROPHYSICS AND COSMOLOGY

N. DERUELLE. Much Ado about Geminga 238
R. FABBRI. The 3K Background Radiation : Observational and Theoretical Status 249
C. VANDERRIEST. Close-up on Gravitational Lensing : the Gravitational Mirages 265
F. HAMMER. Amplification of Light by Gravitational Lens : Dynamics and Thick Lens Effects 281
D. PAVON and J.M. RUBI. Thermodynamical Fluctuations of Massive Black Holes 286
B. BARBERIS and D. GALLETTO. Newtonian and Relativistic Bianchi I Models of the Universe 290
A. BLANCHARD and F.X. DESERT. The Cosmological Constant 294
R. HAKIM. The Inflationary Universe : a Primer 302

List of Participants 333

DEVELOPMENTS AND PREDICTIONS

Joshua N. Goldberg
Laboratoire de Physique Théorique
Université P. et M. Curie
Unité Associée au C.N.R.S. (769)
INSTITUT HENRI POINCARE
11, rue P. et M. Curie
75231 Paris Cedex 05

I - Introduction

In preparing this review of research in general relativity over the past 35 years, I have been impressed by how much in fact has been accomplished. As a result I have had to make a severe selection of material in order to avoid being entirely trivial. Some of you undoubtedly would have made other choices. My remarks are divided into five sections which are titled : Gravitational Radiation, Conservation Laws, Blackholes, Quantum Gravity, and Predictions. The uneven emphasis of these areas results in part from my own experience and in part from what I believe have been important accomplishments.

II - Gravitational Radiation

It is perhaps surprising to most people in the audience to realize that as late as 1957, at the Chapel Hill Conference, H. Bondi and T. Gold argued that gravitational radiation could not exist. Their arguments were tied to the steady state cosmology which at that time still had a few more years of life. What is particularly interesting is that within a year Bondi, Pirani, and I. Robinson[1] published their historic paper giving an exact plane wave solution and within a second year Bondi was lecturing about gravitational radiation in asymptotically flat space-times although the detailed paper[2] was not published until 1962. The plane wave solution which is based on earlier work by Einstein and Rosen[3], may be written in the form

$$ds^2 = e^{2A(u)}\, du\,(du + 2dx) - u^2 \left(e^{2B(u)} dy^2 + e^{-2B(u)} dz^2 \right) . \tag{1}$$

Satisfaction of the Einstein equations implies

$$2A' = uB'^2 \tag{2}$$

while the vanishing of the Riemann tensor implies

$$B'' + 2u^{-1}B' - uB'^3 = 0 . \tag{3}$$

If one attempts to cover the manifold with a single coordinate system, the metric of Eq. (1) exhibits a nasty singularity at u = 0. However, following the very important work of Lichnerowicz and the people around him[4], Bondi, Pirani and Robinson define a non-singular solution using three coordinate patches (Fig.1). The flatness condition, Eq. (3) is satisfied everywhere except in the cross-hatched region where Eq. (2) holds, but not Eq. (3). In region II and III, which overlap with I,

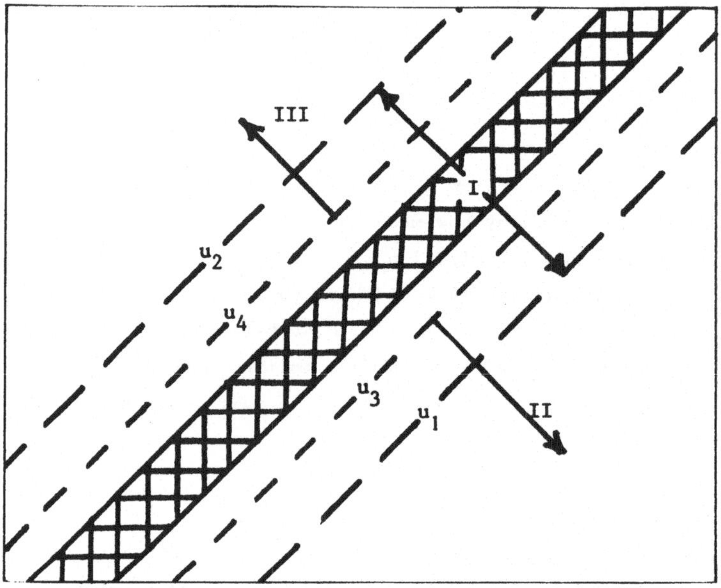

Fig. 1. The three coordinate patches for the plane wave solution. The curvature tensor R_{abcd} is different from zero only in the cross hatched region in patch I : $(0 < u_1 < u < u_2)$. The metric takes the standard Minkowski form in patch II $(u < u_3)$ and patch III $(u > u_4)$.

coordinates can be chosen so that $g_{ab} = \eta_{ab}$, the Minkowski metric. This solution has a five parameter symmetry group : rigid translations in the y-z plane ; rigid translation along x, and a two parameter group of null rotations which leaves du unchanged.

I have presented this example in order to emphasiez two things :

1) At that time there were very few exact solutions known and

2) the methods for looking for exact solutions were relatively undeveloped.

The development of methods for studying exact solutions come from three directions :

1) The Petrov classification introduced to physicists by Pirani[5],

2) The analysis of the geometry of timelike and null congruences by Ehlers, Kundt, and Sachs[6], and

3) The systemitizing of the use of symmetries begun by Ehlers and Kundt[6].

Now, of course, there is a whole book devoted to exact solutions[7] and in England MacCallum is creating a catalog of exact solutions on a computer.

The Bondi analysis[2] of asymptotically flat space-times did not attempt to construct exact solutions. Rather he proceeded physically to ask whether the structure of the Einstein equations was such that they allowed behavior far from matter which one could identify with the radiation of gravitational energy. Indeed, he assumed that in future null directions far from matter space-time was sufficiently close to Minkowski space that there exist nice null surfaces which are like null cones. He introduced an asymptotic coordinate system based on these outgoing "null cones", the null geodesics generate the cones and a foliation of sphere-like 2-surfaces :

$$ds^2 = e^{2b}du^2 + 2e^{2b}du\,dr - r^2 h_{AB}(dx^A - U^A du)(dx^B - U^B du) ,$$

$$\phi = 1 - 2m(u,\theta,\varphi)/r ,$$

$$h_{AB} = \begin{pmatrix} 1 + (\sigma^\circ + \bar{\sigma}^\circ)/2r & \sin\theta\,(\sigma^\circ - \bar{\sigma}^\circ)/2ir \\ \sin\theta\,(\sigma^\circ - \bar{\sigma}^\circ)/2ir & 1 - (\sigma^\circ + \bar{\sigma}^\circ)/2r \end{pmatrix} + \cdots . \tag{4}$$

Note that the 1/r part of h_{AB} has the form of a shear tensor. The important result which Bondi found is that the total mass

$$M = \frac{1}{4\pi}\oint m(u,\theta,\varphi)\,\sin\theta\,d\theta\,d\varphi \tag{5}$$

is a non-increasing function of time :

$$\dot{M} = -\frac{1}{4\pi}\oint \dot{\sigma}^\circ\dot{\bar{\sigma}}^\circ\,\sin\theta\,d\theta\,d\varphi . \tag{6}$$

His analysis was limited to axial symmetry. It was extended and made more rigorous by Sachs[8] and Newman-Penrose[9] by use of tetrad components and spin coefficients.

Anyone who has read the Sachs paper of 1962 knows that it required a major effort to analyze the Einstein equation $G_{ab} = 0$. Newman and Penrose had the brilliant idea that by considering the components of the Weyl tensor as independent field variables, the Bianchi identities become field equations. This yields a quasi-linear system of equations which can be studied in a transparent fashion. Another important technical innovation is the spinor analysis developed by Roger

Penrose[10,11]. In a natural way, the introduction of a spinor basis splits the Weyl tensor and the rotation coefficients into self-dual and anti-self-dual parts, and this gives one better control of the calculation. The same decomposition, of course, can be carried out with tetrads alone, but later I shall discuss a more fundamental use of spinors.

The analysis by Bondi, Sachs and Newman-Penrose depended on taking limits $r \to \infty$. While their results are physically and intuitively satisfying, it was not clear to what extent they depended on the specific coordinate system adopted. It was not easy to study how energy momentum or angular momentum depended on the particular foliation of null surfaces u = constant. Also one had an asymptotic geometry, but one did not have a geometry at infinity ; thus, one could have an asymptotic symmetry group, but not a symmetry group. In other words, the geometrical structures which might have physical content could not be easily studied because they had no home.

The difficulty was over come by Penrose[11] with the introduction of a space-time with boundary $(\hat{M}, \hat{g}_{ab})$ with the properties :

1) in the interior of $\hat{M}$, $\hat{g}_{ab} = \Omega^2 g_{ab}$;

2) on the boundary, $\partial\hat{M}$, $\Omega = 0$, $\nabla_a \Omega = n_a \neq 0$;

3) on $\Omega = 0$, $K_{abcd} = \Omega^{-1} C_{abcd}$ exists ;

4) $R_{ab} = 0 \Rightarrow (\nabla_a \Omega)(\nabla_b \Omega)\, g^{ab}\big|_{\Omega=0} = 0$.

Property (3) implies that $\Omega \sim 1/r$ and (4) tells us that the boundary $\mathscr{I}$ is a null surface. Therefore future null infinity $\mathscr{I}^+$ has a singular induced metric q_{mn} such that $q_{mn} n^n = 0$. The restriction to $\mathscr{I}^+$ of $n^a = g^{ab}\nabla_b \Omega$ is tangent to the null generators of $\mathscr{I}^+$.

One can show that for all asymptotically flat space-times q_{mn} and n^n define a universal structure which is independent of the particular physical space-time as long as the conditions of asymptotic flatness are satisfied[12]. The asymptotic symmetries of the physical space-time can be defined in terms of this universal structure :

$$\mathcal{L}_{\xi^a} q_{mn} = 2k\, q_{mn} , \qquad \mathcal{L}_{\xi^a} n^m = -k\, n^m . \tag{7}$$

Under action of the mapping, $\mathscr{I}^+$ undergoes a conformal transformation which, because q_{mn} is singular, is a six parameter group isomorphic to the Lorentz transformations. In addition there is an infinite dimensional abelian group, the supertranslations. These are non-rigid translations along the generators of

$\mathscr{I}^+$. The translations are constant along each generator but may vary continuously and differentiably from one generator to the next. Thus the symmetry group G is a semi-direct product of the conformal transformations and the supertranslations. The supertranslations form an invariant subgroup and the factor group is isomorphic to the Lorentz group. There is a four parameter invariant subgroup which defines the rigid translations. These are important in defining the Bondi energy-momentum.

If one only has a universal structure on $\mathscr{I}^+$, where does the physics come in ? First of all, the requirement that $\Omega^{-1} C_{abcd}$ has a limit gives rise to the Sachs peeling theorem[13]. That is, it tells us that in the physical space the radiative part of the Weyl tensor falls off as 1/r, the component associated with the mass falls off as $1/r^3$, and that associated with the quadrupole moment, as $1/r^5$. Furthermore, one observes that because the metric becomes singular, the connection on $\mathscr{I}^+$ is not uniquely defined by the universal structure. However, the connection in the physical space-time induces a connection on $\mathscr{I}^+$. The difference between the induced connection and the connection defined only by $D_j g_{mn} = 0$ is of the form $\Delta \Gamma_{mn}{}^j = \gamma_{mn} n^j$ where γ_{mn} is a symmetric tensor. One can show that γ_{mn} contains a part N_{mn} which is conformally invariant and satisfies the algebraic conditions $N_{mn} n^n = N_{mn} g^{mn} = 0$ where g^{mn} is any quasi-inverse of g_{mn} $(g_{mj} g^{jk} g_{kn} = g_{mn})$. N_{mn} depends only on $\dot{\sigma}^0$ (see Eq.(4)) and therefore is the rate of change of shear tensor.

II - Conservation Laws

From Noether's theorem, we know the diffeomorphisms of general relativity lead to differential identities among the field equations which in turn lead to conservation laws. Very early in the history of general relativity one understood that there are problems with energy. For example, there is a coordinate system for the Schwarzschild solution in which the Einstein pseudo-tensor vanishes everywhere, yet there exists a surface integral which defines the total energy as the mass. A local energy density is still elusive, but in asymptotically flat spaces-times invariant expressions for the energy-momentum and, in part, for the angular momentum have been constructed[15].

In a manner similar in spirit to the conformal completion at null infinity, one can study the geometry and structures defined on the hyperboloid of space-like directions at space-like infinity. If one assumes that the "magnetic" part of the Weyl tensor, that is, that part which results from rotational motions of the mass, falls off as $1/r^4$ instead of $1/r^3$ as is true of the "electric" part, then the asymptotic symmetry group is just the Poincaré group - the super translations can be eliminated. Therefore, one can write down invariant integrals for energy-momentum and angular momentum which are constants of the motion and have the usual properties of such quantities in Lorentz covariant theories[16]. At null infinity, the situation is not as good. First of all, when gravitational radiation is present one cannot have constants of the motion. None the less, using the Komar expression of the conservation laws in terms of a vector field ξ^{μ},

$$U^{\mu} = 2\sqrt{-g}\,\nabla_{\nu}(\nabla^{\nu}\xi^{\mu} - \nabla^{\mu}\xi^{\nu}) \, , \quad U^{\mu}{}_{,\mu} \equiv 0 \tag{8}$$

(with the additional condition $\nabla_{\nu}\xi^{\nu} = 0$ one can use the translation subgroup of the supertranslations to define energy-momentum as a 2-dimensional surface integral on $\mathcal{I}^{+}$ and a flux integral to define the change in energy-momentum if the surface of integration is distorted[15]. However, a similar construction for angular momentum has not been constructed. One can write down an angular momentum integral but its behaviour under distorsions of the two-surface in general will not vanish even in Minkowski space. It is my understanding that some progress has been made in terms of a suggestion by Roger Penrose[17], but I do not know the details.

Perhaps the most important question which has only fairly recently been settled is the question of the positivity of energy in general relativity. Actually there are two related questions : Given $T_{\mu\nu}t^{\mu}t^{\nu} \geqslant 0$ for all time-like vectors t^{μ} , $t^{\rho}t_{\rho} > 0$,

1) is ADM energy defined at spatial infinity necessarily non-negative, and

2) in the presence of gravitational radiation is the Bondi energy defined on $\mathcal{I}^{+}$ non-negative ?

The answer to both questions is in the affirmative.

The first question was answered definitively in 1979 by Schoen and Yau[18] who used rather delicate theorems about minimal surfaces to prove the theorem. This was followed by a beautiful, relatively simple proof by Edward Witten[19]. The argument can be put in the following form :

Define $D_a := \nabla_a - t_a t\cdot\nabla =: D_{AA'}$ where t_a is the unit normal to a space-like 3-surface Σ . Consider $t^{BA'}\bar{\xi}^{B'}(D_{AA'}D_{BB'} - D_{BB'}D_{AA'})\xi^{A}$ and use the Witten equation $D_{AA'}\xi^{A} = 0$.

Then we find

$$-D_m\,(t^{AA'}\bar{\xi}_{A'}\,D^m\xi^{A}) = -\,t^{AA'}(D_m\xi_A)(D^m\bar{\xi}_{A'}) + 4\pi\,T_{ab}\,t^a k^b , \tag{9}$$

where $k^a = \sigma^a{}_{AA'}\,\xi^A\bar{\xi}^{A'}$. One can show that there exists a unique spinor which is a constant at spatial infinity and which satisfies the Witten equation. Then we find that with the positive energy condition, $T_{ab}\,t^a k^b \geqslant 0$, the right hand side is positive or zero. Thus,

$$I = -\oint_{\partial\Sigma} (t^{AA'}\bar{\xi}_{A'}\,D^m\xi_A)\; dS_m \geqslant 0 \; . \tag{10}$$

One can show further that

$$I = P_a k^a \tag{11}$$

where P_a is the ADM four-momentum. This argument can be modified to show that the Bondi mass at null infinity is likewise positive or zero[20,21].

The importance of Witten's proof goes beyond the theorem itself. While spinors have been used extensively in general relativity, in every other case when spinors have been used to discuss the Einstein equations, one could equally well have used tetrad vectors. For the first time spinors have an intrinsic role for which tetrads cannot be substituted.

III - Black holes

Certainly the most fascinating objects of study in general relativity are the black holes. They were very poorly understood until recently. Some of us heard a lecture by K.C. Wali[22] who described Chandrasekhar's difficulty in having his theory of white dwarfs accepted. Although the theory of white dwarfs does not involve general relativity, it very definitely involves gravitation. Eddington clearly understood that the implications of Chandrasekhar's theory was that a sufficiently massive star could collapse to a singularity. He felt this was absurd and thereby delayed acceptance of the theory of white dwarfs among astronomers and astrophysicists.

In 1939 Oppenheimer and Snyder[23] calculated the spherically symmetric collapse of pressure free dust using the Einstein equations. They showed that there is nothing in the Einstein equations which would stop the collapse and the formation of the horizon associated with the Schwarzschild solution. However, this result was not exploited until the 50's when John Wheeler and his students[24] began looking at the collapse of various stellar models. Most of this work used spherically symmetric distributions, but the models did take into account nuclear forces. Their results showed that cold stars - after the completion of nuclear burning - less than 1.4 M could reach equilibrium as white dwarfs. Stars more massive would pass through the white dwarf stage to another equilibrium position for neutron stars. The upper limit for a neutron star depends on the assumptions made concerning the equation of state. It is estimated to be $1.5\, M_\odot \lesssim M \lesssim 5 M_\odot$. Stars much more massive than this are known and while there is no explicit proof that such stars cannot lose enough mass to fall below this limit, there is also no proof then they can and always will. Furthermore, the collapse of matter below the Schwarzschild radius does not require an exotic equation of state. If a globular cluster of $10^8\, M_\odot$ collapses, the mean density in the volume $V_s = (4\pi/3)\, R_s^3$ is that of water, $\rho = 1\,\mathrm{gm/cm^3}$.

In 1963 R.P. Kerr constructed the axi-symmetric stationary solution for a rotating mass. A year and a half later the solution including electric charge was constructed. These solutions also exhibited a horizon inside of which there is no escape to time-like infinity. In 1967 Werner Israel[26] proved that a static space-time with a smooth spherically symmetric horizon was necessarily Schwarzschild and in 1975, after considerable work by Brandon Carter[27] and others, David Robinson[28] completed the proof that the Kerr solution was the unique stationary axi-symmetric solution. Israel himself extended his proof to include charge, but the proof that the charged Kerr or Kerr-Newman solution is unique was published only last year by P. Mazur[29] in Poland and G. Bunting[30] in Australia. However, John Wheeler had been saying since the late 60's that "Black holes have no hair" by which he meant that

1) Kerr-Newman is the unique (physically important) stationary axi-symmetric solution ;

2) A collapsing star will radiate away its quadrupole and higher moments and settle down in an equilibrium state which is Kerr or Kerr-Newman and therefore depends only on the three parameters M, J, Q.

While there was general belief that symmetry was not important in the collapse of a massive star, before 1965 there was no geometrical characterization of the properties of the Schwarzschild (or Kerr) singularity which would allow one to study this question. There was, of course, the Raychaudhuri equation which follows from the Einstein equations with an irrotational perfect fluid as a source :

$$\frac{d\theta}{ds} = -\sigma_{ab}\sigma^{ab} - \frac{1}{3}\theta^2 - 4\pi\kappa(\rho+3p), \tag{12}$$

θ is the divergence and σ_{ab} the shear tensor for the flow lines, while ρ and p are the density and pressure of the fluid, respectively. If $(\rho + 3p) \gtrsim 0$, the right hand side is negative definite and the divergence necessarily decreases. Furthermore, from $\nabla_b T^{ab} = 0$ one finds

$$\frac{d\rho}{ds} = -(\rho + p)\theta \tag{13}$$

so that if θ becomes negative, ρ necessarily increases monotónically. However, this result is local and does not give the global information needed to characterize the horizon.

The first global theorem on singularities is due to R. Penrose[31]. This begins with the missing link - the characterization of the essential property of the horizon : the existence of a trapped surface. Penrose defines a trapped surface to be a closed space-like 2-surface such that the family of orthogonal outgoing null geodesics as well as the family of orthogonal ingoing null geodesics is converging so then one can expect the causal future of the trapped surface to be bounded. That is effectively what the Penrose theorem proves. More precisely, he shows that if

1) There exists a global Cauchy surface,

2) A positive energy condition is satisfied, $R_{ab}k^a k^b \gtrsim 0$ for all null vectors

k^a,

then null geodesic completeness and the existence of a trapped surface are incompatible. More loosely, positive energy and a trapped surface implies a singularity. This form of the theorem has a weakness in the requirement of a global Cauchy surface which in general will not exist in a space-time with collapsing matter. However, further work by Hawking and Penrose separately and together (for a review see reference 32) has substituted for that condition two others :

1) Causality - no closed or almost closed time-like

2) An algebraic generality condition which is a local condition to be satisfied by the Riemann tensor.

Again loosely :

i) Generality, positive energy, and causality implies that geodesic completeness and the existence of a trapped surface are incompatible.

ii) Positive energy and the equation of geodesic deviation imply that neighboring time-like or null geodesics will intersect once they start to converge. These results suggest that trapped surfaces are generic and that singularities cannot be avoided.

Penrose has put forward the conjecture that all such singularities are hidden from view behind a horizon - cosmic censorship, but this has not been proved in spite of considerable effort. The converse has also not been proved and this conjecture remains the most important unsolved problem in this area.

Work done by Hawking, Carter, and others in studying the classical properties of black holes have been summarized in the statement of "Four Laws of Black Hole Physics" corresponding to the laws of thermodynamics[33].

0) For a black hole in equilibrium, i.e. stationary or static, the surface gravity κ is a constant over the horizon;

1) $\Delta M = (\kappa/8\pi)\Delta A + \Omega_H \Delta J + \Phi_H \Delta Q$,

M is the mass of the black hole, A its surface area, Ω_H its angular velocity relative to a non-rotating observer at ∞ , J its angular momentum, Φ_H the electrostatic potential on the horizon, and Q its electric charge;

2) $\Delta A \geq 0$;

3) lim $\kappa \to 0$ cannot be reached.

A classical black hole can absorb energy, hence entropy, and appears to violate the second law of thermodynamics. In an attempt to save the second law and the meaning of entropy for processes taking place in the vicinity of a black hole, Beckenstein[34,35] argued that the four laws of black hole physics should indeed be considered thermodynamic. From the second law he argued that the entropy of a black hole S_{BH} should be proportional to the area, A , and from the first three laws that the temperature should be proportional to κ . Indeed one has (k = Boltzmann's constant) for a spherically symmetric black hole

$$S_{BH} = \frac{1}{4}\left(\frac{c^3}{G\hbar}\right) k A \quad , \quad T_{BH} = \frac{\hbar c^3}{8\pi G k M_\odot} \frac{M_\odot}{M} = 5 \times 10^{-8} \frac{M_\odot}{M} K . \quad (14)$$

This suggestion saved the second law in many classical situations, but it broke down when considering the black hole in a thermal radiation bath at a temperature $T < T_{BH}$.

The curious thing about Beckenstein's work is his introduction of Planck's constant into the classical theory. With hind sight this foreshadowed the Hawking radiation of a black hole. No doubt the discovery by Stephan Hawking[36] that a black hole radiates quanta with a thermal spectrum, where the temperature is determined by the Beckenstein temperature, was the most unexpected result of this period. Together with the inclusion of black hole entropy in thermodynamics, this is also potentially the most profound result. First of all, because of quantum interactions the black hole is no longer simply a negative object in physics. It does not simply absorb energy from the surrounding universe while hiding its structure behind the horizon. Quantum mechanically the black hole interacts with all physical fields. However, the calculations are semi-classical in that quantum gravity and the quantum aspects of the black holes itself have not been taken into account. Is it possible that when these effects are taken into account the internal structure of the black hole will be important ? If so, John Wheeler's "no hair" theorem itself may be transcended.

Because of the importance of the Hawking radiation, let me sketch the origin of the result. Hawking considers the spherically symmetric collapse of a star. He is only interested in the field outside the surface of the star which he assumes to be Schwarzschild. In a conformal diagram the collapse is pictured as follows :

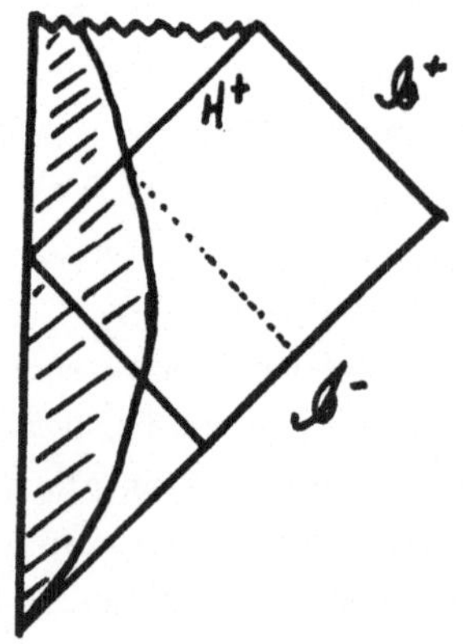

Fig.2. The Penrose diagram for the spherically symmetric collapse of a star. This represents the background for the calculation of Hawking radiation.

On this background one considers the quantum theory of a massless scalar field :

$$\Box \phi = 0 . \tag{15}$$

The field at $-\infty$ is completely determined by its configuration on $\mathcal{J}^-$. Here it has a well defined positive and negative frequency decomposition :

$$\phi = \Sigma (f_i a_i + \bar{f}_i a_i^+) . \tag{16}$$

The f_i are positive frequency solutions of the wave equation and (a_i, a_i^+) are the usual annihilation and creation operators. If there were no collapse and no horizon, then the field could equally well be described by its decomposition into positive and negative frequency on $\mathcal{J}^+$:

$$\phi = \Sigma (g_i b_i + \bar{g}_i b_i^+) . \tag{17}$$

In this case, however, energy and momentum conservation would show us that the vacuum state on $\mathcal{J}^-$ goes to the vacuum state on $\mathcal{J}^+$. With the formation of the horizon the field is not determined only by the configuration on $\mathcal{J}^+$ but one has to take into account the field on H^+, the horizon. Thus, in fact,

$$\phi = \Sigma (g_i b_i + \bar{g}_i b_i^+) + \Sigma (h_i c_i + \bar{h}_i c_i^+) \tag{18}$$

where the g_i are solutions of the scalar wave equation which vanish on H^+ and the h_i are solutions which vanish on $\mathcal{J}^+$. Because the f_i themselves are complete, we have that

$$b_i = \Sigma (\alpha_{ij} a_j + \beta_{ij} a_j^+) \tag{19}$$

$$c_i = \Sigma (\gamma_{ij} a_j + \eta_{ij} a_j^+) \tag{20}$$

and

$$\Sigma \langle 0^- | b_i^+ b_i | 0^- \rangle = \Sigma \beta_{ij} \bar{\beta}_{ij} . \tag{21}$$

A detailed calculation of β_{ij} then shows that the outgoing field at $\mathcal{J}^+$ has a thermal spectral distribution with a temperature given by the Beckenstein temperature. Intuitively one can see that vacuum fluctuations occur in the vicinity of the horizon creating and annihilating pairs of quanta. Occasionally one member of a pair crosses the horizon and ends up in a negative energy state while the other member goes out to $\mathcal{J}^+$.

IV - Quantum gravity

I find it difficult to report on the past and present status of quantum gravity in part because so much has been done and in part because so little has been accomplished. (For a review see reference 37). There is no thread one can follow to gain some understanding. The quantization of the Einstein theory is the most difficult and the most profound problem facing relativists for over 40 years. It raises fundamental questions about the meaning of geometry, of space-time, of manifolds, and of the relationship of gravitation to the rest of physics, particularly now to particle or high energy physics.

The first attempt to quantize gravity was by Leon Rosenfeld in the 30's, but the prolonged current effort began in the early 50's with the work of Dirac[38,39] and Bergmann[40] and it culminated in the canonical formulation which is best known in the manner given by ADM[41]. One assumes a phase space Γ whose points (g_{mn}, p^{mn}) are a positive definite 3-metric and conjugate momenta related to the extrinsic curvature of the surface Σ on which g_{mn} is the metric. To be suitable data for a solution of the Einstein equations these points lie in a subspace $\bar{\Gamma}$ defined by the constraint equations

$$\mathcal{H}_L = \sqrt{g}\,\{{}^3R - g^{-1}(p^{mn}p_{mn} - p^2)\} = 0,$$
$$\mathcal{H}_m = D_n p_m{}^n = 0 . \qquad (22)$$

The Hamiltonian is an combination of these constraints plus a surface term.

$$H = \frac{1}{16\pi}\int_\Sigma \{N\mathcal{H}_L + N^m\mathcal{H}_m\}\, d\tau_{(3)}$$
$$+ \frac{1}{16\pi}\oint_{\partial\Sigma}\{N(g_{km,n} - g_{kn,m})\,g^{km} + 2N^m g^{-\frac{1}{2}} p_{mn}\}\, dS^n \qquad (23)$$

The surface terms are needed to guarantee differentiability of the Hamiltonian and to define the Hamiltonian as the total energy.

Unfortunately, although a tremendous effort has been expended on this problem, its solution has remained elusive. Very recently what I believe is an important step has been made, and I shall describe that briefly later. Over the years there have been other equally unfruitful approaches to quantization : the covariant quantization developed by DeWitt, supergravity which unifies gravity with half-integral spin fields in particular 3/2, and the path integral, or functional integral formulation by Hawking and his co-workers. All of the above work certainly has uncovered interesting properties of the gravitational field, but none is close to what might be called a solution of the problem.

At this point it might be well to ask what one would consider a solution. I think that a conservative, even conventional, point of view would be

1. An identification of the Hilbert space of physical states and the operators which act on that space ;

2. The development of a consistent set of rules which allows one to calculate the evolution of a physical state and the transition between physical states.

In the canonical and covariant formalisms this means

3. An identification of a "correct" configuration space. The Hilbert space then consists of functions on this configuration space and one requires

4. An invariant measure on the configuration space.

And finally

5. Some rules for incorporating fields other than gravity.

6. Agreement with experiments and observations.

By a "correct" configuration space I means a minimal set of dynamical variables, the diffeomorphisms have been factured out, and the constraints either are satisfied trivially or perhaps can be satisfied easily. The functional integral approach is interesting in that it seems to avoid the problem of a "correct" configuration space because it uses all possible classical paths. However, it must have its definition of "correct" built into its measure.

Recently an interesting step has been taken by Abhay Ashtekar[42] towards the solution of step 3 in the above scheme. Recall that the Witten propagation allows one to take a constant spinor at space-like infinity and to propagate it uniquely over a space-like surface Σ . Two independent Witten spinor fields give us a Witten dyad. Combinations of the spinor fields with constants allow us to define Witten vector fields on Σ . If we use the equation leading to the positive energy theorem, we can show that the Hamiltonian can take the form

$$H_W = \frac{1}{16\pi} \int_\Sigma \phi_m{}^A{}_B \; \bar{\phi}^{m}{}^{A'}{}_{B'} \, W^{BB'} \, dS_{AA'} \tag{24}$$

where $W^{BB'}$ is just the Witten vector field defined in terms of the spinor dyad and ($\phi_m{}^A{}_B$, $\bar{\phi}_m{}^{A'}{}_{B'}$) are the self-dual and anti-self-dual spin coefficients defined by the dyad and the Witten projected derivative operator. This suggests that self-dual (anti-self-dual spin coefficients) may be useful in defining an appropriate configuration space on which the Hilbert space state vectors are to act. Indeed, this is what Ashtekar proposes to do. The further development is too elaborate to present here quickly. I mention it in this much detail because I consider this work to be a very important step in the solutions of at least part of the problems we find in quantum gravity.

I shall close here because anything more which I may have to say comes under the heading of predictions (speculation ?) and will be discussed in the next section.

V - Predictions

In a way predictions are easy to make. In the short range everyone knows the direction the field is moving. In the long range, no one knows. My remarks will be divided into several subsections

1. Growth of the Field

There is one prediction I can make with certainty. That concerns the growth of the field. At the meeting celebrating the 50th anniversary of special relativity in Bern, there were about 50 people present and many, like Pauli, though interested, were not active in the field. Two years later in Chapel Hill about 75 active workers in the field gathered in Chapell Hill including a strong representation from France. In Jena and Padova 800-1000 people registered for the GR meetings. I don't expect attendance at meetings to increase, because they become unproductive in such large numbers. But the number of workers is increasing as is shown by the fact that in addition to the conventional broad journals in physics and mathematics which publish articles on general relativity, there are two journals now devoted exclusively to reporting research in general relativity. So it is easy to predict that the number of people working in the field will increase and perhaps at a very rapid rate.

2. Theory

a. Classical

There is a need for exact solutions which describe asymptotically flat space-times containing sources and gravitational radiation. It would be nice to have a solution like the Lienard-Wiechert solutions of electromagnetism. It may not be possible to achieve such solutions in closed form, but good approximate solutions would be welcome. These are important not only to resolve theoretical questions about the existence and structure of $\mathscr{I}$, but also to help in the building of models for sources of gravitational radiation. If we are to be able to interpret the coming observations of gravitational radiation, realistic models will have to be constructed. Therefore, much effort will be going in this direction.

b. Quantum

Recent work by Abhay Ashtekar gives me some confidence that a canonical quantization of general relativity may be within our grasp. The main problem I see is the construction of a measure for the space of functionals describing state vectors. However even if this can be accomplished, it may not be possible to extend this work to include the other fields of particle physics. As formulated at present, general relativity tacks on other fields by minimal coupling of Lagrangians or Hamiltonians. If unification means anything, this must certainly be wrong.

In this respect the functional integral approach does not appear to be better. It too relies on a Lagrangian and minimal coupling between fields. At the very least, it too requires a measure on the space of geometrics.

I don't know where to place my bets as far as the more exotic theories are concerned. Twistor theory, supergravity, Kaluza-Klein theories, and random networks have elements which make each very attractive. At the moment I think it safe to say that 35 years from now none will be important.

3. Experiments

In 1950 no experiments were being done. There were only the 3 classical tests of general relativity. By 1960 the redshift had been measured on the Earth at Harvard and in England. Measurement of the spin orbit interaction - Lense-Thirring and the inertial drag on the spin axis - was proposed and should be carried out in the next few years. The time delay for rays passing near the sun was calculated by Irwin Shapiro and he began systematic measurements on the solar system. Calculations of the gravitational lense effect were done, but I don't think anyone expected such a dramatic observation of the effect as we have recently had with the double image of a very distant galaxy being formed by the convergence of rays by an intervening galaxy acting as the lense.

We have observed the Big Bang in the 3°K background radiation. Quasars by their distance appear to be remnants of very early processes. Perhaps if we get above earth's atmosphere, possibly with an observatory on the Moon or on a space station we may be able to detect clouds which are in a prequasar state. Certainly we should expect to find more pulsars and other double neutron star (or black-hole neutron star) systems. Our understanding of galactic structure and the formation of globular clusters should develop to the point where we may be able to recognize the existence of a black hole in the nucleus of large galaxies and globular clusters. The evidence is marginal at present, but that may change. Also in the 60's gravitational radiation began to be considered seriously and in 1969 Joe Weber announced his results. While his results have not been accepted because they have not been reproduced by others, his work has stimulated research in this area and I believe that by the year 2000 gravitational astronomy will be a recognized field, although still in its infancy.

Other than to predict that there will be perhaps an exponential growth in gravitational experiments, because a large part of this meeting is devoted to that question by people who are more knowledgable than I about that subject, I will not describe experiments which are in the planning state.

4. Role in Future Physics

35 years ago physicists on the whole were uninterested in general relativity or gravitational interactions. Now almost all physicists are aware that the frontier of physics research involves gravitation, conceptually if not directly in experiment. The energies involved in the study of nuclear physics are in the range of MeV to 10's of GeV. The multiplicity of particles and interactions has enforced a search for unifying ideas. The first major step in this direction has been the Weinberg-Salam model based on the invariance group SU(3). It's sucess has triggered the search for a theory of broader extent the so called Grand Unified Theories or

GUT theories. GUT is an acronym but it is also a colloquial word for intestine. When one speaks about a gut reaction one refers to the most basic reaction and the gut of an idea is its core. GUT theories now consider unifying energies of the order of $10^{15} - 10^{16}$ GeV. The Planck energy, the energy associated with the Planck mass is 6×10^{18} GeV. The jump from nuclear to GUT energies is much greater than the jump from GUT to Planck energies. As a result many physicists now believe that fundamental problems of particle physics will not be solved without quantum gravity. Some now study inflationary universes to produce the observed cosmological isotropy and to give time for element formation. But inflationary universes are still fairly conventional. The guts of general relativity, quantum gravity, and black hole physics do not yet play a role and I believe that without them a basic understanding of matter will not be achieved.

5. Final Remarks

If we look back over the past 50 years and ask what in todays gravitational physics is so strange that it could not have been foreseen or was not discussed in some form, we find very little. The existence of neutron stars and black holes were discussed although their properties were not as well known as they are today. Most physicists, if asked, would probably have agreed that gravitational radiation exists but probably not of much importance to physics for a long-long time. There have been many more surprises in particle physics than in general relativity simply because of the ability to do experiments with the strong interactions. The one area which I believe was unpredictable is the Hawking radiation and the associated thermodynamics of black holes. I believe this must have a profound significance and may be crucial in our ultimate understanding of gravitation and the existence of matter. Up to now we have begun with a universe which is filled with matter and we have tried to describe that universe, and to understand the variety of matter which we find in that universe. Physicists have begun to go beyond that point to describe the big bang and the origin of matter itself. I believe that to achieve that understanding the Hawking radiation and the thermodynamics of black holes - perhaps the Statistical Mechanics Black Holes - will play an important role. In any case as research involving general relativity increases in volume, as I am sure it will, we are in for many more surprises over the next 35 years than we found in looking back over the past 35 years.

Some years ago I began a talk with the phrase "Gravity is the organizing force of the universe. I had in mind then the external structure of the universe and went on to describe the role of gravity in the formation of galaxies, stars, pulsars, and black holes. Today I shall close with the statement that Gravity is the unifying force in Physics, but it may take more than the next 35 years for us to understand how that comes about.

REFERENCES

1. H. Bondi, F. Pirani, and I. Robinson, Proc.Roy.Soc.London A215, 519 (1957).
2. H. Bondi, M.G.J. van der Berg, Proc.Roy.Soc. (London) A269, pp. 21-52 (1962).
3. N. Rosen, Phys.Z. Sowjetunion, 12.4, 366 (1937).
4. A. Lichnerowicz, Théories Relativistes de la Gravitation (Masson et cie, Paris, 1955).
5. F.A.E. Pirani, Phys.Rev. 105, 1089 (1957).
6. J. Ehlers and W. Kundt ; "Exact Solutions of the Gravitational Field Equations", in Gravitation, ed. L. Witten (John Wiley and Sons, New York, 1962).
7. D. Kramer, H. Stephani, M. Mac Callum, and E. Herlt, Exact Solutions of Einstein's Field Equations, (Cambridge University Press, 1980) London.
8. R.K. Sachs, Proc.Roy.Soc.London A270, 103 (1962).
9. E.T. Newman and R. Penrose, J.Math.Phys. 3, 566 (1962).
10. R. Penrose, Ann.Phys. 10, 171 (1960).
11. R. Penrose, Proc.Roy.Soc.London, A284, 159 (1965).
12. R. Geroch, "Asymptotic Structure of Space-Time", in Asymptotic Structure of Space-Time, ed. F.P. Esposito and L. Witten (Plenum Press, New York 1977).
13. R.K. Sachs, Proc.Roy.Soc.London A264, 309 (1961).
14. A. Ashtekar, J.Math.Phys. 22, 2885 (1981).
15. R. Geroch and J. Winicour, J.Math.Phys. 22, 803 (1981).
16. A. Ashtekar and A. Magnon-Ashtekar, J.Math.Phys., in press.
17. R. Penrose, Proc.Roy.Soc. London A381, 53 (1982).
18. R. Schoen and S.T. Yau, Phys.Rev.Lett. 43, 1457 (1979).
19. E. Witten, Comm.Math.Phys. 80, 381 (1981).
20. G.T. Horowitz, M.J. Perry, Phys.Rev.Lett. 48, 371 (1982).
21. M. Ludvigsen and J.A.G. Vickers, J.Phys.A. 15, L 67 (1982).
22. K.C. Wali, Phys.Today 35, N°10, 23 (1982).
23. J.R. Oppenheimer and H. Snyder, Phys.Rev. 56, 455 (1939).
24. B.K. Harrison, K.S. Thorne, M. Wakano, and J.A. Wheeler ; Gravitation Theory and Gravitational collapse, (University of Chicago Press, Chicago, 1965).
25. R.P. Kerr, Phys.Rev.Lett. 11, 237 (1963).
26. W. Israel, Phys.Rev. 164, 1776 (1967).
27. B. Carter, "The General Theory of the Mechanical Electromagnetic and Thermodynamic Properties of Black Holes", in General Relativity, ed. S.W. Hawking and W. Israel (Cambridge University Press, London 1979).
28. D.G. Robinson, Phys.Rev.Lett. 34, 905 (1975).
29. P.O. Mazur, J.Phys. A15, 3173 (1982).
30. G. Bunting, "Proof of the Uniqueness Conjecture for Black Holes", Ph.D. Thesis, Department of Mathematics, University of New England, Armidale, N.S.W., Australia (1983).
31. R. Penrose, Phys.Rev.Lett. 14, 57 (1965).
32. F.J. Tipler, C.J.S. Clarke, and G.F.R. Ellis, "Singularities and Horizons - A Review Article", in General Relativity and Gravitation 2, ed. A. Held (Plenum Press, New York 1980).
33. J.M. Bardeen, B. Carter and S.W. Hawking, Comm.Math.Phys. 31, 161 (1973).
34. J.D. Bekenstein, Phys.Rev.D. 9, 3292 (1974).
35. J.D. Bekenstein, Phys.Today, 33, N°1, 24 (1980).
36. S.W. Hawking, Comm. in Math.Phys. 43, 199 (1975).
37. C. Isham, "Quantum Gravity - An Overview" in Quantum Gravity 2, ed. C. Isham, R. Penrose and D. Sciama (Clarendon Press, Oxford, 1981).
38. P.A.M. Dirac, Can.J.Math. 2, 129 (1950).
39. P.A.M. Dirac, Can.J.Math. 3, 1 (1951).
40. P.G. Bergman, Phys.Rev. 75, 680 (1949).
41. R. Arnowitt, S. Deser, and C. Misner, "The Dynamics of General Relativity", in Gravitation, ed. L. Witten (John Wiley and Sons, New York, 1962).
42. A. Ashtekar, Oxford conference on Quantum Gravity 3, March 1984.

RADIATIVE GRAVITATIONAL FIELDS AND RADIATION REACTION FORCES IN GENERAL RELATIVITY.

Luc BLANCHET

Groupe d'Astrophysique Relativiste
Observatoire de Paris-Meudon
92195 Meudon Principal Cedex (France).

We define a Post Minkowskian iteration method for solving Einstein's vacuum equations and we give the general structure of the solution when $r \to o$ or $c \to +\infty$. The method is then used to derive an expression for the radiation reaction force density in the case of a non-relativistic source.

I) INTRODUCTION.

A central problem in Classical General Relativity is to determine the gravitational field generated by an isolated source of matter. Few exact stationary solutions of this problem are known, but the general case where the source does not have any particular symmetry and is non-static (i.e. time varying) is still unsolved. However the hope of soon detecting gravitational waves from astrophysical sources makes it urgent to tackle the problem. Two interesting questions are the following :

Q1) What is the structure of the gravitational field in the exterior vacuum region outside the source ?

Q2) How does the gravitational field react on the source ?

These questions illustrate the two aspects of the coupling between the field and the matter, and we shall give in the following some preliminary answers to these questions.

II) OUTLINE OF THE METHOD.

Let us consider a physical system, playing the role of a gravitational source, with total mass-energy M and finite spatial dimension r_o. The system is constituted by moving fluid masses, with typical velocity v. Let us define the following radii :

$$r_1 := \text{Sup}\left(r_o \, , \, 10 \frac{GM}{c^2}\right) , \tag{1a}$$

$$r_2 := \frac{1}{10} \lambdabar := \frac{1}{10} \frac{r_o c}{v} . \tag{1b}$$

G is Newton's constant, and c is the velocity of light. One may interpret $\lambda = 2\pi\lambdabar$ as a wavelength of the gravitational radiation emitted by the source. In the case where $\frac{v}{c} < \frac{1}{10}$ and $\frac{GM}{c^2} < \frac{r_o}{10}$ (non-relativistic source) one has the situation depicted in fig.1 :

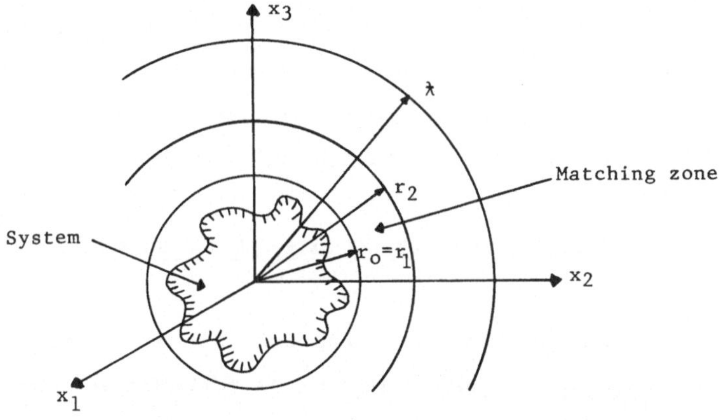

Fig. 1 : a non-relativistic source.

The general method we shall use to determine the gravitational field generated by the system can be divided in three successive steps :

A) to obtain the most general "retarded" vacuum exterior field (in the region $r := \sqrt{x_1^2 + x_2^2 + x_3^2} > r_1$) which admits a Post Minkowskian expansion (see|1|)

$$h^{\alpha\beta} := \sqrt{-\det g_{\gamma\delta}}\, g^{\alpha\beta} - f^{\alpha\beta} = \sum_{n=1}^{+\infty} G^n h^{\alpha\beta}_{(n)} . \tag{2}$$

Here, $f^{\alpha\beta}$ is the Minkowskian metric (signature +2). This field will depend on some unknown retarded functions $X(t-\frac{r}{c})$.

B) To calculate the Post Newtonian expansion of the exterior field, which means to expand formally $h(t,\vec{x},\frac{1}{c})$ when $c \to +\infty$. The first few terms of this expansion are expected to give a good approximation of the exterior field in the region $r_1 < r < r_2$ (when it exists).

C) To compute, from the stress-energy tensor $T^{\alpha\beta}$ of the system, directly in the form of a Post Newtonian expansion, the interior field (in the region $0 \leq r < r_2$). This field is determined up to unknown functions $Y(\vec{x},t)$. Then, in the case of a non-relativistic source, (Fig. 1), to match the exterior and interior Post Newtonian fields in the overlapping region $r_1 < r < r_2$ (matching region). This will give :

- Y as a functional of X,
- X as a functional of the physical caracteristics of the system (density, pressure, etc...).

In this article we present an answer to question Q1) by first defining a Post Minkowskian iteration algorithm for computing $h^{\alpha\beta}_{(n)}$, up to any order n, (step A, see section III) and then, by studying the dependance on $\frac{1}{c}$ of $h^{\alpha\beta}_{(n)}$ (step B, section IV). Next the algorithm is used to derive an explicit expression for the radiation reaction force in the case of a slowly moving source of perfect fluid (step C, section V). This gives us an idea of what an answer to question Q2) should be.

III CONSTRUCTION OF THE EXTERIOR VACUUM FIELD.

Replacing the Post Minkowskian expansion (2) into Einstein's vacuum equations, and using harmonic coordinates, lead to the following equations ($\partial_\beta := \partial/\partial x^\beta$; $\Box := \partial^\beta \partial_\beta = \Delta - c^{-2}\partial_t^2$) :

$$\Box\, h^{\alpha\beta}_{(n)} = \Lambda^{\alpha\beta}_{(n)}\left(h_{(1)}, h_{(2)},\ldots, h_{(n-1)}\right), \tag{3a}$$

$$\partial_\beta\, h^{\alpha\beta}_{(n)} = 0 , \tag{3b}$$

where $\Lambda^{\alpha\beta}_{(n)}$ is a polynomial in $h_{(m)}$ and its first two derivatives ($m<n$, $\Lambda^{\alpha\beta}_{(1)} = 0$). The form of eqs. (3) suggests a recursive construction of $h^{\alpha\beta}_{(n)}$ from $h^{\alpha\beta}_{(1)}$ (linearized approximation). Thorne, |2|, has given the most general $h^{\alpha\beta}_{(1)}$ (see eq. (8-12) of ref. |2|) satisfying (3) together with Fock's "no incoming radiation" condition, |3|, in the form of an infinite sum (multipolar expansion) of "retarded" terms of the type $\partial_P \left(r^{-1} X\left(t - \frac{r}{c}\right)\right)$, where the functions X are either the "mass" multipole M_P or the

"current" multipole S_P or their time-derivatives. Here P is a multi-index $P := (i_1, i_2, \ldots, i_p)$ of order p, and $\partial_P := \partial_{i_1} \partial_{i_2} \ldots \partial_{i_p}$. M_P and S_P are symmetric trace-free euclidean tensors, M (the "mass") and S_i (the "angular momentum") are constant, and M_i (the "mass dipole") is zero (mass centered coordinates). Let us now prove (from ref. |4|), the following theorem, which shows that $h^{\alpha\beta}_{(n)}$ can be constructed up to an arbitrary order n.

Theorem : Starting from a finite set of functions X(t), supposed to be $C^\infty(\mathbf{R})$ and constant in a fixed neighbourhood of $t = -\infty$, (say $t \leq -T$), a formal expansion (2), solution of Einstein's vacuum field equations, can be recursively constructed, and, for all $n \in \mathbf{N}$, the following property $P_{(n)}$ holds for $h^{\alpha\beta}_{(n)}$: $\forall N \in \mathbf{N}$, there exist some functions $F^{\alpha\beta}_{pjk}(t)$ and a function $R^{\alpha\beta}_N(t,\vec{x})$ so that ($n_P := x_{i_1} \ldots x_{i_p}/r^p$)

$$h^{\alpha\beta}_{(n)}(t,\vec{x}) = \sum_{p,j,k} n_P (\log r)^j \, r^k \, F^{\alpha\beta}_{pjk}(t) + R^{\alpha\beta}_N(t,\vec{x}), \tag{4}$$

with : a) $p, j \in \mathbf{N}$, $k \in \mathbf{Z}$, $o \leq p \leq p_{max}$, $o \leq j \leq n-1$ and $k_{min} \leq k \leq N$,

b) $F^{\alpha\beta}_{pjk}(t)$ is constant if $t \leq -T$ and $C^\infty(\mathbf{R})$,

c) $R^{\alpha\beta}_N(t,\vec{x})$ is zero if $t \leq -T$ and both $C^N(\mathbf{R}^4)$ and of order $O(r^N)$ when $r \to o$ (t being fixed).

In other words, the property $P_{(n)}$ is that $h^{\alpha\beta}_{(n)}$ admits, when $r \to o$, an asymptotic expansion on the scale functions $(\log r)^j \, r^k$ ($j \in \mathbf{N}$, $k \in \mathbf{Z}$). We use the name $P_{(n)}$- expansion for that expansion. Note that the only place where the property $P_{(n)}$ depends on the integer n is in the maximum value j_{max} of the power of log r which is : $j_{max} = n-1$.

Main outlines of the demonstration :

1°/ $P_{(1)}$ is true for $h^{\alpha\beta}_{(1)}$ thanks to the Taylor formula at order N, applied to each of the functions $X(t - \frac{r}{c})$ in a neighbourhood of $r = o$.

2°/ Suppose that $h^{\alpha\beta}_{(2)}, h^{\alpha\beta}_{(3)}, \ldots, h^{\alpha\beta}_{(n-1)}$ have been constructed so as to satisfy $P_{(2)}, P_{(3)}, \ldots, P_{(n-1)}$, and let $N \in \mathbf{N}$. The first step of the demonstration is to prove that $\Lambda^{\alpha\beta}_{(n)}$ (see eq. (3a)), which is calculated from $h^{\alpha\beta}_{(1)}, \ldots, h^{\alpha\beta}_{(n-1)}$, satisfies $P_{(n-1)}$, that is to say admits an expansion, up to order N, similar to (4), with $j_{max} = n-2$. Then in order to find a solution of eq. (3a), one is tempted to simply employ the usual retarded integral, defined by

$$\Box^{-1}_{\mathrm{Ret}} \psi(\vec{x},t) := \frac{-1}{4\pi} \iiint_{\mathbf{R}^3} \frac{d^3\vec{y}}{|\vec{x}-\vec{y}|} \psi\left(\vec{y}, t - \frac{1}{c}|\vec{x}-\vec{y}|\right), \tag{5}$$

and to take for ψ the previously calculated $P_{(n-1)}$-expansion of $\Lambda^{\alpha\beta}_{(n)}$. Unfortunately, this leads to divergent integrals because of the singular behaviour of $\Lambda^{\alpha\beta}_{(n)}$ near $r = 0$ (the neighbourhood of $r = +\infty$ does not create any problems because of our assumption of stationarity of the multipole moments in the past $t \leq -T$). Instead we shall use a complex analytic continuation procedure. Let B be a complex number. The second step of the demonstration is to show that the quantities $\Box^{-1}_{\text{Ret}}(r^B \Lambda^{\alpha\beta}_{(n)})$ and $\Box^{-1}_{\text{Ret}}(r^{B-1} n_i \Lambda^{\alpha i}_{(n)})$, ($n_i = x_i/r$), are well defined for B pertaining to some strip $a < \text{Re}\, B < b$ in the complex plane **C**, and can be uniquely analytically continued all over **C**, except for integer values of B (including the value $B = 0$) where the quantities admit in general a multiple pole. Let $\underset{B=o}{\text{Finite Part}}\, Z_B$ and $\underset{B=o}{\text{Residue}}\, Z_B$ be the coefficients of B^o and B^{-1} in the Laurent expansion of Z_B near $B = 0$ and define

$$u^{\alpha\beta}_{(n)} := \underset{B=o}{\text{Finite Part}} \;\Box^{-1}_{\text{Ret}}\; (r^B \Lambda^{\alpha\beta}_{(n)}). \tag{6}$$

One easily finds

$$\Box\, u^{\alpha\beta}_{(n)} = \Lambda^{\alpha\beta}_{(n)}, \tag{7a}$$

$$\partial_\beta u^{\alpha\beta}_{(n)} =: w^{\alpha}_{(n)} = \underset{B=o}{\text{Residue}} \;\Box^{-1}_{\text{Ret}}\; (r^{B-1}\, n_i\; \Lambda^{\alpha i}_{(n)}). \tag{7b}$$

The vector $w^{\alpha}_{(n)}$ is a "retarded" (in Fock's sense) solution of $\Box w = o$, and thus can be written in a form similar to $h^{\alpha\beta}_{(1)}$:

$$w^{\alpha}_{(n)} = \sum_{p\geq o} \partial_P \left(\frac{W^{\alpha}_P\,(t - \frac{r}{c})}{r}\right) . \tag{8}$$

Let us now introduce an algorithmic operator H acting on vectors w^α of the type (8) and computing from $w^{\alpha}_{(n)}$ a tensor $v^{\alpha\beta}_{(n)}$ of the same type (with "retarded" functions $V^{\alpha\beta}_Q(t - \frac{r}{c})$) so that

$$\Box\; v^{\alpha\beta}_{(n)} = 0\ , \tag{9a}$$

$$\partial_\beta v^{\alpha\beta}_{(n)} = -\; w^{\alpha}_{(n)}. \tag{9b}$$

Functions $V^{\alpha\beta}_Q$ are uniquely defined as being some derivatives or some integrals of functions W^{α}_P (see |4| for the definition of H). It follows immediately from eqs. (7) and (9), that

$$h^{\alpha\beta}_{(n)} := u^{\alpha\beta}_{(n)} + v^{\alpha\beta}_{(n)} \tag{10}$$

satisfies Einstein's vacuum equations (3). The next step of the demonstration consists in proving that property $P_{(n)}$ holds for $h^{\alpha\beta}_{(n)}$ defined in (10). We consider firstly the formula

$$\begin{aligned}\Box^{-1}_{\text{Ret}}\left(\hat{n}_p(\log r)^j\, r^{B+k}\, F(t)\right) &= \frac{\partial^j}{\partial B^j}\, \Delta^{-1}\left(\hat{n}_p\, r^{B+k}\, F(t)\right) + \\ &+ \Box^{-1}_{\text{Ret}}\left\{\left(\frac{1}{c^2}\frac{\partial^2}{\partial t^2}\right)\frac{\partial^j}{\partial B^j}\left(\Delta^{-1}\left(\hat{n}_p\, r^{B+k}\, F(t)\right)\right)\right\},\end{aligned} \tag{11}$$

where Δ^{-1} denotes the usual Poisson operator, and $\hat{n}_p$ is a trace-free product of p unit vectors n_i. We apply eq. (11), in an iterative manner, to each term of the $P_{(n-1)}$- expansion of $r^B\ \Lambda^{\alpha\beta}_{(n)}$. This allows one to express $u^{\alpha\beta}_{(n)}$ (eq. (6)) as

$$\begin{aligned}u^{\alpha\beta}_{(n)} &= \underset{B=o}{\text{Finite Part}} \sum_{m\geq o} \left(\frac{1}{c^2}\frac{\partial^2}{\partial t^2}\right)^m \Delta^{-m-1}\left(r^B\ \overline{\Lambda}^{\alpha\beta}{}_{(n)}\right) + \\ &+ \Box^{-1}_{\text{Ret}}\ S^{\alpha\beta}_N(\vec{x},t).\end{aligned} \tag{12}$$

The bar over $\Lambda^{\alpha\beta}_{(n)}$ denotes the formal series of the $P_{(n-1)}$-expansion of $\Lambda^{\alpha\beta}_{(n)}$, $S^{\alpha\beta}_N(\vec{x},t)$ being the $\left(C^N(\mathbf{R}^4)\text{ and } O(r^N)\right)$- function in this $P_{(n-1)}$-expansion (see eq. (4)). Secondly we use the Taylor formula and show that there exist some $C^\infty(\mathbf{R})$-functions $F^{\alpha\beta}_{pi}(t)$ and another $\left(C^N(\mathbf{R}^4)\text{ and } O(r^N)\right)$-function $R^{\alpha\beta}_N(\vec{x},t)$ so that

$$\Box^{-1}_{\text{Ret}}\ S^{\alpha\beta}_N(\vec{x},t) = \sum_{p+2i\leq N} \hat{n}_p\, r^{p+2i}\, F^{\alpha\beta}_{pi}(t) + R^{\alpha\beta}_N(\vec{x},t). \tag{13}$$

We compute now the iterated Poisson operators Δ^{-m-1} acting on $r^B\ \overline{\Lambda}^{\alpha\beta}_{(n)}$ in (12), and take the finite part in B = 0 of the resulting expression. The maximum value j_{max} of j is then increased by one unit, starting from n-2 for $\Lambda^{\alpha\beta}_{(n)}$ to n-1 for $u^{\alpha\beta}_{(n)}$. Doing this, and using (13), leads to the conclusion that $P_{(n)}$ is satisfied by $u^{\alpha\beta}_{(n)}$. The same happens for $v^{\alpha\beta}_{(n)}$ since $v^{\alpha\beta}_{(n)}$ has a structure of the type (8) simular to $h^{\alpha\beta}_{(1)}$ and finally $h^{\alpha\beta}_{(n)}$ admits itself a $P_{(n)}$-expansion. This completes our proof.

Remark 1 : The a priori validity of our construction is not limited in an asymptotic region $r \gg \lambda\!\!\!^{-}$, but extends in all the region $r > r_1$.

Remark 2 : We expect that this construction yields, when formally extended to infinite multipolar series, the most general power series solution of the vacuum

Einstein equations which is constant in the past (modulo an arbitrary coordinate transformation).

IV) POST NEWTONIAN EXPANSION OF THE EXTERIOR FIELD.

The structure of the Post Newtonian expansion of $h^{\alpha\beta}_{(n)}$ follows immediately from the $P_{(n)}$-expansion (4). Indeed, if we consider the multipole moments with their usual physical dimensions (i.e. $(M_P) = (\text{mass}) \times (\text{length})^P$ and $(S_P) = (\text{mass}) \times (\text{time})^{-1} \times (\text{length})^{P+1}$), $h^{\alpha\beta}_{(n)}$ becomes a function of c, (velocity of light), and (4) may be written in terms of c as an asymptotic expansion on the scale functions $(\log c)^j/c^k$ ($j \in \mathbf{N}$, $k \in \mathbf{N}$) :

$$h^{\alpha\beta}_{(n)} = \sum_{\substack{0\leq j\leq n-1 \\ 0\leq k\leq N}} \frac{(\log c)^j}{c^k} + O\left(\frac{1}{c^N}\right) . \tag{14}$$

This result invalidates the usual Post Newtonian approximation methods, which suppose an expansion on the scale functions c^{-k} only.

Let us explore the Post Newtonian expansion of $h^{\alpha\beta}_{(n)}$ in the following framework : we suppose, very generally, that the low order multipole moments are constant, and that the first "radiating" (i.e. non-constant) multipole moments are $M_L(t)$ and / or $S_{L-1}(t)$, where $L = (i_1 \dots i_\ell)$ is a multi-index with ℓ indices ($\ell \geq 2$). When $\ell = 2$ we recover the usual case. With this definition of ℓ , we have shown that the right-hand-side of eq. (13) is of order $O(1/c^{2n+2\ell-i})$, (when $c \to +\infty$), i being the number of spatial indices among α and β . Thus eq. (12) gives us the Post Newtonian expansion of $u^{\alpha\beta}_{(n)}$ as being

$$\bar{u}^{\alpha\beta}_{(n)} = \underset{B=0}{\text{Finite Part}} \sum_{m\geq 0} \left(\frac{1}{c^2}\frac{\partial^2}{\partial t^2}\right)^m \Delta^{-m-1}\left((r/c)^B \bar{\Lambda}^{\alpha\beta}_{(n)}\right) + O\left(\frac{1}{c^{2n+2\ell-i}}\right) . \tag{15}$$

We introduce now the "odd" part of the Post Newtonian exterior metric : this part consists of terms which are the coefficients of c^{-k} with $i+k$ odd. Thanks to eq. (15) together with a related equation for $v^{\alpha\beta}_{(n)}$, we have proven that the "odd" part of the exterior metric, when compared with the "even" part, is quite tiny (when $c \to +\infty$), that is precisely to say

$$(h^{\alpha\beta}_{(n)})_{\text{even}} = O\left(\frac{(\log c)^{n-1}}{c^{2n+\omega}}\right) , \qquad (16a)$$

$$(h^{\alpha\beta}_{(n)})_{\text{odd}} = O\left(\frac{(\log c)^{n-1}}{c^{2n+2\ell-\omega-1}}\right) , \qquad (16b)$$

where $\omega = 0$ if $i = 0$ or 2 and $\omega = 1$ if $i = 1$. The reason why a distinction is made between odd and even parts of the exterior metric is the following : the odd part changes sign upon time reversal, and the corresponding part in the interior metric is thought to be responsible for secular (or irreversible) effects, such as losses of energy, in the source. Radiation reaction forces are generally expected to derive exclusively from the odd terms in the interior metric (|5|, |6|). In section V, we adopt a definition for the reaction force in keeping with this expectation.

V) THE RADIATION REACTION FORCE.

Consider a non-relativistic perfect fluid with matter density ρ (see fig.1). We will give an example of how the matched asymptotic expansions method works (|6|, |7|) in calculating the dominant odd terms ("resistive" terms) in the interior metric of the fluid. Suppose that the interior field admits a Post Newtonian expansion satisfying (16) and plug this expansion into Einstein's equations (with a perfect fluid stress energy tensor $T^{\alpha\beta}$). Then identify the coefficients of c^{-k} in both sides of the equations (note that in the interior region we have : $\Box = \Delta + O(\frac{1}{c^2})$) to obtain (going back to $g_{\alpha\beta}$ instead of $h^{\alpha\beta}$)

$$\Delta \underset{2}{g}{}^{\text{int}}_{oo} = -8\pi\rho , \qquad (17a)$$

$$\Delta \underset{2\ell+3}{g}{}^{\text{int}}_{oo} = \underset{2\ell+1}{g}{}^{\text{int}}_{jk} \, \partial^2_{jk} \, \underset{2}{g}{}^{\text{int}}_{oo} , \qquad (17b)$$

$$\Delta \underset{2\ell+2}{g}{}^{\text{int}}_{oj} = 0 , \qquad (17c)$$

$$\Delta \underset{2\ell+1}{g}{}^{\text{int}}_{jk} = 0 . \qquad (17d)$$

($\underset{k}{g}_{\alpha\beta}$ being the coefficient of c^{-k} in $g_{\alpha\beta}$). Eqs. (17) determine $g^{\text{int}}_{\alpha\beta}$ modulo some functions $Y(\vec{x},t)$ (regular in $r = 0$) which are solutions of $\Delta Y = 0$. On the other hand, thanks to eq. (15), and after having performed a suitable coordinate transformation, we have proven, |8|, that the corresponding exterior terms are given by

$$\underset{2}{g}^{ext}_{oo} = 2 \sum_{p\geq o} \frac{(-)^P}{p!} \left(\partial_P r^{-1}\right) M_P(t) \ , \tag{18a}$$

$$\underset{2\ell+3}{g}^{ext}_{oo} = \frac{(-)^{\ell+1} 2^{\ell+1} (\ell+1)(\ell+2)}{\ell(\ell-1)(2\ell+1)!} x_L \overset{(2\ell+1)}{M_L}(t) + 2 \sum_{p\geq 1} \frac{(-)^P}{p!} \left(\partial_P r^{-1}\right) Q_P(t) \ , \tag{18b}$$

$$\underset{2\ell+2}{g}^{ext}_{oj} = \frac{(-)^{\ell+1} 2^{\ell+1} (\ell-1)(\ell+1)}{\ell(\ell-2)(2\ell-1)!} \varepsilon_{jab}\, x_{aL-2} \overset{(2\ell-1)}{S_{bL-2}}(t) \ , \tag{18c}$$

$$\underset{2\ell+1}{g}^{ext}_{jk} = 0 \ , \tag{18d}$$

where the tensors $Q_P(t)$, which are functions of the time alone, arise from non-linear (quadratic) contributions to the metric. Now the key point is that both $g^{int}_{\alpha\beta}$ and $g^{ext}_{\alpha\beta}$ must coincide since the interior and exterior approximation schemes have a common domain of validity, namely the matching zone $r_1 < r < r_2$ of fig. 1. One easily finds that the matching is fulfilled if

1) We link the functions $M_P(t)$ to the mass multipole moments of the fluid by equations of the type

$$M_P(t) + \frac{1}{c^{2\ell+1}} Q_P(t) = \int_{Fluid} d^3x\, \rho \hat{x}_P + O\left(\frac{1}{c^{2\ell+3}}\right) + \text{"even"} \ , \tag{19}$$

($\hat{x}_P = r^P \hat{n}_P$ = trace free product of $x_{i_1}, \ldots, x_{i_p}$). (We assume that the $S_P(t)$ are similar related to the current multipole moments of the fluid).

2) We determine the Y-functions by taking for $g_{\alpha\beta}$, all over the region $0 \leq r < r_2$, the following values :

$$\underset{2}{g}_{oo} = 2U = 2 \int_{Fluid} \frac{d^3\vec{y}}{|\vec{x}-\vec{y}|} \rho(\vec{y},t) \ , \tag{20a}$$

$$\underset{2\ell+3}{g}_{oo} = -2\, V_{React} = \frac{(-)^{\ell+1} 2^{\ell+1} (\ell+1)(\ell+2)}{\ell(\ell-1)(2\ell+1)!} x_L \overset{(2\ell+1)}{M_L}(t) \ , \tag{20b}$$

$$\underset{2\ell+2}{g}_{oj} = A^j_{React} = \frac{(-)^{\ell+1} 2^{\ell+1} (\ell-1)(\ell+1)}{\ell(\ell-2)(2\ell-1)!} \varepsilon_{jab}\, x_{aL-2} \overset{(2\ell-1)}{S_{bL-2}}(t) \ , \tag{20c}$$

$$\underset{2\ell+1}{g}_{jk} = 0 \ . \tag{20d}$$

Note that U is merely the Newtonian potential of the fluid. We calculate from (20) the radiation reaction force density according to a well-known formula :

$$\begin{aligned} \underset{2\ell+1}{F}{}^{i}_{React} = {} & \frac{1}{2} \rho\, \partial_i \underset{2\ell+3}{g}_{oo} - \rho\, \partial_o \underset{2\ell+2}{g}_{oi} + \rho\, v_j \left(\partial_i \underset{2\ell+2}{g}_{oj} - \partial_j \underset{2\ell+2}{g}_{oi} \right) - \\ & - \rho\, \partial_o \left(v_j \underset{2\ell+1}{g}_{ij} \right) + \rho\, v_j v_k \left(\frac{1}{2} \partial_i \underset{2\ell+1}{g}_{jk} - \partial_j \underset{2\ell+1}{g}_{ik} \right) , \end{aligned} \tag{21}$$

and we find

$$\vec{F}_{React} = \frac{1}{c^{2\ell+1}} \rho \left(\vec{E}_{React} + \vec{v} \times \vec{B}_{React} \right) , \tag{22a}$$

$$\vec{E}_{React} = - \vec{\nabla} V_{React} - \partial_o \vec{A}_{React} , \tag{22b}$$

$$\vec{B}_{React} = \vec{\nabla} \times \vec{A}_{React} . \tag{22c}$$

Therefore, V_{React} and $\vec{A}_{React}$, (given by eqs. (20b,c)), appear as scalar and vector reaction potentials. If $\ell = 2$, we recover the usual reaction force density (see |6|, |9|, |10|). But here, contrarily to the latter references, the full non-linear structure of the Einstein field equations had to and has been taken into account in the derivation of (22). Note that $\vec{F}_{React}$ causes secular decreases of energy and angular momentum, exactly opposite to those computed with the usual flux-integrals in the linearized theory, |2|.

VI CONCLUSION.

The method, on its whole, is promising, as a theory of generation of gravitational waves, mainly because the exterior field determined in section III is valid both in a region near the system ($r \ll \lambda\!\!\!^{-}$) and in the asymptotic region ($r \gg \lambda\!\!\!^{-}$). As a consequence, the link between the source and the field near infinity might be studied by our method. Moreover, the dynamics of the system can be investigated properly. For instance, our derivation of the radiation reaction force in section V, is certainly just one example of the effects that are within the reach of formula (15). Other effects will be the subject of further work.

ACKNOWLEDGEMENTS

I am very grateful to Thibaut Damour for advice and many fruitful discussions on all the points raised in this article.

REFERENCES

|1| K. Westpfahl and H. Hoyler, Lett. Nuov. Cim. 27 (1981) 581 ; L. Bel, T. Damour, N. Deruelle, J. Ibañez and J. Martin, Gen. Rel. Grav. 13 (1981) 963 ; T. Damour, in "Gravitational Radiation", eds. N. Deruelle and T. Piran, North-Holland, Amsterdam (1983) 59.

|2| K.S. Thorne, Rev. Mod. Phys. 52 (1980) 299.

|3| V.A. Fock, "Theory of Space, Time and Gravitation", Pergamon, London (1959) p.365.

|4| L. Blanchet and T. Damour, C.R. Acad. Sc. Paris, Série II, 298 (1984), 431 ; L. Blanchet, thèse de 3ème cycle (non publiée), Université P. et M. Curie, Paris VI.

|5| S. Chandrasekhar and F.P. Esposito, Ap. J. 160 (1970) 153.

|6| W.L. Burke, J. Math. Phys. 12 (1971) 401.

|7| K. Thorne, Ap. J. 158 (1969) 997.

|8| L. Blanchet and T. Damour, (1984), submitted to Phys. Lett. A.

|9| C. W. Misner, K.S. Thorne, J.A. Wheeler, "Gravitation", Freeman, San Francisco (1973) p. 1001.

|10| R.E. Kates, Phys. Rev. 22 (1980), 1871.

MULTIPOLES PARTICLES IN GENERAL RELATIVITY: THE WEYL AND KERR METRICS.

J. Martín, E. Ruiz and M.J. Senosiaín
Departamento de Física Teórica
Universidad de Salamanca (Spain).

1.- INTRODUCTION

In recent articles a method was described for obtaining approximate solutions of the Einstein equations relative to a multipole particle[1]. The reliability of the method was demostrated in the construction of a source model which permits one to reproduce the Kerr metric up to an approximation called quadrupole-postminkowskian. At the same time refinements were made in the interpretation of this metric, due to the restriction which arose in relation to the structure of the quadrupole moment of the source.

The aim of the present work is to make another step in the demostration of the reliability of the above-mentioned method. In order to do so, the Einstein equations will firstly be resolved with a greater degree of approximation. Secondly, the result will not only be applied to the Kerr metric, but also, its reliability will be checked in the reproduction and interpretation of the Weyl metrics (static axysimmetric vacuum solutions).

Section 2 will give a succinct account of the method for obtaining the approximate solution corresponding to a pole-dipole-quadrupole particle and will give the results concerning to the model of source used up to the order of approximation in question. In Section 3, the Kerr and Weyl metrics are written out in a certain system of approximate harmonic coordinates, which will allow us to make the appropiate comparisons with the previous results. Finally, Section 4 comments briefly on the conclusions derived from the other Sections.

2.- FIELD CREATED BY A POLE-DIPOLE-QUADRUPOLE PARTICLE

The goal of this section is to resolve, whitin a certain degree of approximation, Einstein's gravitational field equations[2] relative to a pole-dipole-quadrupole point-like source. We understand by such a source one that is defined by an energy-momentum tensor of the following kind[3]:

$$T^{\mu\nu}(x) = T_m^{\mu\nu}(x) + T_d^{\mu\nu}(x) + T_q^{\mu\nu}(x) \qquad (2.1)$$

where

$$T^{\mu\nu}_{m}(x) \equiv \frac{1}{2}\int_{\mathbb{R}} d\tau \cdot M^{\mu\nu}(\tau)\cdot\bar{\delta}(x^{\rho}-z^{\rho}) \tag{2.2.a}$$

$$T^{\mu\nu}_{d}(x) \equiv -\frac{1}{2}\nabla_{\sigma}\int_{\mathbb{R}} d\tau \cdot M^{\sigma,\mu\nu}(\tau)\cdot\bar{\delta}(x^{\rho}-z^{\rho}) \tag{2.2.b}$$

$$T^{\mu\nu}_{q}(x) \equiv \frac{1}{2}\nabla_{\rho}\nabla_{\sigma}\int_{\mathbb{R}} d\tau \cdot M^{\rho\sigma,\mu\nu}(\tau)\cdot\bar{\delta}(x^{\rho}-z^{\rho}) \quad , \tag{2.2.c}$$

where $z^{\rho} = \varphi^{\rho}(\tau)$ is the parametric curve which locates the source;the quantities $M^{\mu\nu}$, $M^{\sigma,\mu\nu}$ and $M^{\rho\sigma,\mu\nu}$ are three tensors defined on that curve; and $\bar{\delta}$ is the Dirac scalar function, that is, $\bar{\delta} \equiv (-g)^{-1/2}\,\delta$, where g is the determinant of the metric tensor. Expressions (2.2.a),(2.2.b)and (2.2.c) will be called, respectively, the monopole, dipole and quadrupole parts of the energy-momentum tensor (or of the source).

Let us now consider the Einstein equations for the field created by the source (2.1). These equations may be written in harmonic coordinates as follows[4]:

$$\mathfrak{g}^{\rho\sigma}\partial_{\rho\sigma}\mathfrak{g}^{\mu\nu} = \partial_{\rho}\mathfrak{g}^{\mu\sigma}\partial_{\sigma}\mathfrak{g}^{\nu\rho} + Q^{\mu\nu} + 16\pi G\,\mathfrak{T}^{\mu\nu} \tag{2.3.a}$$

$$\partial_{\mu}\mathfrak{g}^{\mu\nu} = 0 \tag{2.3.b}$$

where $\mathfrak{g}^{\mu\nu} \equiv (-g)^{1/2} g^{\mu\nu}$ is the metric density, $\mathfrak{T}^{\mu\nu} \equiv (-g) T^{\mu\nu}$, and $Q^{\mu\nu}$ is the Einstein-Landau pseudotensor which, as is known, depends quadraticaly on the first derivatives of the metric density. The problem consists in determining the solution $\mathfrak{g}^{\mu\nu}(x^{\rho};L)$ of equations (2.3); this is a solution which, as has been indicated symbolically, will be a function of the point (x^{ρ}) where the field is calculated and a functional of the evolution L of the source. Simultaneously it is necessary to determine the differential equations of such evolution.

We shall assume that the solution $\mathfrak{g}^{\mu\nu}$ of (2.3) will admit a formal expansion in a power series of the gravitational constant G. In this sense we shall write:

$$h^{\mu\nu} \equiv \mathfrak{g}^{\mu\nu} - \eta^{\mu\nu} = G\overset{1}{h}{}^{\mu\nu} + G^2\overset{2}{h}{}^{\mu\nu} + G^3\overset{3}{h}{}^{\mu\nu} + O(G^4) \tag{2.4}$$

where the deviation $h^{\mu\nu}$ of the metric density $\mathfrak{g}^{\mu\nu}$ with respect to the Minkowski metric has been introduced.

A).- <u>1-PM APPROXIMATION</u>: Let us begin by writing the Einstein equations (2.3) in the post-minkowskian approximation; that is, overlooking terms of order G^2. Taking into account (2.4), we have:

$$\square h^{\mu\nu} = 16\pi G \overset{\circ}{\mathfrak{T}}{}^{\mu\nu} + O(G^2) \tag{2.5.a}$$

$$\partial_\mu h^{\mu\nu} = 0 \tag{2.5.b}$$

where $\square$ is the flat D'Alambert operator and where $\overset{\circ}{\mathfrak{T}}{}^{\mu\nu}$ represents the zero-order part of the tensor density $\mathfrak{T}^{\mu\nu}$. Now, if we bear in mind the condition of harmonicity (2.5.b) (or what amounts to the same, if we impose the conservation condition to the density $\overset{\circ}{\mathfrak{T}}{}^{\mu\nu}$), we obtain the following:

$$\overset{\circ}{\mathfrak{T}}{}^{\mu\nu}(x) = M\int_{\mathbb{R}} d\tau\, u^\mu u^\nu\, \delta(x^\rho - z^\rho) + \frac{1}{2}\int_{\mathbb{R}} d\tau \left(u^\mu S^{\nu\sigma} + u^\nu S^{\mu\sigma}\right)\partial_\sigma \delta(x^\rho - z^\rho)$$
$$+ \frac{1}{2}\int_{\mathbb{R}} d\tau\, N^{\mu\nu,\rho\sigma}(\tau)\cdot \partial_{\rho\sigma}\delta(x^\rho - z^\rho) \tag{2.6.a}$$

$$\dot{u}^\mu = O(G) \quad , \quad \dot{S}^{\mu\nu} = O(G) \tag{2.6.b}$$

where M represents the mass of the source, u^μ the 4-velocity, $S^{\mu\sigma}$ is the angular momentum tensor and $N^{\mu\nu,\rho\sigma}$ is a tensor defined on the curve $\varphi^\mu(\tau)$ which exhibits the following symmetries:

$$N^{\mu\nu,\rho\sigma} = N^{\nu\mu,\rho\sigma} = N^{\mu\nu,\sigma\rho} \quad , \quad N^{\mu(\nu,\rho\sigma)} = 0 \tag{2.7}$$

(round brackets stand for symmetrization) and represents what we call the quadrupole moment of the source. Finally, the dot over a quantity stands for derivative with respect to the parameter τ, which we assume to be identified with the minkowskian proper time. Furthermore we assume that the Dixon ortogonality condition between the angular momentum tensor and the linear momentum vector of the source is fulfilled[6].

The integration of (2.5.a) by means of the flat retarded propagator, bearing in mind (2.6), yields the following result:

$$h^{\mu\nu}(x;L) = -4GM\hat{\tau}^{-1}\hat{u}^\mu u^\nu + 2G\hat{\tau}^{-3}\left(\hat{u}^\mu \hat{S}^{\nu\sigma} + \hat{u}^\nu S^{\mu\sigma}\right)\hat{\ell}_\sigma +$$
$$+ 2G\hat{\tau}^{-3}\hat{N}^{\mu\nu,\rho\sigma}\left(\eta_{\rho\sigma} - 2\hat{u}_\rho\hat{u}_\sigma + 6\hat{\tau}^{-1}\hat{u}_\rho\hat{\ell}_\sigma - 3\hat{\tau}^{-2}\hat{\ell}_\rho\hat{\ell}_\sigma\right) +$$
$$+ 2G\hat{\tau}^{-2}\hat{\dot{N}}^{\mu\nu,\rho\sigma}\left(\eta_{\rho\sigma} + 4\hat{\tau}^{-1}\hat{u}_\rho\hat{\ell}_\sigma - 3\hat{\tau}^{-2}\hat{\ell}_\rho\hat{\ell}_\sigma\right) - 2G\hat{\tau}^{-3}\hat{\ddot{N}}^{\mu\nu,\rho\sigma}\hat{\ell}_\rho\hat{\ell}_\sigma + O(G^2), \tag{2.8}$$

where

$$\hat{\ell}^\rho \equiv x^\rho - \varphi^\rho(\hat{\tau}) \quad , \quad \hat{\ell}^\rho \hat{\ell}_\rho = 0 \quad , \quad \hat{\ell}^0 > 0 \tag{2.9.a}$$

$$\hat{u}^\mu \equiv \dot{\varphi}^\mu(\tau) \quad , \quad \hat{S}^{\mu\nu} \equiv S^{\mu\nu}(\hat{\tau}) \quad , \quad \hat{N}^{\mu\nu,\rho\sigma} \equiv N^{\mu\nu,\rho\sigma}(\hat{\tau}) \tag{2.9.b}$$

$$\hat{\tau} \equiv -\hat{\ell}_\rho \hat{u}^\rho > 0 \tag{2.9.c}$$

and where the indices are moved with the minkowskian metric (from now onwards this will always be the case).

Let us note that we have not obtained, at the order considered, any information about the evolution of the quadrupole moment $N^{\mu\nu,\rho\sigma}$. In this sense, we shall <u>assume</u> that this moment is stationary and exhibits axial symmetry; this is an assumption which is born out by the following structures:

$$t^{\mu\nu,\rho\sigma} = \bar{A}\{\Delta^{\mu\nu}\Delta^{\rho\sigma} - \Delta^{\mu(\rho}\Delta^{\sigma)\nu}\} + \bar{B}\{\Delta^{\mu\nu}e^\rho e^\sigma + \Delta^{\rho\sigma}e^\mu e^\nu - 2e^{(\mu}\Delta^{\nu)(\rho}e^{\sigma)}\} \tag{2.10.a}$$

$$v^{\mu\nu\rho} = \bar{C}\{\Delta^{\mu\nu}e^\rho - e^{(\mu}\Delta^{\nu)\rho}\} + \bar{D}\, e^{(\mu}s^{\nu)\rho} \tag{2.10.b}$$

$$m^{\mu\nu} = \bar{E}\,\Delta^{\mu\nu} + \bar{F}\, e^\mu e^\nu \tag{2.10.c}$$

where the right hand side of these expressions originate in the following decomposition of the quadrupole moment[7]:

$$N^{\mu\nu,\rho\sigma} = t^{\mu\nu,\rho\sigma} + 2\{v^{\mu\nu(\rho}u^{\sigma)} + v^{\rho\sigma(\mu}u^{\nu)}\} +$$

$$+ u^\mu u^\nu m^{\rho\sigma} + u^\rho u^\sigma m^{\mu\nu} - 2u^{(\mu}m^{\nu)(\rho}u^{\sigma)} \tag{2.11}$$

and where:

$$\Delta^{\mu\nu} \equiv \eta^{\mu\nu} + u^\mu u^\nu \quad , \quad s^{\mu\nu} \equiv S^{-1}S^{\mu\nu} \quad , \quad e_\rho \equiv \frac{-1}{2S}\eta_{\mu\nu\rho\sigma}u^\mu S^{\nu\sigma} \tag{2.12}$$

where S is the length of the angular momentum vector and $\eta_{\mu\nu\rho\sigma}$ is the Levi-Civita symbol($\eta_{0123} = +1$); finally, $\bar{A}$, $\bar{B}$, $\bar{C}$, $\bar{D}$, $\bar{E}$ and $\bar{F}$ are arbitrary constants.

According to these considerations and taking account (2.6.b) and (2.8), the following expression appears for the <u>metric tensor</u>:

$$g_{\mu\nu}(x;L) = \eta_{\mu\nu} + G\left\{ \overset{1}{g}{}^{(m)}_{\mu\nu} + \overset{1}{g}{}^{(d)}_{\mu\nu} + \overset{1}{g}{}^{(q)}_{\mu\nu} \right\} + O(G^2) \tag{2.13}$$

with

$$\overset{1}{g}{}^{(m)}_{\mu\nu} = 2M\hat{r}^{-1}(\eta_{\mu\nu} + 2\hat{u}_\mu \hat{u}_\nu) \tag{2.14.a}$$

$$\overset{1}{g}{}^{(d)}_{\mu\nu} = 2S\,\hat{r}^{-2}(\hat{u}_\mu \hat{m}_\nu + \hat{u}_\nu \hat{m}_\mu) \tag{2.14.b}$$

$$\overset{1}{g}{}^{(q)}_{\mu\nu} = \overset{1}{g}{}^{(q_1)}_{\mu\nu} + \overset{1}{g}{}^{(q_2)}_{\mu\nu} + \overset{1}{g}{}^{(q_3)}_{\mu\nu} \tag{2.14.c}$$

where the indices (m), (d) and (q) indicate, respectively, the contributions to the metric of the monopole, dipole and quadrupole parts of the energy-momentum tensor. Furthermore, the quadrupole contribution has in turn decomposed into the respective contributions of (2.10):

$$\overset{1}{g}{}^{(q_1)}_{\mu\nu} = 2\bar{A}\,\hat{r}^{-3}(\eta_{\mu\nu} + \hat{u}_\mu\hat{u}_\nu - 3\hat{n}_\mu\hat{n}_\nu) +$$

$$+ \bar{B}\,\hat{r}^{-3}\left\{(3\hat{\omega}^2-1)(\eta_{\mu\nu}+2\hat{u}_\mu\hat{u}_\nu) + 4\hat{e}_\mu\hat{e}_\nu - 12\hat{\omega}\,\hat{n}_{(\mu}\hat{e}_{\nu)}\right\} \tag{2.15.a}$$

$$\overset{1}{g}{}^{(q_2)}_{\mu\nu} = -4\bar{C}\,\hat{r}^{-3}\left\{3\hat{\omega}\,\hat{u}_{(\mu}\hat{n}_{\nu)} - \hat{u}_{(\mu}\hat{e}_{\nu)}\right\} + 12\bar{D}\,\hat{r}^{-3}\hat{\omega}\,\hat{u}_{(\mu}\hat{m}_{\nu)} \tag{2.15.b}$$

$$\overset{1}{g}{}^{(q_3)}_{\mu\nu} = \bar{F}\,\hat{r}^{-3}(3\hat{\omega}^2-1)(\eta_{\mu\nu} + 2\hat{u}_\mu\hat{u}_\nu) \tag{2.15.c}$$

where:

$$\hat{m}_\mu \equiv \eta_{\mu\nu\rho\sigma}\hat{e}^\nu\hat{u}^\rho\hat{n}^\sigma \;,\quad \hat{n}^\sigma \equiv \hat{r}^{-1}\hat{\ell}^\sigma - \hat{u}^\sigma \;,\quad \hat{\omega} \equiv \hat{e}^\rho\hat{n}_\rho \tag{2.16}$$

B).- <u>(2-PM)-(4-POLE) APPROXIMATION</u>: By overlooking terms of order G^3 the Einstein equations may be written as follows:

$$\Box h^{\mu\nu} = -G^2\overset{1}{h}{}^{\rho\sigma}\partial_{\rho\sigma}\overset{1}{h}{}^{\mu\nu} + G^2\overset{2}{Q}{}^{\mu\nu} + 16\pi G(\overset{0}{\tau}{}^{\mu\nu} + G\overset{1}{\bar{\tau}}{}^{\mu\nu}) + O(G^3) \tag{2.17.a}$$

$$\partial_\mu h^{\mu\nu} = 0 \tag{2.17.b}$$

where $\overset{2}{Q}{}^{\mu\nu}$ is quadratic in the derivatives of $\overset{1}{h}{}^{\mu\nu}$ and where $\overset{1}{\bar{\tau}}{}^{\mu\nu}$ represents the regularized first order of $\tau^{\mu\nu}$, which identically proves to be zero[1].

We shall <u>assume</u> that $\bar{A} = \bar{B} = \bar{C} = \bar{D} = 0$ and at the same time shall overlook terms of order $\hat{\tau}^{-5}$. According to this, the integration of (2.17) by the flat retarded propagator, and after a lengtly calculation which includes a regularization process[8], leads to the following result with obvious notations:

$$\dot{u}^{\mu} = O(G^2) \quad , \quad \dot{S}^{\mu\nu} = O(G^2) \tag{2.18.a}$$

$$g_{\mu\nu} = \eta_{\mu\nu} + G\left\{ \overset{1}{g}{}^{(m)}_{\mu\nu} + \overset{1}{g}{}^{(d)}_{\mu\nu} + \overset{1}{g}{}^{(q_3)}_{\mu\nu} \right\}$$

$$+ G^2\left\{ \overset{2}{g}{}^{(m^2)}_{\mu\nu} + \overset{2}{g}{}^{(md)}_{\mu\nu} + \overset{2}{g}{}^{(d^2)}_{\mu\nu} + \overset{2}{g}{}^{(mq_3)}_{\mu\nu} + O(\hat{\tau}^{-5}) \right\} + O(G^3) \tag{2.18.b}$$

where:

$$\overset{2}{g}{}^{(m^2)}_{\mu\nu} = M^2 \hat{\tau}^{-2} (\eta_{\mu\nu} - \hat{u}_\mu \hat{u}_\nu + \hat{n}_\mu \hat{n}_\nu) \tag{2.19.a}$$

$$\overset{2}{g}{}^{(md)}_{\mu\nu} = -2MS\hat{\tau}^{-3} (\hat{u}_\mu \hat{m}_\nu + \hat{u}_\nu \hat{m}_\mu) \tag{2.19.b}$$

$$\overset{2}{g}{}^{(d^2)}_{\mu\nu} = \frac{1}{2} S^2 \hat{\tau}^{-4} \{ (9\hat{\omega}^2 - 7)\eta_{\mu\nu} + (13\hat{\omega}^2 - 7)\hat{u}_\mu \hat{u}_\nu +$$

$$+ (9\hat{\omega}^2 + 5)\hat{n}_\mu \hat{n}_\nu + 8\hat{e}_\mu \hat{e}_\nu - 20\hat{\omega}\hat{n}_{(\mu}\hat{e}_{\nu)} \} \tag{2.19.c}$$

$$\overset{2}{g}{}^{(mq_3)}_{\mu\nu} = \frac{1}{2} M\bar{F} \hat{\tau}^{-4} \{ (7\hat{\omega}^2 - 3)\eta_{\mu\nu} - (5\hat{\omega}^2 - 1)\hat{u}_\mu \hat{u}_\nu +$$

$$+ (15\hat{\omega}^2 - 1)\hat{n}_\mu \hat{n}_\nu + 2\hat{e}_\mu \hat{e}_\nu - 12\hat{\omega}\hat{n}_{(\mu}\hat{e}_{\nu)} \} \tag{2.19.d}$$

C.- <u>(3-PM)-(2-POLE) AND (4-PM)-(1-POLE) APPROXIMATIONS</u>: Following this, we shall limit ourselves to writing the results relative to the approximations of order G^3 and G^4, overlooking terms of order $\hat{\tau}^{-5}$:

$$\overset{3}{g}{}^{(m^3)}_{\mu\nu} = 2M^3 \hat{\tau}^{-3} (\hat{u}_\mu \hat{u}_\nu + \hat{n}_\mu \hat{n}_\nu) \tag{2.20.a}$$

$$\overset{3}{g}{}^{(m^2 d)}_{\mu\nu} = 2M^2 S \hat{\tau}^{-4} (\hat{u}_\mu \hat{m}_\nu + \hat{u}_\nu \hat{m}_\mu) \tag{2.20.b}$$

$$\overset{4}{g}{}^{(m^4)}_{\mu\nu} = -2M^4 \hat{\tau}^{-4} (\hat{u}_\mu \hat{u}_\nu - \hat{n}_\mu \hat{n}_\nu) \tag{2.21}$$

Concluding, we have obtained the following approximate metric:

$$g_{\mu\nu} = \eta_{\mu\nu} + G\left\{ \overset{1}{g}{}^{(m)}_{\mu\nu} + \overset{1}{g}{}^{(d)}_{\mu\nu} + \overset{1}{g}{}^{(q)}_{\mu\nu} \right\} +$$

$$+ G^2\left\{ \overset{2}{g}{}^{(m^2)}_{\mu\nu} + \overset{2}{g}{}^{(md)}_{\mu\nu} + \overset{2}{g}{}^{(d^2)}_{\mu\nu} + \overset{2}{g}{}^{(mq_3)}_{\mu\nu} \right\} +$$

$$+ G^3\left\{ \overset{3}{g}{}^{(m^3)}_{\mu\nu} + \overset{3}{g}{}^{(m^2d)}_{\mu\nu} \right\} + G^4\, \overset{4}{g}{}^{(m^4)}_{\mu\nu} + O(\hat{r}^{-5}) \quad . \qquad (2.22)$$

However, it should be pointed out that we have not covered all the approximation in $\hat{r}^{-4}$, since the inclusion of a possible octopole part in the energy-momentum tensor (2.1) would give rise to a term of the kind $\overset{1}{g}{}^{(o)}_{\mu\nu}$ which would be of order $\hat{r}^{-4}$.

To finish, it should be noted that the results (2.6.b) and (2.18.a) are maintained at the orders studied.

3.- WEYL AND KERR METRICS IN APPROXIMATE HARMONIC COORDINATES

A) Weyl's metrics are static axysimmetric vacuum solutions of the Einstein equations. These metrics are written in spherical coordinates $\{t, R, \Theta, \phi\}$ associated to the cylindrical coordinates of Weyl, as follows:

$$ds^2 = -e^{2U} dt^2 + e^{-2U}\left\{ e^{2V}\left(dR^2 + R^2 d\Theta^2\right) + R^2 \sin^2\Theta \, d\phi^2 \right\} \qquad (3.1)$$

with

$$U(R,\Theta) \equiv \sum_{\ell=0}^{\infty} \frac{a_\ell}{R^{\ell+1}} P_\ell(\cos\Theta) \qquad (3.2.a)$$

$$V(R,\Theta) \equiv -\sum_{\ell,k=0}^{\infty} \frac{(\ell+1)(k+1)}{\ell+k+2} \cdot \frac{a_\ell a_k}{R^{\ell+k+2}} \left(P_\ell P_k - P_{\ell+1} P_{k+1}\right) \qquad (3.2.b)$$

where a_ℓ are arbitrary parameters and P_ℓ the Legendre polynomials.

When we wish to write (3.1) in harmonic coordinates $\{t, x, y, z\}$ we find that t is already harmonic and ϕ is an associated azimuthal coordinate. A short calculation reveals then that the problem is reduced to finding two functions $f(R,\Theta) \equiv z$ and $g(R,\Theta) \equiv \sqrt{x^2+y^2}$ which are solutions of the following differential equations:

$$\partial_R\left(R^2 \sin\Theta \cdot \partial_R f\right) + \partial_\Theta\left(\sin\Theta \cdot \partial_\Theta f\right) = 0 \qquad (3.3.a)$$

$$\partial_R\left(R^2 \sin\Theta \cdot \partial_R g\right) + \partial_\Theta\left(\sin\Theta \cdot \partial_\Theta g\right) = \frac{1}{\sin\Theta} e^{2V} \cdot g \qquad (3.3.b)$$

which have the following, respectively exact and approximate, acceptable solutions[9]:

$$f(R,\Theta) = h + R\left\{\Omega + \frac{C_0}{R^2} + \Omega\frac{C_1}{R^3} + \frac{1}{2}(3\Omega^2-1)\frac{C_2}{R^4}\right\} \tag{3.4.a}$$

$$g(R,\Theta) = R\sin\Theta\left\{1 + \frac{1}{2}a_0^2\frac{1}{R^2} + (\beta + a_0 a_1 \Omega)\frac{1}{R^3} + \left[\gamma\Omega + \frac{1}{24}a_0^4(2\Omega^2-1) + \right.\right.$$

$$\left.\left. + \frac{1}{4}a_1^2(3\Omega^2-1) + \frac{1}{4}a_0 a_2(5\Omega^2-1)\right]\frac{1}{R^4} + O(R^{-5})\right. \tag{3.4.b}$$

where $\Omega \equiv \cos\Theta$ and h, C_0, C_1, C_2, β and γ are arbitrary constants.

By taking $h = -a_1/a_0$, which avoids the appearance of a fictitious mass dipole moment, and for example $C_0 = C_1 = C_2 = \beta = \gamma = 0$ [10], we have that after a lengthy calculation the metric may be written in these harmonic coordinates as follows:

$$g_{\mu\nu} = A\eta_{\mu\nu} + Bu_\mu u_\nu + Cn_\mu n_\nu + De_\mu e_\nu + 2E\,n_{(\mu}e_{\nu)} \tag{3.5}$$

where the "vectors" u_μ, n_μ, and e_μ are defined as follows:

$$(u^\mu) \equiv (1,0,0,0) \quad , \quad u_\mu \equiv \eta_{\mu\rho}u^\rho \tag{3.6.a}$$

$$n_\mu \equiv r^{-1}(\eta_{\mu\rho} + u_\mu u_\rho)x^\rho \tag{3.6.b}$$

$$(e^\mu) \equiv (0,0,0,1) \quad , \quad e_\mu \equiv \eta_{\mu\rho}e^\rho \tag{3.6.c}$$

and where the coefficients A,B,C,D,E have the following expressions as a function of the polar coordinates $\{r,\theta\}$ associated to the harmonic coordinates:

$$A = 1 - 2a_0\frac{1}{r} + a_0^2\frac{1}{r^2} - \left(a_2^* - \frac{1}{3}a_0^3\right)(3\omega^2-1)\frac{1}{r^3} +$$

$$+ \frac{1}{2}a_0\left(a_2^* - \frac{1}{3}a_0^3\right)(7\omega^2-3)\frac{1}{r^4} - a_3^*\omega(5\omega^2-3)\frac{1}{r^4} + O(r^{-5}) \tag{3.7.a}$$

$$B = -4a_0\frac{1}{r} - a_0^2\frac{1}{r^2} - 2a_0^3\frac{1}{r^3} - 2\left(a_2^* - \frac{1}{3}a_0^3\right)(3\omega^2-1)\frac{1}{r^3} -$$

$$-2a_0^4\frac{1}{r^4} - \frac{1}{2}a_0\left(a_2^* - \frac{1}{3}a_0^3\right)(5\omega^2-1)\frac{1}{r^4} - a_3^*\omega(5\omega^2-3)\frac{1}{r^4} + O(r^{-5}) \tag{3.7.b}$$

$$C = a_0^2\frac{1}{r^2} - 2a_0^3\frac{1}{r^3} + 2a_0^4\frac{1}{r^4} + \frac{1}{2}a_0\left(a_2^* - \frac{1}{3}a_0^3\right)(15\omega^2-1)\frac{1}{r^4} + O(r^{-5}) \tag{3.7.c}$$

$$D = a_0\left(a_2^* - \frac{1}{3}a_0^3\right)\frac{1}{r^4} + O(r^{-5}) \tag{3.7.d}$$

$$E = -3a_0\left(a_2^* - \frac{1}{3}a_0^3\right)\omega\frac{1}{r^4} + O(r^{-5}) \tag{3.7.e}$$

where:

$$\omega \equiv n^k e_\mu \equiv \cos\theta \quad , \quad a_2^* \equiv a_2 - \frac{a_1^2}{a_0} \quad , \quad a_3^* \equiv a_3 - 3\frac{a_1 a_2}{a_0} + 2\frac{a_1^3}{a_0^2} \quad . \tag{3.8}$$

Comparing these results with those of the previous Section and taking into account specifically the equations of motion up to the order we are dealing with (relationships of the (2.18.a) kind), strict coherence may be observed as long as we have:

$$GM = -a_0 \quad , \quad S = 0 \quad , \quad G\bar{F} = -\left(a_2^* - \frac{1}{3}a_0^3\right) \quad . \tag{3.9}$$

These results show, at the order considered, the characteristics of a possible source of the Weyl metrics in terms of its parameters a_ℓ. In particular, the quadrupole moment, does not coincide with the corresponding "newtonian" of the function U, an aspect which has already been pointed out by Geroch[11]. We should also note that the terms of (3.7) which contain a_3^* do not have homologues in the results of the previous Section, due to the absence of an octopole moment in the source model.

B) The Kerr metric (stationary axysimmetric vacuum solution of Einstein's field equations) is written in the Boyer-Lindquist coordinates $\{t, R, \Theta, \phi\}$ as follows:

$$ds^2 = -\left(1 - \frac{2\alpha R}{\rho^2}\right)dt^2 + \frac{\rho^2}{\Delta}dR^2 + \rho^2 d\Theta^2 +$$

$$+\left(R^2 + a^2 + \frac{2\alpha R}{\rho^2}a^2\sin^2\Theta\right)\sin^2\Theta\, d\phi^2 + \frac{4\alpha R}{\rho^2}a\sin^2\Theta\, dt\, d\phi \tag{3.10}$$

where α and a are arbitrary parameters and where:

$$\rho^2 \equiv R^2 + a^2\cos^2\Theta \quad , \quad \Delta \equiv R^2 + a^2 - 2\alpha R \tag{3.11}$$

By following an almost identical process to that of the Weyl metrics and using the same notations, we find that the Kerr metric is written in the following way in a certain system of harmonic coordinates:

$$g_{\mu\nu} = A\eta_{\mu\nu} + Bu_\mu u_\nu + Cn_\mu n_\nu + De_\mu e_\nu + 2E\, n_{(\mu}e_{\nu)} + 2F\, u_{(\mu}m_{\nu)} \qquad (3.12)$$

with:

$$m_\mu \equiv \eta_{\mu\nu\rho\sigma} e^\nu u^\rho n^\sigma \quad , \quad n^\sigma \equiv \eta^{\sigma\rho} n_\rho \qquad (3.13)$$

and where now:

$$A = 1 + 2\alpha\frac{1}{r} + \alpha^2\frac{1}{r^2} - \alpha a^2(3\omega^2-1)\frac{1}{r^3} + \alpha^2 a^2(\omega^2-2)\frac{1}{r^4} + O(r^{-5}) \qquad (3.14.a)$$

$$B = 4\alpha\frac{1}{r} - \alpha^2\frac{1}{r^2} + 2\alpha^3\frac{1}{r^3} - 2\alpha a^2(3\omega^2-1)\frac{1}{r^3} - 2\alpha^4\frac{1}{r^4} + \alpha^2 a^2(9\omega^2-4)\frac{1}{r^4} + O(r^{-5}) \qquad (3.14.b)$$

$$C = \alpha^2\frac{1}{r^2} + 2\alpha^3\frac{1}{r^3} + 2\alpha^4\frac{1}{r^4} - 3\alpha^2 a^2(\omega^2-1)\frac{1}{r^4} + O(r^{-5}) \qquad (3.14.c)$$

$$D = 3\alpha^2 a^2\frac{1}{r^4} + O(r^{-5}) \quad , \quad E = -2\alpha^2 a^2\omega\frac{1}{r^4} + O(r^{-5}) \qquad (3.14.d)$$

$$F = -2\alpha a\frac{1}{r^2} + 2\alpha^2 a\frac{1}{r^3} + \alpha a^3(5\omega^2-1)\frac{1}{r^4} - 2\alpha^3 a\frac{1}{r^4} + O(r^{-5}) \qquad (3.14.e)$$

Once again, strict coherence with the results of the previous Section may be observed if we impose:

$$GM = \alpha \quad , \quad \frac{S}{M} = -a \quad , \quad \frac{\bar{F}}{M} = -a^2 \ . \qquad (3.15)$$

These are relationships which yield the already known interpretation. We should note that the term of (3.14.e) which contains αa^3 does not have a homologue in the results of the previous Section due to the absence of an octopole component in the energy-momentum tensor, as it occurred in the Weyl metrics case.

4.- CONCLUSIONS

A procedure has been described for obtaining <u>approximate</u> vacuum solutions of the Einstein field equations. In order to do so, an energy-momentum distribution tensor has been used with support on a time-like curve and having some definitive multipole structure. The reliability of the method has been perfectly demonstrated since on imposing stationarity and axial symmetry, it allows us to reproduce the Weyl and Kerr metrics at the order considered. Furthermore, this reliability allows us to adventure a refined interpretation of these metrics in terms of extended sources, which should be lacking, according to the results obtained, in stress quadrupole moment and flow quadrupole moment. However, it is necessary to point out that this latter statement should be revised in the light of an analysis of harmonic coordinate transformations. This analysis will appear in a forthcoming work in which certain aspects of the present one are to be broadened.

REFERENCES

1 .- J. Martín, E. Ruiz & M.J.Senosiaín: "Actas de los E.R.E. 1983"; Universitat de Palma de Mallorca, Spain (1984). "Multipoles particles and the Kerr metric"; submited to Phys. Rev. D.

2 .- The linear signature of space-time should be taken as equal to +2 and the speed of light in vacuum as equal to one. The Greek indices run from 0 to 3, where the first refers to time.

3 .- A straightforward generalization of the energy-momentum tensor of a pole-dipole-particle, W. Tulczyjew. Acta Phys. Polon. 18, 393 (1959) .

4 .- C.W. Misner, K.S. Thorne and J.A. Wheeler, "Gravitation" (Freeman, San Francisco, 1973).

5 .- L. Landau and L. Lifshitz, "Theorie du Champ" (Mir, Moscou, 1966).

6 .- This condition defines center of mass world-line of an extended body W.G. Dixon, in "Isolated Gravitating Systems in General Relativity", proceedings of the International School of Physics "Enrico Fermi", ed. J. Ehlers (North Holland, Amsterdam, 1979) .

7 .- Similar to what is done in J. Ehlers and E. Rudolph, Gen. Rel. Grav. 8, 197 (1977).

8 .- We use the regularization procedure described in L. Bel, T. Damour, N. Deruelle, J. Ibañez and J. Martín, Gen.Rel. Grav. 13, 963 (1981).

9 .- By acceptable, we understand that they do not have singularities except on R = 0 and, moreover, that $\lim_{R\to\infty} f(R,\Theta) = R\cos\Theta$ and $\lim_{R\to\infty} g(R,\Theta) = R\sin\Theta$.

10 .- A different choice from this one will be analysed in a forthcoming work.

11 .- R. Geroch, J. Math. Phys. 11, 2580 (1970).

UNCONSTRAINED DEGREES OF FREEDOM OF GRAVITATIONAL FIELD AND THE POSITIVITY OF GRAVITATIONAL ENERGY

Jerzy Kijowski

Institute for Theoretical Physics
Polish Academy of Sciences
Aleja Lotników 32/46
02-668 WARSAW, Poland

ABSTRACT: The space of Cauchy data for Einstein equations is effectively reduced with respect to Gauss-Codazzi constraints. The mixed initial value - boundary value problem is analysed. The role of boundary degrees of freedom is discussed. The energy-positivity is obtained as a simple consequence of the construction used.

1. Symplectic structure of the space of Cauchy data

The goal of this paper is to present the construction of unconstrained degrees of freedom of gravitational field together with the analysis of the notion of gravitational energy as a hamiltonian of the system. Traditionally, the "canonical" formalisms used in General Relativity were based on: 1)integration by parts and 2) the hope, that under sufficiently strong asymptotic conditions all the "inconvenient" boundary integrals will wanish. Therefore, all the results were valid "modulo surface integrals" (e.g. we have been taught that the hamiltonian of the gravitational field is equal to zero - modulo surface integrals). Recently, the important role of surface phenomena has been stressed by many authors (see e.g. [2] and [8]). A coherent description of these phenomena can be given in terms of the theory of symplectic relations as proposed by W.M.Tulczyjew (see e.g. [9] and [7]). In this theory both volume integrals and boundary integrals have equally legal status.

The analysis of the gravitational field in terms of the theory of symplectic relations leads to the so called affine formulation of General Relativity (see e.g. [5] , [7] and [4]). In the paper [6], based on the affine formulation, the dynamics of the gravitational field within a finite space-time region Σ with boundary $\partial\Sigma$ was analysed (at the end of our considerations $\partial\Sigma$ can be shifted to space-infinity or to null-infinity and the limits of the corresponding boundary terms can be calcutated). The present paper is a straightforward continuation of [6].

Let Σ be a compact, smooth 3-dimensional manifold with boundary $\partial\Sigma$. In the present paper we limit ourselves to the simplest topological situation i.e. we assume that Σ is diffeomorphic to the 3-disc $K(0,R) \subset \mathbb{R}^3$. Let $\Lambda = \partial\Sigma \times \mathbb{R}^1$ and let $\Sigma_t = \Sigma \times \{t\}$.

The space $V = \Sigma \times \mathbb{R}^1$ will be the interior of our space-time tube and the boundary $\Lambda = \partial V$ will be a 1-time-like and 2-space-like surface in our space-time. Points of Σ are observers, each of them having its own method of moving in space-time (e.g. each of them is equipped with a pre-programmed jet-engine and a clock). The change of coordinates in V ("passive gauge") is irrelevant since we can give at the very beginning the names to all the observers and to equip them with clocs. Let us therefore fix a coordinate chart (x^μ) on V such that $x^0 = t$, $x^3 = \log r$, where r is the radial coordinate (i.e. the value of x^3 lies within the half-line $]-\infty, \log R\,]$), and (x^1, x^2) is a coordinate chart on $\partial\Sigma = S^2$ (e.g. spherical angles ϕ and θ). We use the following convention for indices: greek indices μ, ν run from 0 to 3; k,l are coordinates on Σ and run from 1 to 3; i,j are coordinates on Λ and run from 0 to 2; A, B are coordinates on $\partial\Sigma$ and run from 1 to 2. We have $\Sigma_t = \{x^0 = t\}$; $\Lambda = \{x^3 = \log R\}$.

We will consider pseudo-riemannian geometries on V. Intuitively, transformations of the geometry due to the simple "reorganization" of the observation (change of clocks and of programs of the jet-engines) should be considered as "gauge transformations". However, we are not free to decide which transformations can be called gauge. The gravitational field inside V is a generalized (constrained) hamiltonian system and gauge transformations are precisely those which correspond to the degeneracy of the symplectic form on the constraint manifold. The usual description of the symplectic structure is given by the so called A.D.M. symplectic form (see [1]):

$$\omega_{ADM} = \frac{1}{2\kappa} \int_\Sigma dP^{kl} \wedge dg_{kl} \tag{1}$$

where $\kappa = 8\pi G$ is the gravitational constant, g_{kl} is a Riemannian 3-metric on Σ and

$$P^{kl} = \sqrt{\det g}\,(\tilde{g}^{kl}\,\mathrm{Tr}\,K - K^{kl}) \tag{2}$$

By $\tilde{g}$ we denote the contravariant metric inverse to g and K is the extrinsic curvature of Σ with respect to the 4-metric $g_{\mu\nu}$ which we are looking for. The phase space $\mathcal{P}_\Sigma$ is therefore described by the two objects g and P (12 functions on Σ). Real physical situations correspond to fields which fulfill 4 constraint conditions induced by Gauss-Codazzi equations:

$$P^{kl}{}_{|l} = 0 \tag{3}$$

(the covariant divergence of P with respect to g) which we call the vector-constraint, and the following "scalar constraint":

$$(\det g)\,\overset{3}{R}(g) - P^{k}{}_{l}P^{l}{}_{k} + \frac{1}{2}\,(P^{k}{}_{k})^{2} = 0 \tag{4}$$

The form ω_{ADM} is degenerate on the constraint space $\bar{P}_{\Sigma} \subset P_{\Sigma}$. Gauge transformations are generated by vector fields tangent to this degeneracy. There is a very disappointing result about the A.D.M. form (1) : not all the transformations of $\bar{P}_{\Sigma}$ which can be implemented by boundary preserving diffeomorphisms of V belong to this class. A simple computation shows that the value of the form (1) is modified by a boundary term when subjected to the one-parameter group of diffeomorphisms generated by a vector field X on V. The boundary term arises due to integration by parts and is composed of two terms: one term is proportional to the value of X on $\partial V = \Lambda$ and the other term is proportional to derivatives of X on Λ. For Λ - preserving diffeomorphisms we have $X|\Lambda=0$ but the second term does not vanish in general (the symplectic structure is invariant with respect to gauge transformations. This is true only for transformations generated by vector fields X which vanish on Λ together with first derivatives).

Using the theory of symplectic relations it was proved in [6] that the correct sympletic structure in the space of Cauchy data differs from the A.D.M. structure by a surface term and is equal to

$$\omega = \frac{1}{2\kappa}\int_{\Sigma} dP^{kl} \wedge dg_{kl} + \frac{1}{\kappa}\int_{\partial\Sigma} d\lambda \wedge d\alpha \tag{5}$$

where

$$\lambda = \sqrt{\det g_{AB}} \tag{6}$$

is the 2-dimensional volume density on $\partial\Sigma$ and

$$\alpha = \operatorname{arsh} \frac{g^{03}}{\sqrt{|g^{00}|g^{33}}} \tag{7}$$

is the "hyperbolic angle" between the hypersurfaces Σ and Λ. Now, the phase space is the direct sum

$$P = P_{\Sigma} \oplus P_{\partial\Sigma} \tag{8}$$

where $P_{\partial\Sigma}$ is described by "2 functions on the boundary" (actually λ is a scalar density on $\partial\Sigma$ and α is a scalar function on $\partial\Sigma$). The constraint manifold $\bar{P} \subset P$ is defined by those objects which fulfill not only (3) and (4) but also the compatibility condition between λ and the restriction of g to $\partial\Sigma$. The following theorem can be proved:

Theorem. Gauge transformations for the pair $(\bar{P}, \omega)$ are precisely the transformations which can be implemented by Λ - preserving diffeomorphisms.

The theorem is a consequence of the fact that the modification of the "volume term" in (5) due to the derivatives of X on Λ is cancelled by the corresponding modification of the boundary term.

2. Gauge conditions

The reduced phase space $\tilde{P} = \bar{P}/\sim$ is the quotient space, where $\sim$ denotes the gauge-equivalence relation. To describe effectively the quotient we impose 4 gauge conditions which enable us to pick up a representant within each gauge-equivalence class. The conditions are:

$$P^{kl} g_{kl} = 0 \tag{9}$$

(i.e. Σ is a maximal surface) and

$$g_{kl} = f \, \gamma_{kl} \tag{10}$$

where γ is the metric which satisfies 4 conditions:

$$\gamma_{33} = 1 \; ; \qquad \gamma_{3A} = 0 \; ; \qquad \sqrt{\det \gamma_{AB}} = \overset{\circ}{\lambda} \, . \tag{11}$$

Here $\overset{\circ}{\lambda}$ the standard 2-volume density on the unit sphere $S^2 = \partial K(0,1) \subset \mathbb{R}^3$. If (x^1, x^2) are spherical angles ϕ and θ then

$$\overset{\circ}{\lambda} = \sin\theta \tag{12}$$

However, there is no global coordinate chart on $\partial\Sigma \cong S^2$ and therefore we keep the condition (11) in the form which does not depend of the particular choice of coordinates on Σ. As an example of the metric which has a representation satisfying (10)-(11) we can take the flat metric on $\mathbb{R}^3$ which can be written in the form

$$ds^2 = r^2 \left[(d \log r)^2 + d\Omega^2 \right] \tag{13}$$

where $d\Omega^2$ is the standard metric on S^2. Thus

$$f = r^2 = \exp(2x^3) \tag{14}$$

in this case. The possibility of finding for a given metric g the representant fulfilling (10)-(11) is equivalent to the following 2-nd order equation for the function $\rho = x^3$:

$$\left(\tilde{g}^{kl} + \frac{\nabla^k \rho}{|\nabla\rho|} \frac{\nabla^l \rho}{|\nabla\rho|} \right) \nabla_k \nabla_l \rho = 0 \tag{15}$$

The solution has to satisfy the boundary conditions: $\rho|\partial\Sigma = \log R$ and $\rho \to -\infty$ at a given point $\underset{o}{x}$ inside Σ. The choice of the point $\underset{o}{x}$ is also a gauge condition. As an example of the function which satisfies (15) we can take $\rho=\log r$ in the flat space.

For a given solution ρ of (15) we "extend" coordinates (x^1,x^2) from $\partial\Sigma$ to the interior of Σ in such a way that the vector field ∂_3 is orthogonal to the family of surfaces $\{x^3=\text{const}\}$. It can be easily checked that the coordinate chart obtained this way satisfies the conditions (10)-(11). The problem wheather or not equation (15) together with boundary conditions admits always a (unique) solution has not yet been fully solved. There are however partial results which are very favorable for the positive answer.

The equation (15) is written by help of a metric g. However, it is invariant with respect to conformal deformations of g. Therefore, only the conformal structure implied by g is involved. The easiest way to verify this observation is to derive (15) from the variational principle

$$\frac{\delta L}{\delta \rho} = 0 \tag{16}$$

where the lagrangean

$$L = \sqrt{\det g}\ [\ \tilde{g}^{kl}(\partial_k\rho)(\partial_l\rho)\ ]^{3/2} \tag{17}$$

is manifestly invariant with respect to conformal modifications of the metric. The equation (15) is therefore the conformal generalization of the Laplace equation and its solutions will be called conformally-harmonic functions. Corresponding 2-surfaces $\{\rho=\text{const}\}$ will be called conformally-harmonic surfaces.

Our gauge conditions correspond thus to the 3+1 maximal slicing of V and morover to the 2+1 slicing of each Σ_t which is conformally-harmonic:

$$\Sigma - \{\underset{o}{x}\} = \chi \times\ ]-\infty, \log R\ ] \tag{18}$$

where $\chi \cong \partial\Sigma \cong S^2$.

3. Reduction

The following formula is easy to check

$$dP^{kl} \wedge dg_{kl} = d(fP^{kl}) \wedge d\gamma_{kl} - df \wedge d(P^{kl}\gamma_{kl}) \tag{19}$$

Therefore, due to (9)-(11) we have:

$$dP^{kl} \wedge dg_{kl} = d(fP^{AB}) \wedge d\gamma_{AB} \tag{20}$$

We define the following object:

$$p^{AB} = fP^{AB} + \frac{1}{2}\gamma^{AB}P^3{}_3 \tag{21}$$

where γ^{AB} is the contravariant 2-metric on χ inverse to γ_{AB}. We have

$$d(\frac{1}{2}\gamma^{AB}P^3{}_3)\wedge d\gamma_{AB} = dP^3{}_3\wedge\frac{1}{2}\gamma^{AB}d\gamma_{AB} + \frac{1}{2}P^3{}_3\, d\gamma^{AB}\wedge d\gamma_{AB} = 0 \tag{22}$$

because

$$\frac{1}{2}\gamma^{AB}d\gamma_{AB} = d\log\sqrt{\det\gamma} = d\log\overset{o}{\lambda} = 0 \tag{23}$$

due to condition (11). Finally our symplectic structure (5) reduces to

$$\omega = \frac{1}{2\kappa}\int_\Sigma dp^{AB}\wedge d\gamma_{AB} + \frac{1}{\kappa}\int_{\partial\Sigma} d\lambda\wedge d\alpha \tag{24}$$

The number of independent components of γ_{AB} is 2 (because of the symmetry and the unimodularity condition (11)). The number of independent components of p^{AB} is also 2 since equation (9) implies: $p^{AB}\gamma_{AB} = g_{AB}P^{AB} + P^3{}_3 = 0$. The object γ_{AB} can be called a unimodular metric on χ. The reduced phase space $\tilde{P}$ is thus the collection of "2 degrees of freedom per point in Σ" (unimodular metric and its momentum) and "1 degree of freedom per point in $\partial\Sigma$" (the hyperbolic angle α and its momentum λ). For each collection $(p^{AB},\gamma_{AB},\alpha,\lambda)$ of reduced Cauchy data we can reconstruct the complete data $(P^{kl},g_{kl},\alpha,\lambda)$ by solving the four constraint equations. It can be easily verified, that due to our 2+1 -conformally-harmonic decomposition of Σ the vector constraint (3) reduces to a single second order linear elliptic equation

$$q,_{33} + \frac{1}{2}\overset{2}{\Delta} q = \text{function of } p^{AB} \text{ and } \gamma_{AB} \tag{25}$$

for the unknown function

$$q(x^1,x^2,x^3) = \frac{1}{\overset{o}{\lambda}}\int_{-\infty}^{x^3} P^3{}_3(x^1,x^2,\rho)\, d\rho \tag{26}$$

together with the condition:

$$P^3{}_A = \overset{o}{\lambda}\, q,_A + \text{function of } p^{AB} \text{ and } \gamma_{AB} \tag{27}$$

(symbol $\overset{2}{\Delta}$ denotes the 2-dimensional laplacian defined by γ). To solve uniquely the linear equation (25) we do not need any boundary condition for q except the continuity condition for P^{kl} at the center point $\underset{o}{x}$. This condition implies that $q \to 0$ exponentially for $\rho \to -\infty$ and it is sufficient to find uniquely the solution. The only lacking component of the complete Cauchy data is now the conformal factor f which can be found by solving the scalar constraint (4). As it is known (see e.g. [3]) it is a second order, elliptic equation for the function f. To find uniquely the solution we use the value $\lambda = \overset{\circ}{\lambda} f$ on Σ which gives us the Dirichlet data for the equation.

4. Time evolution

In the paper [6] the fundamental formula has been proved:

$$- dH = \frac{1}{2\kappa} \int_{\Sigma} (\dot{P}^{kl} \, dg_{kl} - \dot{g}_{kl} \, dP^{kl}) + \frac{1}{\kappa} \int_{\partial\Sigma} (\dot{\lambda} \, d\alpha - \dot{\alpha} \, d\lambda) + \frac{1}{2\kappa} \int_{\partial\Sigma} (\gamma \, dQ^{00} + \lambda \, ds - 2Q^{0}{}_{A} \, dn^{A} - \sigma^{AB} \, d\gamma_{AB}) \tag{28}$$

where the hamiltonian H is given by the formula

$$H = \frac{1}{2\kappa} \int_{\partial\Sigma} (Q^{00}\gamma + 2Q^{0}{}_{A} \, n^{A} - s \, \lambda) \tag{29}$$

The following notation has been used: dots denote ∂_0 (time derivative); Q^{ij} is the A.D.M. "momentum" for the hypersurface Λ (Q is defined in terms of extrinsic curvature of Λ in the same way as P was defined in terms of extrinsic curvature of Σ);

$$n^{A} = \tilde{\tilde{g}}^{AB} g_{0B} \tag{30}$$

where $\tilde{\tilde{g}}$ is the inverse of the 2-metric g_{AB}

$$\gamma = g_{00} - n^{A} n^{B} g_{AB} \tag{31}$$

$$s = \frac{1}{\lambda} g_{AB} \, S^{AB} \tag{32}$$

where S is the orthogonal projection of Q onto $\partial\Sigma$ i.e.:

$$S^{AB} = Q^{AB} + Q^{0A} n^{B} + Q^{0B} n^{A} + Q^{00} n^{A} n^{B} \tag{33}$$

and finally σ is given by the traceless part of S:

$$\sigma^{AB} = \frac{\lambda}{\overset{\circ}{\lambda}} (S^{AB} - \frac{1}{2} \tilde{\tilde{g}}^{AB} s \, \lambda) \tag{34}$$

The formula (28) is an analog of a definition of a hamiltonian vector field in classical mechanics, written in terms of symplectic relations:

$$- d \, H \, (p,q) = \dot{p} \, dq - \dot{q} \, dp \tag{35}$$

which is equivalent to

$$\dot{p} = - \frac{\partial H}{\partial q} \quad ; \quad \dot{q} = \frac{\partial H}{\partial p} \tag{36}$$

In field theory an additional non-evolutional boundary term is always present. This corresponds to the possibility of controlling not only Cauchy data but also boundary value of the field. For example, the corresponding formula for the scalar field theory would be:

$$- dH(\pi,\phi) = \int_{\Sigma} (\dot{\pi}\, d\phi - \dot{\phi}\, d\pi) + \int_{\partial\Sigma} p^3\, d\phi \tag{37}$$

where $\pi = p^0$ and p^3 are components of the momentum p^μ canonically conjugate to the scalar field ϕ (see [6], [7]). The formulae like (28), (35) or (37) have a precise mathematical meaning and can be written without coordinates (see e.g. [7]).

Intuitively, formula (37) gives rise to the hamiltonian system (π, ϕ) provided the last term is "killed". This can be done by imposing the boundary value of ϕ on Λ, i.e. by restricting the class of the unknown fields to the subclass of those which fulfill the boundary conditions. Within this subclass we have $d\phi = 0$. This enables us (due to integration by parts) to write the formulae

$$\dot{\pi} = -\frac{\delta H}{\delta \phi} \quad ; \quad \dot{\phi} = \frac{\delta H}{\delta \pi} \tag{38}$$

analogous to (36), i.e. to translate the field-evolution problem into the language of hamiltonian systems.

Physically, the choice of the boundary conditions means that we close the system composed of the field within Σ by imposing at the boundary $\partial\Sigma$ an insulation from the outside world. Mathematically, this means that we choose an appropriate functional space (generally a Sobolev space) of functions satisfying boundary conditions as the configuration space of our system. The phase space will be defined as its cotangent bundle. This way we obtain a rigorously defined hamiltonian system.

Let us come back to the problem of gravitational field. The formula (28) implies the following procedure of solving Einstein equations:

i) "Kill" the last term in (28) by imposing 6 boundary conditions on Λ. The particular choice of the first 4 of them, i.e. Q^{00}, s, n^A, is the choice of the tube Λ in space-time (the choice of the family of boundary observers). The evolution will now be referred to this family. For any such a boundary reference we fix also the remaining two boundary conditions i.e. $\gamma_{AB}|\partial\Sigma$. The regularity conditions for the metric g at the center $\underset{\circ}{x}$ give us also boundary conditions for γ at the opposite end of the x^3-axis : $\partial_3\gamma_{AB}$ has to vanish exponentially for $x^3 \to -\infty$. These boundary conditions have to be included into the definition of the functional space which will be our configuration space (the phase space $\tilde{P}$ will be its cotangent bundle).

ii) Choose an element $(p^{AB}, \gamma_{AB}, \lambda, \alpha)$ in $\tilde{P}$ as an initial value of the field.

iii) The hamiltonian (29) has to be expressed in terms of legal variables. We use for this goal the identities

$$Q^0{}_A = P^3{}_A + \lambda\, \alpha,_A \tag{39}$$

and, for q=sh α,

$$\partial_3 \lambda = \frac{1}{\sqrt{1+q^2}} \left(q\, P^3{}_3 + \sqrt{g_{33}}\, \sqrt{|\gamma|}\, Q^{00} \right) \tag{40}$$

which the reader can easily verify. Finally, the first (most difficult one) term of the hamiltonian (29) becomes:

$$Q^{00}\gamma = -\frac{1}{f Q^{00}} \left[\mathring{\lambda} \sqrt{1+\mathrm{sh}^2\alpha}\; \partial_3 f - P^3{}_3\, \mathrm{sh}\alpha \right]^2 \tag{41}$$

The value of the hamiltonian depends on Cauchy data $(p^{AB}, \gamma_{AB}, \lambda, \alpha)$ and on control parameters (Q^{00}, s, n^A) at the boundary. If the latter are not constant in time the hamiltonian is explicitely time-dependent. The non-conservation of the energy is due to the interaction of our system with the outside world.

iv) Solve the initial value problem for the unconstrained hamiltonian system $(\tilde{P}, \omega)$ with the hamiltonian found above. This way we find the values of p^{AB}, γ_{AB}, λ and α on different surfaces Σ_t (possibly within a "thin sandwich $|t| < \varepsilon$ only).

v) Solve constraint equations (25) and (4) on each Σ_t separately. Use the value of λ as the boundary condition for f. This way you know the complete Cauchy data on each Σ_t.

vi) To find lacking four components of the 4-metric $g_{\mu\nu}$ use gauge-conservation conditions. The reader can easily check that the condition $\partial_0 (P^{kl} g_{kl}) = 0$ is equivalent to the elliptic equation for the lapse function $N = |g^{00}|^{-1/2}$:

$$\Delta N = NR \tag{42}$$

where R is the scalar curvature of the 3-metric g. The boundary value of N on $\partial\Sigma_t$ (necessary to find the solution) can be easily calculated from equation (41).

Due to our 2+1 conformally harmonic decomposition of Σ the conservation conditions for the gauge (10)-(11) reduce to the single elliptic linear equation for the radial component N^3 of the shift vector $N^k = \tilde{g}^{kl} g_{0l}$:

$$(f N^3{}_{,3})_{,3} + \frac{1}{2} \mathring{\Delta} N^3 = \text{function of } P^{kl} \text{ and } g_{kl}. \tag{43}$$

plus the condition

$$N^A{}_{,3} = -\gamma^{AB} N^3{}_{,B} + \text{function of } P^{kl} \text{ and } g_{kl}. \tag{44}$$

To solve uniquely (43) we use the boundary value for N^3 on $\partial\Sigma_t$ which is given by the already known function α. The value of N^3 for $x^3 \to -\infty$ is also necessary. The choice of the latter means that we prescribe the space-time direction of the trajectory of the center $\overset{o}{x}$. The simplest way to do it is to decide that $\overset{o}{x}$ moves orthogonally with respect to Σ_t. This corresponds to homogeneous (vanishing) boundary condition at $-\infty$. However, all other choices are equally possible.

To integrate condition (44) from the boundary towards the center we use the boundary value of the 2-dimensional vector field $N^A = n^A$ which we fixed on the whole Λ at the beginning. This completes the information about $g_{\mu\nu}$.

5. Energy positivity

Due to our 2+1 conformally harmonic decomposition of Σ the scalar constraint equation (4) can be rewritten in the following way:

$$- \frac{1}{2\kappa} \partial_r (r^2 \overset{o}{\lambda} \phi ,_r) - \frac{1}{2\kappa} \partial_A (\overset{o}{\lambda} \gamma^{AB} \phi ,_B) + \frac{\overset{o}{\lambda}}{4\kappa} \overset{2}{R} (\gamma) - \frac{\overset{o}{\lambda}}{2\kappa} =$$

$$= \frac{1}{4\kappa f^2} P^k{}_l P^l{}_k + \frac{\overset{o}{\lambda}}{8\kappa} \gamma^{kl} \phi ,_k \phi ,_l + \frac{\overset{o}{\lambda}}{16\kappa} \gamma_{AC,3} \gamma_{BD,3} \gamma^{AB} \gamma^{CD} \tag{45}$$

where $\phi = \log \frac{f}{r^2}$; $r = \exp x^3$ and $\overset{2}{R}(\gamma)$ is the scalar curvature of the 2-metric γ. It turns out, that in the asymptotically flat case the total energy (i.e. the limit of the expression (29) for $R \to \infty$) equals:

$$M = \lim_{R \to \infty} (- \frac{1}{2\kappa} \int_{\partial K(0,R)} \overset{o}{\lambda} r^2 \phi ,_r) \tag{46}$$

The above quantity is, by the way, equal to the so called A.D.M. mass of the system. Integrating the left hand side of the equation (45) over asymptotically flat, non-compact Cauchy surface we obtain exactly the value of M since the second term gives no contribution whereas the third and the fourth ones cancel each other due to the Gauss-Bonnet theorem. Therefore, M is equal to the integral of the manifestly positive right-hand side of (45).

In the case of non-empty space the right-hand side of (45) is modified by the matter energy density which is again positive. Therefore, our simple argument for the positivity of the energy remains valid.

Similar argument can also be used in topologically non-trivial case (the radial variable will run outside the horizon only!) but we do not discuss it in the present paper.

6. Final remarks

There are also different ways of controlling the boundary parameters. For example, one can perform a partial legendre transformation in (28) exchanging γ with Q^{00}, λ with s but also $\dot{\lambda}$ with α. This way we obtain a hamiltonian system which is different from the one discussed here. The 6 boundary control parameters of this system will be the 3-metric g_{ij} on Λ. The energy of this system defined again as the hamiltonian differs from the energy (29) by the term coming from the Legendre transformation. The properties of such a system will be discussed in a forthcoming paper. Also the limit $R \to \infty$ will be discussed for both systems.

REFERENCES

[1] Arnowitt R., Deser S., Misner.: The dynamics of general relativity. In: Gravitation - an introduction to current research (Witten L. ed.) N.Y. 1962, Wiley

[2] Ashtekar A.: Asymptotic properties of isolated systems: Recent developments. To appear in the Proceedings of GR 10, Padova, July 83

[3] Choquet-Bruhat Y., York J.W.: TheCauchy problem. In: General Relativity and Gravitation (Held A. ed.), N.Y. 1980, Plenum Press

[4] Ferraris M., Kijowski J.: Gen. Rel. Grav. 14 (1982) p.165

[5] Kijowski J.: Gen. Rel. Grav. 9 (1978) p.857

[6] Kijowski J.: Asymptotic Degrees of Freedom and Gravitational Energy. To appear in the Proceedings of Journées Relativistes, Torino, May 83

[7] Kijowski J., Tulczyjew W.M.: A Symplectic Framework for Field Theories. Springer Lecture Notes in Physics, vol 107 (1979)

[8] Regge T., Teitelboim C.: Ann. Phys. 88 (1974) p.286

[9] Tulczyjew W.M.: Symposia Mathematica 14 (1974) p. 247

A METHOD FOR GENERATING EXACT SOLUTIONS OF EINSTEIN'S FIELD EQUATIONS

Joseph Hajj-Boutros
Lebanese University
Faculty of Sciences II
Mansourieh Meten
LEBANON

The exact analytical solutions of the field equations for a spherical star in mechanical and thermodynamical equilibrium have an obvious interest, hence we propose a method for generating exact analytical solutions from existing ones. In the case of a perfect fluid, the energy-momentum tensor is :

$$T_{ij} = (\rho + P)U_i U_j + P g_{ij} \tag{1}$$

ρ being the density and P the pressure.

We take the Schwarzschild (1) coordinates defined by :

$$ds^2 = e^{\lambda} dr^2 + r^2 d\theta^2 + \sin^2\theta \, d\varphi^2 - e^{\nu} dt^2 \tag{2}$$

where λ and ν are two functions of r only.

In a comoving frame of reference U^i is defined by

$$U^i = (0, 0, 0, e^{\nu/2}) \tag{3}$$

Hence the field equations

$$R_{ij} - \frac{1}{2} R g_{ij} = 8\pi T_{ij}$$

become

$$8\pi P = e^{-\lambda}\left[\nu'/r + 1/r^2\right] - \frac{1}{r^2} \tag{4}$$

$$8\pi P = e^{-\lambda}\left[\nu''/r - (\lambda'\nu')/4 + (\nu')^2/4 - (\nu'-\lambda')/2r\right] \tag{5}$$

$$8\pi\rho = e^{-\lambda}\left[\lambda'/r - 1/r^2\right] + \frac{1}{r^2} \tag{6}$$

The pressure has been assumed isotropic. By the substitution

$$\frac{\nu'}{2r} = u \qquad \text{and} \qquad e^{-\lambda} = v \tag{7}$$

D.N PANT and SAH obtain from (5) and (6) :

$$\frac{dv}{dr} + 2r^2\,\frac{\frac{du}{dr} + u^2 r - \frac{1}{r^3}}{1+ur^2}\,v = \frac{-2}{r(1+ur^2)} \tag{8}$$

(8) has the solution

$$v(r) = v_0(r) + C\,\exp\left[\int \frac{2r^2\left(\frac{du}{dr} + u^2 r - \frac{1}{r^3}\right)}{1+ur^2}\,dr\right] \tag{9}$$

This result has been already obtained by Heintzman (4).

We have made the substitution $W = ur^2$ in (8), and we have obtained the following Ricatti equation :

$$2rv\frac{dW}{dr} + W\left[r\frac{dv}{dr} - 4v\right] + 2vW^2 + r\frac{dv}{dr} - 2v + 2 = 0 \tag{10}$$

By making the substitution $W = W_0 + \frac{1}{Z}$ in (10) (W_0 being a known solution of (10)) we get :

$$\frac{dZ}{dr} + Z\left[-\frac{1}{2v}\frac{dv}{dr} + \frac{2}{r} - \frac{d\nu}{dr}\right] = \frac{1}{r} \tag{11}$$

By quadrature we obtain :

$$Z(r) = \frac{1}{r^2}\sqrt{v}\,e^{\nu}\left[\int \frac{r\,dr}{\sqrt{v}\,e^{\nu}} + C_1\right] \tag{12}$$

C_1 being a constant.

Hence from a particular solution $u_0(r)$, (12) gives a new solution u(r). We have applied this method to the solution of R.C. Adams ; M. Cohen, R. Adler and C. Schiffield (5) in which the pressure cannot vanish at finite distance. The line element of R. Adams et al. is :

$$ds^2 = B^2 dr^2 + r^2 (d\theta^2 + \sin^2\theta\, d\varphi^2) - C^2 r^{\frac{4\alpha}{1+\alpha}} dt^2 \tag{13}$$

Once again (12) yields

$$u(r) = \frac{2\alpha}{r^2(1+\alpha)} + \left[\frac{1+\alpha}{2(1-\alpha)} r^2 + C_1 r^{\frac{4\alpha}{1+\alpha}} \right]^{-1} \tag{14}$$

The new line element is :

$$\begin{aligned} ds^2 = {} & B^2 dr^2 + r^2 (d\theta^2 + \sin^2\theta\, d\varphi^2) \\ & - C^2 r^{\frac{4\alpha}{1-\alpha}} \exp\left[\frac{2}{C_1} \operatorname{Ln} r + 2 r^{\frac{2\alpha-2}{1+\alpha}} \operatorname{Ln} C_1 \right] \end{aligned} \tag{15}$$

The pressure becomes :

$$8\pi P = \frac{1}{B^2 r^2} \left[\frac{4\alpha}{1+\alpha} + \frac{2}{C_1 + C r^{\frac{3\alpha-2}{1+\alpha}}} + 1 \right] - \frac{1}{r^2} \tag{16}$$

and hence P can vanish at a finite distance $r = r_0$; which can be taken as the radius of the body

$$r_0 = \left[\frac{1}{C} \cdot \frac{2}{B^2 - (1 + \frac{4\alpha}{1+\alpha})} - C_1 \right]^{\frac{1+\alpha}{2\alpha-2}} \tag{17}$$

References

1. D. Kramer et al., in Exact Solutions of Einstein's Field Equations (VEB, Deutscher Verlag der Wissenschaften, Berlin 1980) p. 1963.
2. Misner, C.W., Thorne, K.S., and Wheller, J.A. (1973) Freeman, San Francisco (Gravitation).
3. D.N. PANT and A. SAH 1982, Phys.Rev.D. 26, 1254.
4. H. Heintzmann, 1969, Z.Physik 228, 489.
5. R.C. Adams et al., 1973, Phys.Rev.D. Vol 8, N°6, p. 1652.

CAUSAL RELATIVISTIC THERMODYNAMICS OF TRANSITORY PROCESSES IN ELECTROMAGNETIC CONTINUOUS MEDIA

C. Barrabès
Département de Physique, Faculté des Sciences
37200 Tours, France

In the standard treatment of the transport phenomena in continuous media the heat conduction and the viscosity are described by the law of Fourier and Navier-Stokes or by their relativistic version proposed by Eckart[1]. It is well known that these laws suffer from the two following drawbacks : 1°) thermal and viscous disturbances propagate acausally 2°) there exist generic short wave-length secular instabilities[2].

The usual way to evade such difficulties is to introduce relaxation terms in the transport equations. A justification to the existence of the relaxation terms was proposed by Müller [3] in classical mechanics. It was later rediscovered by Israel[4] in the relativistic theory and shown to be in agreement with the study of fluid properties by means of the relativistic kinetic theory[5]. According to Müller ans Israel, a causal description of the transient thermodynamics is available if all the dissipative effects are introduced, up to the second order, in the expression of the entropy.

This work is devoted to a study in General Relativity of the constitutive equations of electromagnetic media, when submitted to transitory processes. We have followed the causal thermod namics of Müller-Israel together with an axiomatic approach to the constitutive equations of electromagnetic continuous media[6]. Our formalism is then general enough to include electromagnetic deformable solids as well as electromagnetic fluids, while a recent work of Israel ans Stewart[7] only dealt with electromagnetic fluids. Applications of this work may for instance concern the study of manetospheres, pulsars, black-hole accretion rings and the early eras of the cosmological evolution.

In can be shown[6,8] that the balance law of the internal energy ε for a spin less medium may be written :

$$\rho\dot{\varepsilon} = \frac{1}{2}\,(t - E\otimes P - B\otimes M).[g]^{\cdot} - \text{div}\, q + q.\dot{U} + E.j + P.[E]^{\cdot} + M.\,[B]^{\cdot}$$

where t is the stress tensor, q the heat current vector, U the 4-velocity, j the electric conduction current, E and B (P and M) the electric and magnetic fields (polarization). The symbol $[\ ]^{\cdot}$ stands for the convective derivative, $\frac{1}{2}\,[g]^{\cdot}$ is equal to the strain rate tensor.

Any process is admissible if it satisfies the entropy principle, div S $\geqslant$ 0. In a reversible or in an irreversible quasistationnary process the entropy current S is written :

$$S = \rho\,\eta\,U + \frac{q}{\Theta}$$

where η is the entropy density and Θ the temperature. Moreover a functionnal relation gives the internal energy ε in terms of the entropy density η and a set of mechanical and electromagnetic variables which are characteristic of the medium.

According to Müller and Israel, the entropy current which describes a transitory process , following a small departure from an equilibrium state, has to be written :

$$S = \rho\,(\eta+\eta') + \frac{q}{\Theta} + s'$$

where the perturbative terms, η' and s', are of the second order in all the dissipative effects (heat, electric conduction, viscosity, polarization). Furthermore, these terms have scales of variation which are much smaller than the ones of the quasistationnary terms (η, q), and their space-time derivatives will therefore take non negligible values.

By applying these considerations in the inequality of the entropy principle and by using the balance law of energy and the functionnal relation which defines the internal energy, it gives the transport equations of the medium. These equations constitute a set of coupled partial differential equations of the first order having a complicated form in the general case. In the case of a dielectric fluid they have the following form :

$$t + A_1.\dot{t} + A_2.\dot{P} = A_3.[g]^{\cdot} + A_4.\nabla q + A_5.\nabla P$$

$$P + \Sigma_1.\dot{P} = \Sigma_2.E + \Sigma_3.\overset{*}{\Theta} + \Sigma_4.\dot{E} + \Sigma_5.\dot{q}$$

$$q + K_1.\dot{q} = K_2.\overset{*}{\Theta} + K_3.E + K_4.\dot{E} + K_5.\dot{P}$$

where $\overset{*}{\Theta}$ is the relativistic temperature gradient.

The transport equations for an electromagnetic deformable solid have similar forms. The axiom of rheological invariance[6,8,9] implies that the convected derivative has to be used in place of the covariant derivative in the direction of the flow vector U. In the case of a magneto-elastic body without heat and electric conduction, the transport equations will be :

$$t + A_1 \cdot [t]^{\cdot} = A_2 \cdot e + A_3 \cdot [g]^{\cdot} + A_4 \cdot \nabla M$$

$$M + \mu_1 \cdot [M]^{\cdot} = \mu_2 \cdot B + \mu_3 \cdot [B]^{\cdot} + \mu_4 \cdot \mathrm{div}\, t$$

where e is the strain tensor.

The tensorial coefficients A_i, Σ_i, K_i, μ_i which appear in the foregoing two sets of transport equations depend on the equilibrium parameters of the medium and take simple forms when the medium admit symmetry properties. The transport equations make evident the relaxations terms thanks to which the propagation equation will be hyperbolic and some generic instabilities may not occur[2].

1 C.Eckart, Phys. Rev. 58, 919 (1940).

2 L. Lindblom and W.A. Hiscock, Astrophys. J. 267, 383 (1983) ; Ann. Phys. (N.Y.) 151, 466 (1983).

3 I. Müller, Z. Phys. 198, 329 (1967).

4 W. Israel, Ann. Phys. (N.Y.) 100, 310 (1976).

5 W. Israel and J.M. Stewart, Ann. Phys. (N.Y.) 118, 341 (1979).

6 G.A. Maugin, J. Math. Phys. 19, 1198, 1206, 1212, 1220 (1978).

7 W. Israel and J.M. Stewart, General Relativity and Gravitation (Plenum, 1980).

8 C. Barrabès, J. Math. Phys. to be published.

9 B. Carter, Proc. Roy. Soc. A 372, 169 (1980).

La relativité générale : une théorie sans problème(s) ?

J. Eisenstaedt
Equipe de Recherche Associée au C.N.R.S. n° 533
Laboratoire de Physique Théorique, Institut Henri Poincaré
11, rue P. et M. Curie, 75231 Paris Cedex 05, France

A l'origine de ce travail, l'étonnement de trouver, dans la littérature scientifique concernant la relativité générale, un grand nombre de jugements d'ordre idéologique : critique ou dithyrambique. Aussi bien est-il clair à chacun qu'entre 1925 et 1955 - grosso modo - la théorie marque le pas ; le seul fait du "renouveau" le prouve amplement. Ces éléments sont assez frappants pour qu'on s'y arrête.

Il ne s'agit pas ici de faire un bilan des résultats de la théorie ou quelque revue [1], ni d'opposer aux critiques de certains physiciens, à l'inquiétude de quelques relativistes, des réponses ; la théorie se défend fort bien grâce à ses propres mérites ... et aux travaux de ses spécialistes. Il s'agit de décrire grâce aux documents disponibles [2] cet état de fait et de tenter d'en comprendre les raisons à partir des structures du champ et de la discipline. Il s'agit en particulier de montrer qu'il ne suffit pas qu'une théorie soit "juste" pour qu'elle s'insère aisément dans le champ institutionnel.

"Je me souviens que pendant ma lune de miel en 1913, j'avais dans mes bagages quelques exemplaires des articles d'Einstein qui, au grand dam de mon épouse, ont absorbé mon attention pendant des heures. Ces papiers me semblaient fascinants, mais difficiles et presque effrayants. Lorsque j'ai rencontré Einstein à Berlin en 1915, la théorie était très perfectionnée et couronnée par l'explication de l'anomalie du périhélie de Mercure, découverte par Leverrier. Je l'ai comprise, non seulement grâce aux publications mais aussi grâce à de nombreuses discussions avec Einstein, - ce qui eut pour effet que je décidai de ne jamais entreprendre aucun travail dans ce champ. Les fondations de la relativité générale m'apparaissaient alors, et encore aujourd'hui, comme le plus grand exploit de la pensée humaine quant à la Nature, la plus stupéfiante association de pénétration philosophique, d'intuition physique et d'habileté mathématique. Mais ses liens à l'expérience étaient ténus. Cela me séduisait comme une grande oeuvre d'art que l'on doit apprécier et admirer à distance" [3].

Telle est la manière dont Max Born; lors du congrès de Berne en 1955 évoquait ses rapports à la relativité générale. Mieux qu'un long préambule, ce texte me permet de situer mon propos ; il pose en effet une question essentielle, celle de la vraie place de la relativité générale en tant que théorie physique dans l'institution scientifique entre le début des années vingt, moment où la théorie est reconnue et son renouveau que l'on peut situer, symboliquement, en 1955 date de la mort d'Einstein.

Mais malgré les événements qui vont jalonner l'histoire de la confirmation de la théorie et sur lesquels je ne m'étendrai pas ici [4], il n'y aura guère de raisons après 1915, mais surtout dès les années vingt, de douter de la relativité générale qui va subir avec plus de succès que toute autre théorie de la gravitation - et d'abord celle de Newton - les rares tests rendus possibles par la technique, à la précision qu'elle permet, et qui sur le plan de sa structure ne pose aucun problème de fond.

Ainsi, est-ce au-delà de sa validité, au-delà de la logique scientifique mise en forme par K. Popper [5] qu'est questionnée la théorie d'Einstein ; d'un point de vue productiviste [6].Il ne s'agit pas tant de savoir si la théorie est juste, on s'y accorde très généralement, mais ce qu'elle apporte de plus, ce qu'elle rapporte. On fait donc le bilan de la théorie, on oppose l'actif au passif sans d'ailleurs mettre nécessairement la même chose dans les plateaux de la balance ... Bref, c'est "d'économie relativiste" qu'il est ici question.

Ainsi, loin des arguments physiques, les considérations avancées seront de nature philosophique, épistémologique ou esthétique ; mais elles seront généralement présentées comme secondaires, destinées à expliquer l'attrait - l'intérêt - qu'exerce la théorie sur ses spécialistes plutôt qu'à en conforter l'assise scientifique. Aussi bien, ce sont des arguments défensifs qui viennent compenser le peu de moyens dont dispose la relativité générale au plan empirique.

"La théorie de la relativité a un attrait particulier à cause de sa consistance interne et de la simplicité logique de ses axiomes" écrit Einstein dans la préface au livre de P.G. Bergmann [7]. Simplicité logique, tel est en effet l'un des mots-clef de la question, qui, selon Popper, doit être rapportée à la rareté des paramètres, impliquant la haute improbabilité a priori de la théorie ou encore sa réfutabilité. En effet, la relativité générale, parce qu'elle ne possède pas de paramètre libre - sinon Λ la constante cosmologique - n'a a priori que peu de chance de faire face à de nombreux tests expérimentaux ou observationnels. Elle est donc remarquablement rigide et hautement réfutable ce qui a contrario explique, entre autres éléments, le grand intérêt dont jouiront les "théories alternatives".

"La magie de cette théorie est telle qu'à peu près personne ne peut y échapper pourvu qu'il l'ait bien comprise" [8]. Après Einstein qui, sous l'emprise de sa découverte présente ainsi sa théorie en novembre 1915, bien des relativistes se montreront sensibles à cette architecture étrange et la portent au crédit de la relativité générale couramment citée comme modèle de théorie physique. Ainsi sera-t-elle considérée par Bergmann comme "le plus parfait exemple de théorie des champs jusqu'alors connue" [9] mais l'on ne peut manquer de citer ici Paul Langevin pour lequel "Nous n'avons rien actuellement qui puisse lui être comparé au point de vue [physique], pas plus qu'au point de vue de la beauté intérieure, de la nécessité logique et de la fidélité à ce que doit être toute physique, une construction théorique sur une base exclusivement expérimentale" [10]. Mais, il faut aussi rappeler l'admiration de H. Weyl "un des plus grands exemples de la pensée spéculative" [11], celle d'Eddington bien sûr et la "force d'intime conviction" que M. von Laue souhaite qu'elle exerce sur ses lecteurs ... On pourrait multiplier les exemples à l'envie.

Les relativistes n'oublient pourtant pas l'essentiel : "Ni l'harmonie interne, ni la satisfaction logique qu'offre une telle théorie ne peut être un critère de sa validité. Il s'agit seulement de savoir qu'elles sont les conséquences que l'on peut en tirer pour l'observation et comment ces conséquences peuvent être vérifiées par l'expérience. La théorie de la relativité générale ne joue pas à ce propos un rôle différent que n'importe quelle autre théorie" [12].

Ainsi Lanczos distingue-t-il à juste titre validité de la théorie et intime conviction, ici la satisfaction logique, l'harmonie interne. On relève pourtant parfois un glissement de sens à ce niveau. Ainsi, P.G. Bergmann introduisant dans son manuel le chapitre sur les "tests expérimentaux" écrit-il : "Les arguments les plus convaincants en faveur de la théorie générale de la relativité, restent, néanmoins, jusqu'à présent théoriques" [13].

C'est là, une affirmation défensive qui vient implicitement compenser le manque d'arguments dont dispose la relativité générale au plan empirique, qui tend sinon à donner à la structure interne un rôle premier du moins à placer les éléments théoriques "plus convaincants" avant les arguments empiriques.

Le thème de la structure logique de la théorie d'Einstein est inépuisable, un modèle porté aux nues ou récusé suivant l'orientation philosophique mais dont personne ne conteste la qualité. C'est un thème qui par un glissement de sens courant dérive souvent vers celui de l'esthétique de la relativité générale. Pourtant, la solidité du bâtiment a-t-elle quelque chose à voir avec son élégance ? Ce passage d'un thème à l'autre naît probablement de celui de "simplicité", un concept cher à Einstein comme on le sait tandis qu'il s'est dit étranger à celui d'élégance "qu'il faut laisser au tailleur et au cordonnier" [14]. C'est pourtant un thème que l'on rencontre souvent chez les relativistes eux-mêmes, qui s'en émerveillent, mais bientôt aussi chez certains de leurs collègues, quanticiens pour la plupart qui le leur retournent avec une toute autre connotation. Ce thème, dès lors péjoratif de l'esthétisme est opposé à celui de l'expérience et plus précisément à son manque, formant les deux termes d'un faux conflit épistémologique.

Elégance, harmonie, beauté intérieure, incomparable esthétique, c'est là l'expression de la séduction qu'exerce la théorie sur ses spécialistes qui ne cachent pas le plaisir que leur procure une théorie bien tournée. Que l'on s'en réjouisse ou que l'on s'en afflige, ce n'est pas pour rien que tant d'images de l'ordre esthétique sont utilisées pour qualifier la relativité générale. Indéniablement, elle a plus d'un point commun avec une oeuvre d'art abstraite ; quant à l'exigence de la structure qu'elle s'impose, des matériaux qu'elle utilise, quant à la distance qu'elle met entre l'image première du phénomène et la représentation qu'elle en donne, quant au caractère révolutionnaire, radical de l'image du monde qu'elle inaugure, des points sur lesquels bien des scientifiques ont insisté, que ce soit pour l'admirer ou l'en blâmer.

Si les relativistes reviennent ainsi si fréquemment à ce thème esthétique, c'est sans doute bien sûr parce que la relativité générale est vraiment belle et qu'ils y sont sensibles. Mais c'est aussi, à mon sens, par compensation car très souvent ce thème s'articule à celui du manque expérimental. Le plaisir esthétique leur est une raison supplémentaire d'y travailler que ne justifierait pas suffisamment le peu de résultats concrets - effets physiques, satisfactions institutionnelles - que la théorie leur apporte. Ce thème prend donc le sens d'un argument d'ordre économique.

Comme l'a fort justement fait remarquer S. Chandrasekhar dans un article récent sur l'histoire de la discipline, "la description du travail d'Einstein comme oeuvre d'art est souvent le masque sous lequel les physiciens désavouent la pertinence de la relativité générale quant à l'avance de la physique" [15]. Une constatation qui s'appuie entre autre sur les propos de Rutherford :

"Au-delà de sa validité, la théorie de la relativité générale ne peut être considérée que comme une magnifique oeuvre d'art" [16].

Ainsi, l'argument esthétique est-il retourné à leurs auteurs sous une forme péjorative. Et c'est bien sur le plan économique, au niveau de sa fécondité, "au-delà de sa validité" que la relativité générale est condamnée. Car, si la relativité générale n'est "qu'une oeuvre d'art" c'est que ses spécialistes ne sont que des artistes qui produisent des idées, magnifiques certes, mais peu utiles : luxueuses. Et c'est là un thème qui renvoie au procès que feront certains physiciens à leurs collègues accusés d'être avant tout des mathématiciens qui plus que d'autres scientifiques sont censés être particulièrement sensibles à l'esthétique et y puiser leur inspiration plutôt que dans l'expérience.

Comme un tableau (trop) abstrait, la théorie d'Einstein sera considérée par de nombreux scientifiques comme étant d'un accès difficile. Au-delà de la boutade bien connue d'Eddington contée par Chandrasekhar [17], des physiciens aussi sérieux que M. Born, P. Ehrenfest, M. von Laue, J.J. Thompson expliciteront ce point, d'où découle la réputation d'incompréhensibilité dont elle jouira auprès du public cultivé. Et il faut remarquer avec Born qu'il s'agissait "d'une théorie neuve, révolutionnaire. Un effort était nécessaire pour l'assimiler". Un effort que "tout le monde ne pouvait pas ou ne voulait pas faire" [18].

Une remarque qui renvoie en partie à l'isolement de la relativité générale qui développera durant ces années infiniment peu de liens avec les autres théories physiques. Il faut dire aussi que sur le front de ses développements les plus neufs elle posera - comme toute autre théorie - à ses spécialistes des problèmes difficiles et bien des relativistes se plaindront de ce fait. Un fait qu'il faut rapporter au concept de fécondité, face à un champ observationnel que nous avons décrit rapidement, face à l'image de la théorie de Newton qui avait eu la chance de trouver un champ d'action quasiment vierge, face aussi à la mécanique quantique et à la relativité restreinte qui s'étaient trouvées dans des situations infiniment plus enviables. Mais bien sûr les difficultés qu'affronte la théorie n'ont pas grand chose à voir avec son incompréhensibilité prétendue. Il semble que J.J. Thomson en soit à l'origine que Chandrasekhar et Franck citent : "Je dois confesser que nul n'a encore réussi à mettre en langage clair ce qu'est en réalité la théorie d'Einstein". Un trait empoisonné qui vise la clôture de la théorie et de ses spécialistes enfermés dans un langage hermétique. Ainsi le thème de la difficulté est-il utilisé d'une manière analogue à celui de l'esthétique. Et tandis que l'esthétique devient esthétisme, la difficulté devient incompréhensibilité : un double enfermement.

Ainsi, l'incompréhensibilité supposée de la théorie d'Einstein n'est que le revers de son manque de fertilité, le coup bas porté par ceux qui, n'ayant pas eu le loisir, le désir de s'y investir, n'y ayant aucun intérêt propre et qui, complexés de ne la comprendre réellement pas - et pour cause ! - pour se justifier accusent : "c'est une théorie incompréhensible !". Traduisez : "c'est une théorie dont *l'intérêt* m'est incompréhensible".

Ainsi, le peu de liens que la théorie d'Einstein propose avec les "vrais" problèmes de la physique la rejette, pour un temps, du côté de l'art pour l'art.

Pourtant, il faut redire avant tout que, tout au long de son histoire, la relativité générale n'a jamais été sérieusement mise en défaut et qu'elle parvient plus qu'honorablement et mieux que toute autre théorie concurrente à rendre compte du champ observationnel qu'elle tendait à couvrir, tel qu'il se présente en 1915 puis tel qu'elle le restructure ; un champ mince certes, limité comme toujours par les techniques disponibles aussi bien que par l'état de la prospective théorique ; mais dans ces bornes banales, un champ dont elle rend fort bien compte au niveau qualitatif, ce qui ne signifie pas qu'elle le couvre parfaitement au niveau quantitatif.

Mais les spécialistes sont unanimes pour déplorer les difficultés spécifiques quant à l'observation des effets propres à la théorie. D'autant qu'il s'agit toujours d'observation - et non d'expérimentation - dont par nature on est loin de posséder toutes les données, dont on ne peut manipuler aucun paramètre. Il s'agit aussi - en particulier pour les éclipses - de difficultés d'un tout autre ordre : la guerre, les nuages, le matériel ... les récits des observateurs fourmillent d'anecdotes à ce sujet [19].

Entre l'éclipse de 1919 et l'expérience de Pound et Rebka en 1960, si l'on met à part le domaine cosmologique, malgré les timides espoirs nourris et aussitôt déçus concernant diverses questions astronomiques, malgré quelques autres effets plus ou moins mûrement calculés, le statut empirique de la théorie, toujours limité à ses trois tests classiques - périhélie de Mercure, déviation des rayons lumineux, déplacement des raies spectrales - s'est plutôt rétréci, ce qui a constitué le prix à payer des trop belles certitudes des années vingt [20]. Mais, aussi bien, c'est là un manque d'abord lié à l'étonnante proximité de la théorie de Newton qui, après plus de deux siècles d'hégémonie, ne laisse à toute théorie concurrente qu'une marge infime pour se déployer empiriquement ; un manque "compensé" par l'extraordinaire architecture de la théorie, deux thèmes que l'on oppose indéfiniment.

"Il en va beaucoup plus mal encore qu'avec la relativité restreinte" s'écrie Hermann Weyl dans la première édition de son "Raum, Zeit, Materie", opposant ensuite le peu de phénomènes observables au "bouleversement que la théorie apporte" [21]. C'est là un thème qui revient constamment dans la littérature relativiste, une préoccupation lancinante qui est aussi celle d'Einstein [22]. Et face aux "trois tests classiques", d'assez nombreuses tentatives théoriques ont été faites, avant le renouveau des années soixante pour appliquer la relativité générale à d'autres problèmes ; qu'il s'agisse de l'accélération séculaire de la lune, du déplacement de l'orbite de Mars, du niveau atomique ou plus réalistes concernant l'effet de lentille gravitationnelle, le mouvement du périhélie de la terre, faisant appel à un disque tournant ou à un gyroscope, sans parler bien sûr du champ cosmologique. Mais, il faut dire qu'il s'agit parfois quasiment d'expériences de pensée tant on est loin de pouvoir atteindre techniquement la précision requise pour que des effets spécifiques soient décelables. C'est alors un quatrième test introuvable ! [23].

Mais au-delà de ces résultats décevants, ce qui est stupéfiant c'est bien qu'Einstein ait embrassé, avant même que sa théorie fût complète, ce qu'il faut bien appeler pour près de cinquante ans, l'ensemble du champ empirique de sa théorie ; qu'il ait fallu attendre les années soixante pour que se renouvelle quelque peu le maigre stock des tests de la théorie. Ce fait historique n'eût guère étonné dans un siècle moins dynamique quant à l'innovation technologique. Il prend, à cause d'un effet d'optique lié au véritable réseau expérimental dont dispose la mécanique quantique, des allures de désaveu. Il indique plus prosaïquement, que la théorie de Newton était encore plus juste qu'on ne le croyait - qu'on ne l'espérait. Il montre donc que le champ expérimenté de la gravitation n'a guère évolué et n'est alors guère différent de celui sur lequel s'appuie la théorie de Newton : la banlieue solaire. Et en ce sens la relativité générale est une théorie révolutionnaire sur un champ classique, qui bouleverse le cadre de la gravitation sans disposer d'un véritable champ propre accessible. C'est ce décalage qui mettra la théorie dans une position extrêmement inconfortable, à la fois point de mire conceptuel et point aveugle de la physique, référence obligée des épistémologues mais repoussoirs des (vrais) physiciens.

Lors du congrès de Berne, consacré en 1955 à la relativité générale, sur trente-quatre conférences une seule, celle de R.J. Trumpler sera consacrée aux résultats observationnels. Dans l'hommage qu'il écrit dans la Review of Modern Physics lors de la mort d'Einstein, J.R. Oppenheimer note :

"Dans les quarante ans qui se sont écoulés [ces trois tests] sont restés le principal et, à une exception près, le seul lien entre la relativité générale et l'expérience. L'exception repose dans le champ de la cosmologie" [24].

Et l'on ne peut ici passer sous silence l'opinion de R.H. Dicke qui vient à la fin des années cinquante aux théories relativistes de la gravitation avec la ferme intention de remettre la théorie d'Einstein dans le droit chemin expérimental et dénoncera avec vigueur "l'indigence de la preuve expérimentale" et comme "une chose affligeante [---] le manque de contact avec l'observation et les faits expérimentaux" [25]. Et plus d'un spécialiste remarquera que, sur le plan empirique, la théorie d'Einstein était loin d'être d'une "pressante nécessité". Il s'agit là d'un point essentiel à plus d'un titre. Parce que plus une théorie est corroborée, plus elle assure de liens entre des champs divers, expérimentaux ou théoriques, plus elle sert d'outil à d'autres théories, plus grande est la confiance qu'on lui porte aussi bien sur le plan épistémologique que technique. C'est cette relation dialectique, cette spirale des investissements et des profits intellectuels qui va cruellement manquer à la théorie d'Einstein, rejetée du côté du spéculatif.

Je n'en veux pour preuve que le choc - et l'espoir - que va représenter en 1960 l'expérience de Pound et Rebka qui, grâce à l'effet Mossbauer récemment découvert, viennent de vérifier - à 1% ! - le troisième test. Une bonne nouvelle que A. Schild annonce dans l'American Journal of Physics sur un ton biblique :

"Voici des jours excitants : la théorie de la gravitation d'Einstein, sa théorie générale de la relativité de 1915, est passée du royaume des mathématiques à celui de la physique. Après 40 ans de contrôles astronomiques maigrement parsemés, de nouvelles expériences terrestres sont possibles et sont projetées" [26].

Durant la période de réception, on a pris acte de l'étonnante proximité - sur le champ expérimentable - des théories newtonienne et einsteinienne de la gravitation ; une proximité qui bientôt pourtant inquiète certains spécialistes ; ainsi, dès 1916, J. Droste, élève de Lorentz, qui est probablement le premier à s'exprimer à ce sujet, pointera "un résultat <u>pour une fois</u> différent de tout ce que prévoit la théorie de Newton" [27]. Sans doute son résultat est-il loin d'être achevé théoriquement et encore plus loin de pouvoir être observé. Il n'en demeure pas moins qu'il a mis en évidence un champ d'action spécifiquement relativiste et flairé le danger que représente la trop grande proximité de la théorie de Newton [28].

Bien loin de ces visées alors purement spéculatives, un travail indispensable de développement, de justification sera accompli tout au long de ces années. Ainsi, s'agit-il, bien souvent dans le cadre d'approximations post-newtoniennes, de trouver de nouveaux effets spécifiques ou de calculer les contributions relativistes à des effets d'origine newtonienne. T. Levi-Civita a su très concrètement exprimer cette nécessité : "le mouvement des corps célestes dans des conditions ordinaires diffère si peu de sa représentation newtonienne que, pour les besoins astronomiques, les effets relativistes peuvent être traités comme des perturbations de premier ordre" [29] écrit-il au détour d'un article technique. C'est là une démarche nécessaire, indispensable, mais qui reste par construction dans la dépendance technique et conceptuelle de la théorie de Newton. Une démarche qui domine toute cette période et que H. Bondi défendra malicieusement en 1962 dans une conversation avec Synge, évoquant "cette méthode particulière d'approximation" [à la relativité générale] inventée comme chacun sait 250 ans avant la théorie" et qui "sauf pour ce qui concerne quelques points mineurs" [---] "satisfait largement à [ses] propres aspirations à la réalité" [30].

Dans le droit fil de ces approches post-newtoniennes, il faut signaler les nombreux travaux qui dans le cadre de la solution de Schwarzschild s'appuyant sur un système de coordonnées particulier - celui de Droste-Schwarzschild - considéré de fait comme absolu, constitueront de facto une véritable interprétation néo-newtonienne sinon de la théorie du moins de sa solution la plus importante. Mais bien au-delà de cette démarche, il faut aussi penser aux nombreuses interprétations particulières de la théorie, réponses au problème que pose la covariance aux spécialistes.

Bien souvent, ces approches particulières se justifient de la complexité de la structure de la théorie et tout particulièrement des difficultés qu'imposent covariance générale et non-linéarité. Et plus d'un relativiste se plaindra du peu de solutions exactes connues, un fait qui n'est bien sûr pas indépendant de la complexité de la théorie. Ainsi, est-ce à l'occasion d'un travail important concernant la solution de Schwarzschild, dont il est précisément l'un des premiers à repenser l'interprétation traditionnelle - néo-newtonienne - que Synge s'inquiète du manque de fécondité de la relativité générale face à la vigueur de la théorie de Newton. Un manque de fécondité qu'il attribue à la non-linéarité de ses équations de champ ("une formidable difficulté") et à la covariance générale ("embarrassante plutôt qu'avantageuse") [31].

Il n'empêche que les spécialistes de la théorie - et en particulier Synge - sont conscients de la nécessité d'une approche spécifiquement relativiste :

"Au temps où la relativité devait gagner la croyance dans un monde incrédule, il était naturel de lui donner de la respectabilité en l'expliquant autant que possible en termes des vieux concepts, écrit-il. Mais cela a conduit à des concepts confus. Ces jours ont passé et l'on peut entreprendre un nouvel examen du problème de l'introduction des concepts relativistes" [32].

Ainsi, les spécialistes de la relativité générale ont-ils développé deux stratégies complémentaires. Une tactique pragmatique, à travers une vision basse de la théorie, en terme des vieux concepts - un choix qui s'exprime par exemple dans l'interprétation néo-newtonienne - afin de tenter de faire accepter la théorie face aux nombreuses critiques. Soit, faisant front, favoriser une vision décidément relativiste, payante - ce n'est alors qu'un espoir - à long terme, sans hésiter devant les spéculations cosmologiques par exemple, ni les techniques mathématiques sophistiquées mais en restant loin des préoccupations empiriques et donc tout en prêtant le flanc aux accusations de "formalisme". C'est là, évidemment, fort de leur intime conviction, le choix des vrais relativistes.

Entre le début des années vingt, les années d'or de la relativité générale et le milieu des années trente, par rapport au nombre total des publications recensées dans les "Fortschritte der Mathematik" qui est globalement multiplié par trois, la mécanique newtonienne conservera sa place en pourcentage (7%) tandis que la relativité générale verra la sienne se réduire comme peau de chagrin passant de 7% à 2%.

Comment s'étonner de la grande indifférence de la grande majorité des astronomes quant à la relativité générale ? Les astronomes qui semblaient devoir être naturellement les utilisateurs privilégiés d'une nouvelle théorie de la gravitation restent sourds à ses attraits ... et à ses techniques sophistiquées. Sans doute savent-ils que la relativité générale permet d'expliquer l'avance du périhélie de Mercure ; peut-être connaissent-ils la formule du second, voire du troisième test ; guère plus. Entre 1920 et 1960, à part l'ouvrage de J. Chazy, aucun traité d'astronomie n'accordera plus de quelques petites pages à la théorie d'Einstein, souvent moins, parfois rien. Un point qui n'est pas contradictoire avec l'intérêt que quelques astronomes porteront à la théorie jusqu'à en devenir de brillants spécialistes tel Eddington.

Le champ cosmologique possède quant à lui, un statut tout à fait à part et particulièrement intéressant. Même si la "cosmologie relativiste" [33] dépend conceptuellement, techniquement presque entièrement de la relativité générale, elle en reste alors de fait essentiellement distincte. Ainsi est-elle assez fréquemment ignorée des manuels et lors de la conférence de Chapel-Hill en 1957, la seconde conférence internationale consacrée à la relativité générale, P.G. Bergmann excluera explicitement de son rapport la cosmologie qui "est un champ en soi et au moins jusqu'à présent, n'est pas intimement connectée aux autres aspects de la relativité générale ..." [34]. A l'inverse, le rapport de G. Lemaître au Congrès Solvay de 1958 qui concerne "l'état général de la théorie cosmologique" ne fait pas mention de la théorie d'Einstein. De plus, ainsi que le souligne fortement Synge avec bien des auteurs : "de toutes les branches de la science moderne, la théorie cosmologique est la moins liée à l'observation" [35]. Car, sur le plan empirique, la cosmologie n'apporte alors que fort peu de choses à la relativité générale.

De tels faits permettent d'expliquer le peu de confiance sinon les invincibles réticences dont témoignaient beaucoup de relativistes quant au champ cosmologique et les précautions dont s'entouraient ceux qui y travaillaient. De plus, à la trop grande liberté que l'observation laissait à la théorie se conjuguait la proximité épistémologique de la "philosophie" jetant une ombre diabolique sur une spécialité déjà marginale et suspecte.

Pourtant, d'une manière paradoxale et longtemps souterraine, ce sont précisément ces caractéristiques spéculatives qui donneront à la cosmologie une importance manifeste dans le développement de la relativité générale. C'est un cosmologue réputé, pourtant peu suspect d'idéalisme R.C. Tolman qui écrit en 1934 :

"Puisque nous avons basé notre traitement sur une théorie physique acceptable, nous sommes en droit d'attendre du comportement théorique de nos modèles au moins qu'ils nous informent et qu'ils libéralisent notre manière de penser quant aux possibilités conceptuelles du comportement de l'univers réel" [36].

C'est bien là en effet qu'il faut voir tout l'intérêt de la cosmologie pour la relativité générale ou plus précisément pour l'image de la relativité générale que s'en font et que forgent ses propres spécialistes. Car, à l'inverse de toutes les autres applications de la théorie, les effets cosmologiques ne sont pas liés à une vision néo-newtonienne. Et la cosmologie relativiste représente alors une des rares branches de la relativité générale qui dispose réellement, précisément en raison de son caractère spéculatif, de quelque autonomie face à la théorie de Newton ... En fait, jusqu'au début des années soixante, la cosmologie a constitué le seul domaine où la relativité générale a pu être projetée, pensée jusqu'au bout, dans le cadre d'une structure de l'espace-temps nettement dégagée des schémas newtoniens, d'un espace vraiment courbé. Ce n'est certainement pas un hasard s'il se trouve au moins deux cosmologues réputés, G. Lemaître et H.P. Robertson, aux sources de la refonte de l'interprétation de la solution de Schwarzschild et tout particulièrement de sa "singularité". C'est en ce sens que la cosmologie représentera un apport essentiel au développement récent de la théorie. Sans doute, l'exiguïté du champ directement expérimentable est à l'origine des problèmes de la relativité générale. Mais plus encore, c'est la structure même des champs envisagés et envisageables, du domaine dont peuvent s'autoriser les spécialistes qui bloque son évolution. Un domaine qui, pour être pris au sérieux, ne peut se situer en un lieu à jamais inaccessible observationnellement mais qui doit aussi laisser à la théorie la place de s'exprimer ; c'est précisément à cette frontière entre le spéculatif et l'empirique que se situait alors la cosmologie.

Einstein ne s'attardera guère à sa théorie de la relativité générale. Il n'est que peu satisfait de sa création [37] et en particulier de la description des sources de champ, un "pis-aller" notera-t-il dans ses notes autobiographiques ; à tel point qu'il s'étonnera - en 1921 - que C. Lanczos cherche des solutions exactes "à un tel ensemble éphémère "d'équations"" [38]. Pour lui, la relativité générale est d'abord pensée comme prolongement de la relativité restreinte, comme une généralisation du principe de relativité en présence d'un champ de gravitation. Selon cette interprétation, que conforte d'ailleurs tout simplement son nom la relativité générale n'est théorie de la gravitation qu'en second lieu, un statut qui ne dominera son image sans ambiguïté que plus tardivement. Ainsi la relativité générale ne représente-t-elle pour Einstein qu'une halte sur un chemin qui partant de la relativité restreinte aboutirait à une théorie unitaire des interactions gravitationnelles et électromagnétiques. Une halte qui lui permet d'obtenir deux "résultats" auxquels il tient particulièrement, "la covariance des lois de la nature et leur non-linéarité" ainsi qu'il l'exprime très clairement dans la préface du livre de P.G. Bergmann ; des "résultats" d'où il repart à la conquête de nouvelles terres ... Une "quête sans espoir", pour reprendre l'expression d'Abraham Taub [39], que plus d'un relativiste regrettera.

Quant au statut de la relativité restreinte, il se distingue désormais totalement de celui de sa soeur cadette. Sur le plan de l'expérience, bien sûr, mais plus encore sur celui de la construction théorique. Plus qu'une théorie, elle est considérée comme un outil de travail, une "super-loi" selon l'expression de Wigner, la première du bagage de tout physicien théoricien, tandis que la relativité générale, elle, est la discipline d'un petit groupe de théoriciens bien particuliers.

On recense plus d'une vingtaine de théories alternatives à la théorie d'Einstein ce qui est la marque d'un intérêt considérable. Mais pareil projet ne peut se concevoir, se soutenir après 1915 sans une certaine insatisfaction à propos de la théorie d'Einstein, sans arrière-pensées. Whitrow et Morduch, qui ont d'ailleurs travaillé eux-mêmes à une telle théorie, abordent cette question dans l'article de revue qu'ils consacrent en 1965 à ce sujet. Leur "vision critique" de la théorie d'Einstein reprend en substance les reproches qu'affronte la théorie depuis cinquante ans : faiblesse de la preuve empirique, importance "des éléments méthodologiques et esthétiques" [40]. A l'évidence ce dernier point est au centre des motivations de chacun car la faiblesse des résultats empiriques de la relativité générale est une conséquence inéluctable de la proximité des prédictions newtoniennes et des observations ; elle concerne donc également toutes les théories alternatives. C'est que, ainsi que nous l'avons souligné à maintes reprises, dans l'esprit des scientifiques se trame une sorte de bilan qui oppose les investissements consentis aux résultats obtenus ; un bilan qui pourrait être favorable à quelque théorie alternative, en rognant du côté des principes. D'autant plus que la relativité générale, ne disposant d'aucun paramètre arbitraire, a pu apparaître comme relativement fragile, telle une construction très rigide que le moindre événement pouvait déstabiliser et rendre caduque ; une construction qu'il ne devrait pas être si difficile de concurrencer, de remplacer par quelque architecture moins ambitieuse.

Une analyse des objections que soulève la relativité générale parmi ceux qui tentent de construire, entre les deux guerres, une autre théorie de la gravitation, permet de percevoir deux thèmes bien distincts. D'une part des critiques d'ordre épistémologiques sinon philosophique qui interrogent les principes mêmes de la théorie. D'autre part des objections visant le caractère riemanien et non-linéaire de la relativité générale, qui ont pour point de départ la difficulté de la mise en oeuvre technique de la théorie. Mais personne ne met alors en cause la relativité restreinte ni la capacité de la relativité générale à répondre aux questions posées jusqu'alors par l'expérience.

Dans "The Principle of Relativity" publié en 1922 - au sommet de la gloire d'Einstein - Whitehead propose "une version alternative à la théorie de la relativité" : "Ma théorie maintient la vieille division entre physique et géométrie. La physique est la science des relations contingentes de la nature et la géométrie exprime l'uniformité de ses relations" note-t-il dans sa préface après avoir cité J.J. Thomson qui estime que "notre but ultime est de décrire le sensible en terme du sensible". Ainsi, ne s'étonnera-t-on pas de son refus apriorique d'un cadre riemanien, un refus qu'il partage d'ailleurs avec quasiment tous les auteurs de théories alternatives avant 1960.

Aussi bien, est-ce sur le plan idéologique que Milne attaque violemment la relativité générale d'Albert Einstein : "Le mysticisme qu'Einstein a balayé par la porte d'entrée dans sa relativité "restreinte" lorsqu'il insistait sur l'utilisation de nombres observationnellement déterminés pour fixer les événements est rentré par la fenêtre lorsqu'il introduisit les coordonnées générales de "l'espace-temps". La "relativité générale" implique une forme d'atavisme ..." écrit-il en 1940 [41]. Et c'est d'une manière imagée qu'il exprime son sentiment quant à la covariance générale :

"La relativité générale est telle un jardin où les fleurs et les mauvaises herbes croissent ensemble [42]. Dans notre jardin nous essayons de ne cultiver que les fleurs".

Mais aussi bien, au-delà de la covariance, c'est une fois de plus la structure riemanienne qui est visée. Ainsi, aussi bien chez Whitehead que chez Milne, ce ne sont pas tant les résultats théoriques de la relativité générale qui posent problème, pas plus que leur valeur prédictive ni même quelque point faible particulier ; c'est l'appareil épistémologique lui-même qui est rejeté, sans doute fondamentalement à cause de la trop grande distance entre la structure fondant la théorie et les faits empiriques, entre l'espace figuré et l'espace vécu.

Mais on peut aussi se demander si une des raisons à la multiplicité de ces tentatives ne serait pas le manque de familiarité dont certains auteurs font preuve face aux concepts et aux techniques nécessités par la relativité générale. C'est bien ce que laisse entendre la discussion qui, en 1925, suit l'exposé de l'esquisse théorique - restée sans lendemain - de G. Temple, un physicien théoricien britannique. Après avoir perfidement admiré "l'élégance et l'ingéniosité" des méthodes de l'auteur, Eddington remarque : "qu'un tel travail peut seulement être le recours de celui qui a déjà été conduit à croire que la théorie de la relativité est erronée" [43]. Mais, si l'on en croit les autres interventions, c'est bien plus la difficulté à comprendre et à manipuler la théorie d'Einstein qui fait le succès de ce genre de travaux. Ainsi, la tentative de Temple trouve-t-elle un accueil favorable auprès d'un autre intervenant parce qu'elle est "beaucoup plus facile à suivre" tandis qu'un troisième le félicite d'avoir apporté "une théorie plus sympathique aux physiciens que celle d'Einstein". C'est là sans doute l'opinion, naïvement exprimée, du physicien de base. D'ailleurs, la complexité de la structure mathématique de la relativité générale et plus précisément la difficulté d'en manipuler les éléments est un point sur lequel insistent aussi Whitrow et Morduch et qui rejoint par exemple les inquiétudes de Synge qui a d'ailleurs travaillé à la théorie de Whitehead.

Mais, à ces motivations, dont aucune ne peut finalement être vraiment qualifiée d'interne [44], s'ajoute un fait important pour l'avenir de la physique et qui constitue à mon sens la plus sérieuse raison de quelques-unes de ces tentatives, face au scepticisme général et au manque de résultats qu'avaient rencontrées les théories unitaires. C'est que le caractère géométrique de la théorie d'Einstein l'isole quasi-totalement des autres théories physiques qui sont alors des théories de champ bâties sur un espace "plat". A ce propos, Nathan Rosen poursuivant sa tentative d'interprétation de la relativité générale en ces termes se demande "s'il ne serait pas mieux d'abandonner l'approche géométrique de la gravitation dans l'espoir d'obtenir un traitement plus uniforme de tous les différents champs de force que l'on rencontre dans la nature" [45]. Ce sera en particulier dans les années cinquante la raison essentielle à la construction de théories phénoménologiques de la gravitation ainsi que l'expriment en 1957, F.J. Belinfante et J.C. Swihart : "D'un point de vue de simplicité théorique, notre théorie linéaire, par son refus de la covariance générale, a l'avantage d'éviter les complications des "contraintes" qui causent les difficultés de la quantification" faisant allusion aux travaux de Rosenfeld [46]. Ainsi, symboliquement, ce sont les deux "résultats" auxquels Einstein tenait le plus, "la covariance des lois de la nature et leur non-linéarité" qui sont le plus souvent mises en cause. Un point qui, nous l'avons vu, s'explique d'abord par des considérations externes à la gravitation, par la complexité des objets mathématiques impliqués par la relativité générale mais aussi par le désir de sortir la gravitation de l'isolement qui est le sien face à la théorie quantique. Ainsi donc est-ce, fondamentalement, ce "piédestal" géométrique complaisamment décrit par Lanczos [47], si représentatif du splendide isolement de la relativité générale, qui est visé.

Mais nul ne s'est plus appliqué à en saper les bases que R.S. Dicke, dès la fin des années cinquante : "La présomption de la validité de la relativité générale repose en premier lieu sur la beauté et l'élégance de la théorie plus que sur les observations" déplore-t-il [48]. Car il suspecte les relativistes de tenir à leur théorie pour des raisons esthétiques alors "qu'on peut construire autant de théories *ad hoc* que l'on veut" [49]. Et, parmi les arguments qu'il invoque pour trancher entre elles, au-delà de l'observation et des règles communément admises, il préfèrera la fécondité et la simplicité récusant bien sûr l'esthétique mais aussi l'histoire.

Ainsi, à Varenna en 1961, exprimera-t-il très crûment ce point de vue économiste :

"On a souvent soutenu, dans l'intérêt du conservatisme et de l'économie que l'on doit considérer seulement les théories établies comme la relativité générale jusqu'à ce qu'elles soient reconnues fausses par l'expérience. C'est d'habitude un bon principe" [50].

Si l'on peut comprendre que sa préférence aille à une théorie plus "simple" à manipuler et à relier aux autres théories physiques que la relativité générale, pour autant qu'elle rende compte du même champ empirique, il est plus difficile de le suivre lorsqu'il note aussitôt après qu' "il y a un danger qu'une théorie incorrecte soit propagée à cause d'un précédent établi sur un accident historique" (sic), même s'il précise longuement ce qu'il entend par "accident historique" en reconstruisant, en réécrivant une partie de l'histoire de la physique de 1906 à 1919, de la théorie de l'électron de Lorentz à l'éclipse de Sobral en replaçant différemment dans le temps les faits expérimentaux, les théories et les hommes. Sans entrer dans le détail de cette histoire-là, soulignons que l'argument essentiel de Dicke, c'est que Lorentz, placé devant la possibilité d'exprimer ses équations de mouvement de manière géométrique "n'aurait pas souhaité favoriser ce mode d'expression qu'il aurait considéré comme non-physique".

Bref, c'est toute l'architecture de la théorie que Dicke déplore ici, mais au-delà, ce qu'il ne supporte pas, c'est que la science soit elle-même soumise aux aléas de l'histoire à travers le désir d'un homme, fût-il Albert Einstein.

Même si les succès de la relativité générale lui donnent avant tout le statut d'une théorie de la gravitation, telle n'était pas aussi simplement l'ambition, le désir d'Einstein qui y voyait d'abord une théorie des changements de repère en présence d'un champ de gravitation. Ainsi, loin d'être une discipline "verticale" basée sur la gravitation, elle tend sinon à s'imposer, du moins à se faire accepter comme une théorie "horizontale", outil telle la relativité restreinte de (presque) toute la physique. Ainsi, la discipline est-elle mieux définie avant 1960 par l'ensemble des travaux ayant trait à la théorie d'Einstein que par son objet physique, la gravitation. Et certains manuels, pourtant représentatifs de la discipline ne se recouvrent que fort peu, tant son emprise potentielle est considérable. La comparaison des sujets abordés par les ouvrages de J. Chazy "La théorie de la relativité et la mécanique céleste" (1928-1930), de R.C. Tolman "Relativity, Thermodynamics and Cosmology" (1934) et P.G. Bergmann "The theory of Relativity" (1942) est édifiante.

Quant aux préoccupations de ceux qui, avant la seconde guerre mondiale, ont consacré une partie importante de leur temps à la théorie, elles sont à l'évidence à l'image de cette diversité. Beaucoup sont des mathématiciens derrière H. Weyl, plusieurs des astrophysiciens ou des astronomes après S. Eddington et l'on s'étonne de n'y trouver finalement qu'une petite majorité de physiciens-théoriciens dont M. von Laue, est-au-delà d'Einstein - représentatif. Rares sont pourtant ceux qui y consacreront alors tout leur temps et plus d'un s'intéressera un moment à quelque théorie alternative ce qui montre bien que leur réel attachement à la théorie trouve, face à l'austérité de la conjoncture, ses limites.

Mais, à l'inverse, il ne faudrait pas que notre optique, qui répétons-le est de tenter de comprendre le silence relatif de la théorie durant ces années, induise une image désuète des travaux menés. Bien que ce ne soit pas ici le lieu d'en faire l'inventaire ni d'en écrire l'histoire, il n'en demeure pas moins que le renouveau de la discipline dans les années soixante plonge pour partie ses racines dans un certain nombre d'études dont l'intérêt n'est pas nécessairement apparu aussitôt [51]. Aussi bien, un certain nombre de questions théoriques pendantes ont reçu alors des réponses satisfaisantes. Mais dans la mesure où aucune demande pressante n'apparaissait du côté de l'observation, où aucune expérience ne cristallisait les questions théoriques, la relativité générale a sans aucun doute subi, en tant que spécialité un blocage ; d'autant que son absence de dynamisme se traduisait institutionnellement par un manque évident de postes et de crédits induisant par un effet d'entraînement facile à comprendre un moindre intérêt et de moindres résultats.

Tous les témoignages concordent pour indiquer que l'enseignement de la relativité générale a durement subi cette conjoncture . En 1942 Einstein le regrette tandis que S. Chandrasekhar note amèrement qu'entre 1936 et 1961, aucun cours de relativité générale n'a été donné à Chicago [52]. En fait son enseignement est laissé au gré de l'intérêt personnel des universitaires. Ainsi apprend-on dans sa correspondance avec Einstein que Max Born assure un cours à Göttingen en 1929 puis en 1940 à Edinburgh. En France, quelques cours seront donnés au début des années vingt, mais aucun témoignage ne fait état - avant le milieu des années cinquante - d'un enseignement suivi de la théorie. C'est dire le peu d'intérêt qu'elle suscite au niveau institutionnel.

La description que nous a laissée Infeld de l'ambiance qui régnait alors à Princeton résume fort bien la situation :

"En tout cas, dans les années vingt, les scientifiques témoignaient du plus grand intérêt pour la discipline. Mais déjà en 1936, alors que j'étais en contact avec Einstein à Princeton, j'observai que cet intérêt avait presque totalement cessé. Le nombre de physiciens travaillant dans ce champ à Princeton pouvait se compter sur les doigts d'une main. Je me souviens que très peu d'entre nous se rencontraient dans le bureau du défunt Professeur H.P. Robertson et puis même ces rencontres cessèrent. Nous, qui travaillions dans ce champ, étions plutôt regardés de travers par les autres physiciens. Einstein lui-même me faisait souvent remarquer "A Princeton, ils me prennent pour un vieil imbécile : Sie glauben ich bin ein alter Trottel". Cette situation resta à peu près sans changement jusqu'à la mort d'Einstein. La Théorie de la Relativité n'était pas très estimée dans "l'ouest" et mal vue dans "l'est""[53].

Un témoignage qui corrobore celui de Bergmann qui confiait récemment à Abraham Pais que dans ces années-là "Vous n'aviez qu'à savoir ce que faisaient vos six meilleurs amis et vous saviez tout ce qui se passait en relativité générale" [54]. C'est là un témoignage qui souligne le peu de contacts entre les relativistes isolés dans leurs universités mais qu'il ne faut certainement pas prendre au pied de la lettre, sauf à réduire les spécialistes de la discipline aux quelques fidèles proches d'Einstein.

Peu d'ouvrages seront édités entre 1925 et 1955. Fort peu d'articles de revue paraîtront. Ainsi dans le très influent "Handbuch der Physik", la relativité générale, abordée en 1929 sera ignorée jusqu'en 1962, date à laquelle Bergmann y écrira un assez long article. Mais, plus que tout autre le fait que la première conférence internationale consacrée à la relativité générale se tienne quarante ans après la naissance de la théorie pour commémorer, au lendemain de la mort de son inventeur, le cinquantenaire de la théorie de la relativité restreinte, a valeur de symbole.

Mais ces caractéristiques spécifiques de la théorie d'Einstein ne prennent assurément tout leur sens que si on les rapporte aux conditions qui règnent tout au long de cette période dans la discipline voisine, le champ quantique. Là, la structure de production des connaissances y est bien différente : un champ expérimental explosif, un champ théorique extrêmement vivant, une dynamique incomparable. Une rude concurrence pour la relativité générale que le texte de Born [3] illustre parfaitement bien.

On ne peut ici passer sous silence la distance qu'Einstein lui-même oppose aux tenants de l'interprétation de Copenhague qui domine la physique quantique, le point le plus marquant des relations délicates qu'il entretient avec ses collègues quanticiens ; un point qui concerne d'abord bien sûr la position personnelle, l'autorité d'Einstein dans le milieu, mais qui ne sera évidemment pas sans influencer l'image de la théorie qu'il - et qui le - représente ; une prise de position symbolique d'une conception bien définie de la physique théorique.

Aussi bien la personnalité d'Einstein, sa vision du monde, l'image de la physique qu'il projette, son exigence et ses refus, sa solitude, évidemment très proches des caractéristiques de la seconde partie de son oeuvre, tout cela n'est pas sans influencer ceux qui choisissent alors, du milieu des années vingt au sortir de la seconde guerre mondiale, de travailler en relativité générale ; les rares théoriciens qui font ce choix développeront inéluctablement des travaux relativement formels aussi bien parce qu'ils partagent peu ou prou la vision globale d'Einstein, que parce que la théorie les y entraîne et que l'exiguïté de son champ empirique les y contraint.

Exigence théorique, volonté unitaire, intérêt épistémologique affirmé, caractère relativement formel de la production scientifique, refus d'une conception phénoménologique de la construction théorique, large impact de la structure mathématique face à la faiblesse des liens empiriques, caractère artisanal des structures de la recherche, tels sont alors les traits dominants de la production relativiste, des caractéristiques que l'on ne retrouve pas précisément dans le champ quantique. Mais les caractéristiques extraordinairement différentes de ces champs théoriques se révèlent peut-être encore plus nettement quant à la manière de travailler des spécialistes, quant aux structures de la recherche. D'un côté, des théoriciens souvent intégrés dans des laboratoires importants, travaillant en collaboration sur des théories en constante évolution et en liaison avec des expérimentateurs nombreux au service de machines de plus en plus puissantes. De l'autre, des professeurs d'université travaillant le plus souvent isolément sur quelque aspect des équations d'Einstein.

Mais si la spécificité relativiste s'exprime par des éléments techniques et institutionnels, elle ne s'y résume pourtant pas. Une question se pose, que nous n'avons jusqu'alors pas évoquée ; s'il y a tant de raisons, dès 1915, de choisir de travailler en physique quantique - Born nous a éclairés à ce sujet - quelles sont les motivations de ceux qui vont, malgré la logique économique que nous avons esquissée, préférer se plonger dans l'étude de la relativité générale ?

Dans la préface de son "Relativity : the General Theory", un manuel qui paraît en 1960 et qui plus que tout autre représente la somme du travail accompli tout au long de ces années, c'est cette question que Synge pose, décrivant avec beaucoup d'humour l'image qu'il a du relativiste :

"De tous les physiciens, le relativiste est le moins engagé socialement. Il est le grand spécialiste en théorie de la gravitation et la gravitation est socialement signifiante, mais il n'est pas consulté pour la construction d'une tour, d'un pont, d'un bateau, ou d'un avion et même les astronautes peuvent se débrouiller sans lui jusqu'à ce qu'ils se demandent dans quel éther voyagent leurs signaux.

Couper les cheveux en quatre dans une tour d'ivoire n'est pas du goût de tout le monde, et sans aucun doute plus d'un relativiste attend le jour où le gouvernement lui demandera son opinion sur les questions importantes. Mais que signifie "important"? La science a un double but, comprendre la nature et conquérir la nature, mais pour ce qui concerne la vie intellectuelle de l'homme, c'est sûrement la compréhension qui est la chose la plus importante. Alors laissons le relativiste rejoindre sa tour d'ivoire où il a la paix pour chercher à comprendre la théorie d'Einstein aussi longtemps que ce monde mouvementé se satisfera de faire ses affaires sans lui" [55].

Ainsi Synge exprime-t-il ici avec beaucoup de vigueur l'opposition entre deux mondes qui n'ont que peu de points communs sinon celui de partager une même ambition et une même institution : la physique théorique. Les images et les fonctions de ces deux courants sont amplifiés pour donner à penser ; celle du relativiste vu par son collègue physicien, à n'en pas douter spécialiste de la mécanique quantique, celle du quanticien vu par le relativiste. Derrière ces deux mondes qui le plus souvent s'ignorent, derrière ces deux projets - comprendre et conquérir - qui coexistent Synge donne à voir deux cultures qui s'affrontent, deux philosophies qui se heurtent. C'est avant tout une conception artisanale, monastique même de la science et des scientifiques qu'il défend ici ; refus de la physique triomphante aussi bien que de la société aliénante qui n'est pas sans faire penser à celle d'Einstein lui-même ; scepticisme d'un homme idéaliste qui n'espère pas grand-chose de l'évolution de la science moderne.

Mais au-delà de sa position personnelle, sont on ne peut plus clairement posées les questions qui agitent le milieu. Elles trouvent leur source dans l'articulation de la théorie avec l'expérience, la physique et les mathématiques, ainsi que nous y avons insisté plus haut, mais bien plus encore désormais dans l'organisation quasi-industrielle du monde de la physique plus que jamais lié aux pouvoirs économique, militaire, politique. Un monde auquel les relativistes ont jusqu'alors échappé mais, (qu'Einstein nous en préserve !), auquel ils sont de plus en plus confrontés à travers la proximité institutionnelle des théories quantiques dans le cadre de la physique théorique.

Ainsi ce texte de Synge marque-t-il particulièrement bien la frontière entre deux époques que la mort d'Einstein sépare symboliquement et que le congrès de Berne marquera institutionnellement ; celle que nous avons décrite où la relativité générale constitue, à l'intérieur de la physique théorique, un îlot quelque peu suranné à l'abri des grands courants qui agitent les théories quantiques et celle du renouveau que pressent Synge. Une frontière que les organisateurs d'une école d'été consacrée en 1973 aux "Astres Occlus" marquent dans la préface des comptes rendus :

"L'histoire de la transformation prodigieuse de la Relativité Générale pendant ces dix dernières années est chose connue ; d'une baie tranquille où quelques théoriciens poursuivaient leurs recherches, elle est passée aux avant-postes, en pleine effervescence, qui attirent un nombre croissant de jeunes talents, ainsi que de crédits importants destinés aux recherches expérimentales" [56].

Bref, les relativistes vont désormais pouvoir vivre de la relativité générale et non plus seulement pour la théorie d'Einstein.

Références

[1] On se reportera utilement aux articles de revue contenues dans ce volume.
[2] Je serais très reconnaissant à ceux qui voudront bien me signaler les sources - certainement nombreuses - qui ont nécessairement dû m'échapper (lettres, documents d'archives, etc.).
[3] Born (M.), 1955. - Physics and Relativity. - Fünfzig Jahre Relativitätstheorie, Helvetica Physica Acta, Sup. IV, 1956, p. 253.
[4] A ce sujet on se reportera utilement aux travaux suivants :
. Chandrasekhar (S.), 1975. - Verifying the theory of relativity. - Bull. of At. Sc., 31 : 17-22.
. Chandrasekhar (S.), 1979. - Einstein and general relativity : historical perspectives. - Am. J. of Phys., 47(3) : 212-217.
. Crelinsten (J.), 1984. - W.W. Campbell and the Einstein problem. -
. Earman (J.), Glymour (C.), 1980. - Relativity and eclipses : the british eclipse expeditions of 1919 and their predecessors. - Hist. Stu. Phys. Sci., 11(1) : 49-85.
. Earman (J.), Glymour (C.), 1980. - The gravitational red-shift as a test of general relativity : history and analysis. - St. Hist. Phil. Sci., 11 : 175-214.
. Eisenstaedt (J.), 1982. - Histoire et Singularités de la Solution de Schwarzschild (1915-1923). - Arch. for Hist. of Ex. Sci., 27(2) : 157-198.
. Pais (A.), 1982. - Stubtle is the Lord ... . - Oxford Un. Press. N.Y.
. Stachel (J.), 1979. - Einstein's Odyssey. - The Sciences, March 1979 : 14-34.
. Stachel (J.), 1979. - The genesis of general relativity. - Einstein Symposium, Berlin : 428-442.
[5] Popper (K.), 1935. - Logik der Forschung. - J. Springer. On sait l'influence immense d'Einstein et singulièrement de la relativité générale sur Popper.
[6] A ce propos, il faut citer :
. Kuhn (T.S.), 1962. - The structure of scientific revolutions. - Chicago : the University of Chicago Press.
mais aussi :
. Bourdieu (P.), 1976. - Le Champ Scientifique. - Actes de la Recherche en Sciences sociales, 2/3 : 88-104.
[7] Einstein (A.), 1942. in Bergmann (P.G.), 1942. - Introduction to the Theory of Relativity. - New York, Prentice Hall. Preface.
[8] Einstein (A.), 1915. - Zur allgemeinen Relativitätstheorie. - Sitzungber. Berlin, p. 779, 4 Nov. 1915.
[9] Bergmann (P.G.) op. cité p. 211.
[10] Langevin (P.), 1922. - L'aspect général de la théorie de la relativité. - Bull. Sci. des étudiants de Paris, 30 Mars 1922, p. 20.
[11] Weyl (H.), cité par :
Chandrasekhar (S.), 1972. - The increasing role of General Relativity in Astronomy. - Observatory, 92 : p. 160.
[12] Lanczos (C.), 1932. - Stellungder Relativitätstheorie zu anderen physikalischen Theorien. - Naturw., 20(7) : p. 115.
[13] C'est moi qui souligne : Bergmann (P.G.), op cité p. 211.
[14] Einstein (A.), 1916. - Uber die spezielle ... . - Braunschweig, Vieweg, 1917, préface p. V.
[15] Chandrasekhar (S.), op. cité p. 213.
[16] cité par Chandrasekhar, ibid.
[17] op.cité p. 216.
[18] Born (M.), op. cité p. 244.
[19] A ce propos, on lira par exemple l'article de Crelinsten.
[20] Cf. Earman et Glymour, op. cité.
[21] Weyl (H.), 1918. - Raum Zeit Materie. - 1ère éd. Berlin : Springer Verlag, p. 198.
[22] Par exemple dans la préface au livre de Bergmann.
[23] Il est amusant de noter que suivant les époques et les auteurs, ce quatrième test est attribué à des effets différents : l'expérience d'Oetvos, l'effet Hubble, l'effet Shapiro, P.S. R 1913+16 ...

[24] Oppenheimer (J.R.), 1956. - Einstein. - Review of Modern Physics, 28 : p.1.
[25] Dicke (R.H.), 1961. - Mach's principle and equivalence. - Proc. E. Fermi, Varenna, XX : p. 1, Acad. Press New York.
[26] Schild (A.), 1960. - Equivalence principle and Red-shift measurements. - Am. Journal of Physics, 28 : p. 778.
[27] C'est moi qui souligne : Droste (J.), 1916. - Het zwaartekrachtsveld ... . - Leiden : E.J. Brill éd. 1916, p. 26.
[28] A ce propos : Cf. J. Eisenstaedt, op. cité.
[29] Levi-Civita (T.), 1937. - Astronomical consequences of the relativistic two-body problem. - Amer. J. Math., 59 : p. 227.
[30] Bondi (H.), 1962. - A discussion on the present state of relativity. - Proceeding of the R.S. of London (A), 270 : p. 325.
[31] Synge (J.L.), 1950. - The gravitational field of a particle. - Proc. Roy. Irish Soc., 53 : p. 83.
[32] Synge (J.L.), 1970. - Talking about Relativity. - North-Holland Pub., p. 16.
[33] Néologisme qui désigne la cosmologie dans le cadre de la relativité générale.
[34] Bergmann (P.G.), 1957. - Review of Modern Physics, 29 : p. 352.
[35] Synge (J.L.), 1960. - Relativity, the general theory. - Amsterdam : North-Holland Pub., p. 329.
[36] Tolman (R.C.), 1934. - Relativity, Thermodynamics and Cosmology. - Oxford : Oxford Un. Press, p. 445.
[37] A ce propos, on lira : J. Stachel, op. cité.
[38] in Whitrow (G.J.), 1967. - Einstein : the man and his achievement. - British Broadcasting Corp., p. 49.
[39] Whitrow (G.J.), op. cité, p. XII.
[40] Whitrow (G.J.), Morduch (G.E.), 1965. - Relativistic theories of gravitation. - Vistas in Astronomy (6) : 1-67. Oxford : A. Beer ed.
[41] Milne (E.A.), 1940. - Kinematical Relativity. - Oxford : Oxford Un. Press, p. 52. Milne qui aurait bien dû balayer devant sa propre porte, lui qui dans la conclusion de son "Kinematical Relativity" invoquera Dieu (1948, p. 233). Et l'on se demande bien de quel "atavisme" il est vraiment question ... On est en 1940.
[42] Cité par Chandrasekhar, op. cité p. 214.
[43] Temple (G.), 1925. - On mass and energy. - Proc. Phys. Soc., 37 : 269-281.
[44] Car les difficultés techniques n'ont jamais empêché que soit "expliqué" quelque phénomène et vérifiée ou réfutée la théorie.
[45] Rosen (N.), 1940. - General relativity and flat space. - Phys. Rev., 57 : 147-153.
[46] Belinfante (F.J.), Swihart (J.C.), 1957. - Phenomenological linear theory of gravitation. - Annals of Physics, 1 : p. 168.
[47] Lanczos (C.), 1955. - Albert Einstein and the theory of relativity. - Nuovo Cimento, 10(2), Supp. : 1193-1220.
[48] Dicke (R.S.), 1964. - Gravitation and relativity. - New York : Chiu and Hoffmann ed., p. 1.
[49] Dicke (R.S.), 1957. - Gravitation without a principle of equivalence. - Rev. of Mod. Phys., 29 : 363-376.
[50] Dicke (R.S.), 1961. - Mach's principle and equivalence. - in Proceedings "Enrico Fermi", vol. 20 : p. 5. Møller (C.) ed., New York : Academic Press.
[51] Je pense ici tout particulièrement à la cosmologie mais aussi à certains travaux d'ordre mathématique.
[52] Chandrasekhar (S.), 1979. - op. cité p. 214.
[53] Infeld (L.), 1962. - Proceeding on theory of gravitation (G.R.3). - Paris : Gauthier-Villars éd., 1964. p. XV.
[54] Pais (A.), 1982, op. cité p. 268.
[55] Synge (J.L.), 1960. - Relativity : the general theory. - Amsterdam : North-Holland.
[56] DeWitt (C.), DeWitt (B.), 1973. - Black Holes/Les astres occlus. - New York : Gordon and Breach Sc. Pub.

GEOMETRIE ET PHYSIQUE

André Lichnerowicz

A la demande des organisateurs, cet exposé va être un exposé général, aussi peu technique que possible. J'aimerais vous y faire part de réflexions concernant les interactions entre géométrie différentielle et physique théorique, leur passé, leur présent et peut-être certains éléments concernant leur avenir.
Il est bien des manières d'employer nos mathématiques à une meilleure intelligence du concret. Dans certain nombre de cas elles interviennent seulement comme *instrument*. Le scientifique veut parvenir le plus rapidement possible à une comparaison numérique avec l'expérience et assurer ainsi son pouvoir sur les choses.
Mais la mathématique assume souvent un rôle plus ambitieux. Elle se veut aussi *mode de pensée* pour appréhender la réalité et ne prétend à son intelligence que lorsqu'il a été possible de construire, pour l'ensemble des phénomènes étudiés, un modèle mathématique *cohérent*. Les grandes théories physiques de notre temps aboutissent à la création de tels modèles, généralement d'aspect géométrique en un sens large et, ce faisant, elles ont ou bien eu recours à des disciplines mathématiques déjà développées, ou bien ont puissamment contribué au développement de nouvelles structures mathématiques.

L'élément primitif de la géométrie différentielle est la notion de *variété différentiable* et celle-ci provient historiquement de la Mécanique. Le premier exemple de variétés différentiables naturelles à un nombre arbitraire de dimensions que l'homme ait rencontré nous a été donné par la Mécanique Analytique de Lagrange. Dans ses études reposant sur les espaces (ou espace-temps) de configuration des systèmes dynamiques, étaient déjà présents non seulement le concept de variété différentiable mais, plus ou moins implicitement, beaucoup d'autres concepts géométriques décrits en termes de coordonnées locales $\{q^i\}$. Mais nous devons attendre Riemann et sa célèbre Dissertation inaugurale *Sur les hypothèses qui servent de fondement à la géométrie* pour voir partiellement formulé ce qui était précédemment largement implicite. La géométrie des espaces de Riemann était née. L'apparition, due au *cristallographe* Voigt du concept de tenseur, mal dégagé des notions métriques, devait donner à la première et la plus importante des grandes géométries différentielles l'un de ses instruments essentiels. L'analyse tensorielle riemannienne, développée par Christoffel dans le domaine des mathématiques pures était, il faut l'avouer, quelque peu formelle et gratuite, mais dès 1900 avec Ricci et Levi-Civita d'une part elle se géométrise, d'autre part elle donne naissance à de multiples applications à la dynamique classique, à l'élasticité et à la théorie des milieux anisotropes. C'est là qu'à partir de 1912, Einstein devait trouver un cadre tout prêt à accueillir le développement de la théorie relativiste de la gravitation.

A propos de cet exposé, j'ai relu l'extraordinaire mémoire de 1900 dû à Ricci et Levi-Civita et savouré à nouveau sa modernité. Que ce soit du point de vue de la géométrie riemannienne ou de celui des applications physiques, on trouve déjà là, parfaitement développée, bon nombre des concepts de base. Grâce à leur approche et à leurs idées, il devient possible de traduire la mécanique analytique du système mécanique le plus banal (système holonome à contraintes indépendantes du temps) en termes d'une géométrie déjà raffinée. L'étude du mouvement d'un tel système devient l'étude du mouvement d'un point dans un espace de Riemann dont la variété de base est l'espace de configuration et dont la métrique est donnée par l'énergie cinétique. Les applications de ce point de vue, qui est lagrangien dans son essence, sont remarquables par exemple en ce qui concerne certains problèmes de stabilité. Bien entendu la métrique est toujours définie positive.

On peut dire en somme qu'Einstein a eu la possibilité d'élaborer rapidement son admirable et courageuse théorie de 1915 grâce à l'édifice construit préalablement par Ricci et Levi-Civita et il a toujours reconnu ce qu'il devait à ces géomètres.

En fait, quand parurent les travaux de Ricci, ils ne furent accueillis par les mathématiciens de l'époque que par une estime polie. Bien qu'ils aient suscité l'intérêt de quelques spécialistes à travers le monde, leur champ était malheureusement trop loin des centres d'intérêt de la communauté mathématique principalement préoccupée d'ensembles, d'intégration ou de fonctions de variables réelles ou complexes. Quant aux géomètres différentiels, ils polissaient studieusement la théorie des surfaces plongées dans l'espace euclidien ou se préoccupaient de géométrie projective et conforme, ou plus généralement des espaces de Klein venus du programme d'Erlangen et ils trouvaient l'oeuvre de Ricci "quelque peu formelle et artificielle" selon un mot de Darboux. Seul Poincaré envisageait la possibilité d'une "géométrisation globale des problèmes dynamiques".

La situation change complètement avec l'apparition de la théorie de la Relativité générale. A partir de la fin de la première guerre mondiale, cette théorie a fourni pendant quinze ans un champ d'intérêt commun aux mathématiciens et aux physiciens théoriciens. Autour d'elle et stimulés par elle et par sa vogue alors prodigieuse, les savants se mettent à créer de nouvelles structures géométriques et l'on peut dire, je crois, qu'elle fut l'un des moteurs et l'une des sources authentiques de la géométrie différentielle contemporaine.

D'abord vint la découverte presque simultanée par Levi-Civita, Pérès et Schouten de la notion de parallélisme dans un espace de Riemann. Malgré ce qui subsistait encore de "trop analytique" dans ce projet géométrique, on voyait se révéler un instrument, un langage, des intuitions qui devenaient vraiment géométriques.

D'autre part la physique elle-même voulait briser le cadre trop restrictif de la géométrie riemannienne. Les physiciens du temps ne connaissaient que deux champs physiques : le champ électromagnétique et le champ gravitationnel. C'était l'étude du champ électromagnétique qui avait conduit les physiciens à construire la théorie de la Relativité restreinte et à s'intéresser à cet étrange espace-temps plat de Minkowski, avec sa métrique indéfinie. Dans le cadre de la Relativité générale, qui est essentiellement une théorie relativiste de la gravitation, le champ électromagnétique n'intervenait pas naturellement, mais semblait surimposé artificiellement de l'extérieur. Il y avait là scandale, un scandale qui ne nous affecte plus de la même façon, à cause de la multiplicité et de la richesse des champs physiques que nous avons été obligés de prendre au sérieux. Mais nous y reviendrons.

Ainsi la physique théorique s'est sentie dans l'obligation de découvrir une "théorie du champ unifié". Pendant bien des années, elle a cherché à développer une théorie qui puisse apporter l'unification des champs gravitationnel et électromagnétique au sein d'un unique hyperchamp dont la donnée soit équivalente à celle d'une structure géométrique pour l'univers.

Depuis 1919, date où Hermann Weyl développe le premier essai d'une telle théorie, les efforts se multiplient, se révélant tous insatisfaisants physiquement pour une raison ou une autre. A côté des espaces de Weyl, pour lesquels c'est le groupe des similitudes qui joue le rôle fondamental, on vit apparaître l'espace d'Eddington, premier exemple d'un espace à connexion affine. Ces premières ébauches, maladroites du point de vue physique, ont été importantes pour le développement de la géométrie différentielle. Mais il s'agissait de constructions assez isolées les unes des autres et il en était de même des premiers essaisd'Einstein lui-même. Ces constructions isolées furent englobées à partir de 1922 dans la synthèse réalisée par le théoricien des groupes qu'était Elie Cartan. Pour celui-ci, ses "espaces généralisés", comme il les nommait, devaient avoir une importance considérable en Physique Mathématique. Devant le développement présent des théories de jauge, il ne nous appartient pas de le contredire.

On me permettra de citer sur ce sujet un passage remarquable de la Notice d'Elie Cartan qui met en évidence clairement cette interaction entre géométrie et physique qui me préoccupe aujourd'hui.
"Les espaces que j'ai imaginés", dit Elie Cartan", sont une généralisation à laquelle il était impossible de parvenir en suivant les idées directrices de Riemann, Weyl et Eddington. C'est ma conception de la structure des groupes qui m'a guidé et montré la voie".
"On sait que, par la théorie de la relativité générale, Einstein a bouleversé la conception qu'on se faisait jusqu'à lui, d'un espace-temps homogène, préexistant aux phénomènes et dans lequel ces phénomènes viennent s'insérer sans l'altérer... La théorie de la relativité restreinte respectait cette conception et avait tout au plus pour résultat de changer la nature du groupe fondamental de l'univers. Il n'en est plus de même avec la relativité générale pour laquelle les propriétés géométriques et cinématiques de l'espace-temps sont en quelque sorte contingentes et dépendent de la distribution de la matière... Est-ce à dire que la notion de groupe est exclus de la Physique ? Il n'en est rien, car l'hypothèse fondamentale de la théorie d'Einstein est non pas, comme beaucoup de personnes l'ont cru, qu'il est possible de formuler les lois de la Physique dans tout système arbitraire de coordonnées, ce qui serait une simple tautologie, mais que dans toute région suffisamment petite de l'espace-temps, les lois de la relativité restreinte sont vraies en première approximation.
"D'après l'hypothèse d'Einstein, les phénomènes physiques doivent permettre de révéler au physicien le caractère euclidien d'une très petite portion de l'univers. Or ... le caractère euclidien infinitésimal de l'univers se révèlera dans le déplacement euclidien infinitésimal (qui amène le repère orthonormé d'un observateur en coïncidence avec le repère orthonormé d'un observateur infiniment voisin). Par suite, non seulement la distance des deux observateurs, mais encore la comparaison des directions issues de ces deux observateurs sont l'une et l'autre du domaine de l'expérience"
De ces considérations, Elie Cartan déduit que l'espace-temps le plus

général compatible avec ce qu'il nomme l'hypothèse d'Einstein est un espace à connexion "euclidienne" (nous dirions lorentzienne), pouvant comporter une torsion arbitraire et il montre que, dans ses recherches concernant une théorie du champ unitaire, les idées d'Einstein ont évoluée précisément dans cette voie qu'il avait indiquée dès 1922.
"Ce que je viens de dire", ajoute Cartan,"s'applique à tout autre groupe que celui des déplacements euclidiens et conduit ainsi à la création de classes étendues d'espaces qui interviennent en Analyse et Mécanique... J'ai en particulier recherché en 1923 si les phénomènes mécaniques seuls permettraient la comparaison des directions dans l'espace-temps et j'ai montré qu'il en serait ainsi si chaque élément de matière avait un moment cinétique non infiniment petit par rapport à la quantité de mouvement".
On voit clairement, me semble-t-il, sur ce passage combien historiquement géomètrie et physique se sont appuyées l'une sur l'autre, en se communiquant aussi bien leurs exigences propres que leurs intuitions respectives. Physiquement, nous nous trouvons aux sources de la théorie dite d'Einstein-Cartan développée dans un grand mémoire de 1923 et renouvelée récemment en particulier par Trautman. Mathématiquement, Elie Cartan a réussi, en rapprochant les différentes géométries différentielles existantes du programme d'Erlangen, à montrer comment à tout groupe de Lie, il est possible d'associer "une classe d'espaces à connexion de ce groupe". Le programme tracé par Elie Cartan dépassait de beaucoup les réalisations effectivement construites à cette époque. Il est encore largement notre programme.

Ce programme s'énonce aujourd'hui en termes d'espaces fibrés, une structure qu'Elie Cartan semblait appréhender presque intuitivement et dont les fondements et les premières études furent développés par son élève Ehrresmann et par Whitney au cours des années 1937-1939. Le langage des fibrés n'a jamais été explicitement utilisé par Cartan, mais je l'ai entendu souvent parler de "la variété des repères", ces repères pouvant correspondre à différents groupes.
Les espaces fibrés naturels les plus simples sont ceux définis par l'espace des vecteurs ou celui des covecteurs tangents aux différents points d'une variété différentiable, ou bien entendu l'espace des repères linéaires aux différents points de la variété. Dans ce dernier cas la fibre en un point est <u>difféomorphe au groupe structural</u> qui est ici le groupe linéaire. Un tel fibré est appelé un fibré principal et aux espaces à connexion du groupe G de la terminologie de Cartan correspon-

dent dans la terminologie contemporaine les connexions sur un fibré principal de fibre G . Dans le cas du fibré des repères linéaires, on obtient les connexions linéaires ; celles-ci sont souvent incorrectement appelées affines parce qu'à chacune d'elles est canoniquement associée une connexion affine (c'est-à-dire sur le fibré des repères affines) dont la courbure traduit à la fois la courbure de la connexion linéaire et sa torsion. Si la variété est munie d'une métrique, riemannienne ou lorentzienne, au fibré principal des repères orthonormés correspondent les connexions métriques dont une et une seule, la connexion dite riemanienne, est sans torsion. Selon la signature, le groupe structural est soit le groupe orthogonal, soit le groupe de Lorentz. Dans le cas affine, c'est soit le groupe des déplacements, soit le groupe de Poincaré.
Une remarque concernant la Relativité générale. Contrairement à de qui est trop souvent cru, l'interprétation physique ne doit faire intervenir en principe que le fibré des repères orthonormés et les grandeurs qui se déduisent de repères orthonormés. Les cartes locales servent seulement en principe, si j'ose dire, à la cartographie de l'espace-temps, à la détermination de la place d'un évènement, mais non directement aux mesures. Beaucoup de difficultés apparentes sont évitées dans les interprétations physiques, si on utilise systématiquement des repères orthonormés dont le vecteur temporel est donné par le vecteur-vitesse unitaire de l'observateur.

Même en Mécanique Analytique classique, le concept de fibré s'est révélé l'instrument fondamental pour une géométrisation globale des problèmes dynamiques, pour reprendre l'expression de Poincaré. Il s'agit en fait d'assurer à la représentation mathématique une objectivité qui corresponde à l'objectivité même des phénomènes physiques étudiés et c'est, sans doute, cette exigence qui rend nécessaire la liaison entre géométrie et physique pour assurer la pleine intelligence de celle-ci
En formalisme lagrangien, l'espace de base pour l'étude de la mécanique d'un système dynamique holonome est le fibré des directions tangentes à l'espace-temps de configuration du système ; un état dynamique n'est autre qu'un point de cet espace. Si les liaisons sont indépendantes du temps, on peut se réduire à l'étude du fibré tangent à l'espace de configuration. L'équation différentielle du second ordre qui détermine le mouvement n'est pas ici directement liée à la géométrie de l'espace. Il n'en est pas de même en formalisme hamiltonien. Pour un système à liaisons indépendantes du temps, l'espace de base est le fibré cotan-

gent à l'espace de configuration ; ce n'est autre que l'espace de phase des physiciens ou mécaniciens. Il est bien connu qu'un tel fibré admet une structure symplectique naturelle exacte, définie par la 2-forme de Poincaré F qui s'écrit localement $\sum dp_i \wedge dq^i$. L'étude des équations de la Mécanique a montré, depusi Hamilton et Jacobi, qu'il est essentiel de pouvoir effectuer des changement de coordonnées (p_i, q^i) ne respectant pas la structure cotangente. On est ainsi conduit à prendre pour espace de phase une variété symplectique (W, F) qui définit la géométrie du système dynamique.

Un mouvement est alors décrit par une courbe intégrale c(t) d'un champ hamiltonien X_H, automorphisme infinitésimal défini par l'hamiltonien H , de la structure symplectique, t désignant le temps classique Telle est la signification géométrique des classiques équations d'Hamilton. L'étude de l'algèbre de Lie des automorphismes infinitésimaux symplectiques conduit naturellement au crochet de Poisson et à une forme intrinsèque (sans coordonnées) de l'équation de la dynamique. L'étude systématique, dans l'optique de Souriau, de l'espace des mouvements (étude reprise selon lui de Lagrange) rend encore plus manifeste l'importance des structures symplectiques.

L'approche hamiltonienne d'un système dynamique holonome dont les liaisons dépendent du temps est moins connu. Dans ce cas, l'espace des états est le fibré des directions cotangentes à l'espace-temps de configuration. Cette variété W , de dimension (2n+1) admet une structure de contact naturelle π (où π est une 1-forme densité convenable de poids-1/(n+1)).

Un mouvement est alors décrit sur la variété de contact (W,π) par une courbe intégrale c(s) d'un automorphisme infinitésimal de la structure de contact ; le paramètre s n'est autre que l'action à un facteur constant près.

On notera que cet énoncé est indépendant de tout repèrage de l'espace-temps de configuration et englobe à lui seul tous les principes de la mécanique classique. On peut rendre cet énoncé plus maniable en passant à la symplectisation naturelle de la variété de contact, variété qui peut être identifiée ici au fibré cotangent de l'espace-temps de configuration dont la dimension est (2n+2).

Les rapports géométriques entre formalismes hamiltonien et lagrangien ont été récemment approfondis en particulier par Tulczyjev. A partir d'efforts concernant l'analyse, une géométrisation de la dynamique des champs physiques et une présentation rigoureuse des formalismes lagrangien et hamiltonien correspondants est recherchée à partir de la considération de variétés d'applications, variétés qui sont de dimension infinie et comportent des singularités dont certaines présentent de grandes difficultés. Des approches ont été effectuées par Madame Choquet-Bruhat, Marsden ou Alan Weinstein. Des progrès concernent les champs scalaires ou électromagnétiques, de grosse difficultés demeurent pour le champ gravitationnel et le rôle de singularités joué par les isométries.

La présentation conventionnelle de la mécanique quantique à partir du formalisme hamiltonien présente le grave défaut de n'être pas explicitement covariante. La géométrie ne semble y jouer aucun rôle. Au cours des quinze dernières années, deux approches distinctes présentant certaines interactions se sont efforcées de donner à la mécanique quantique un statut géométrique analogue à celui de la mécanique classique. Il s'agit d'une part de la quantification géométrique de Kostant-Souriau, d'autre part de la théorie des star-produits due à Flato, J. Vey et moi-même. La première approche utilise comme espace de phase un espace homogène symplectique G/H, revêtement selon un théorème de Souriau d'une orbite de G correspondant à sa représentation coadjointe selon le point de vue de Kirillov-Kostant-Souriau ; elle met en oeuvre des polarisations réelles ou complexes qui correspondent à des généralisations des structures cotangentes. La seconde approche vise à considérer la mécanique quantique comme déformation non triviale de la mécanique classique au sens de Gerstenhaber, le paramètre de déformation étant $\hbar/2i$. On obtient ainsi un processus de quantification qui conduit aux mêmes résultats que la mécanique quantique conventionnelle, mais qui s'applique dans des conditions plus larges.

Ces deux voies semblent intéressantes et prometteuses, mais elles suscitent encore, sans doute à tort, plus d'intérêt chez les mathématiciens que chez les physiciens. Elles posent d'ailleurs des problèmes mathématiques et, en particulier, cohomologiques d'un intérêt certain. En ce qui concerne les déformations, c'est la cohomologie dite de Hochschild qui joue le rôle principal.

Un objet géométrique remarquable, distinct du tenseur, était né en 1913 des recherches d'Elie Cartan sur les représentations irréductibles du groupe orthogonal. Mais cet objet était resté un peu dans l'ombre, soigneusement caché derrière la transparence des mathématiques pures. Ce n'est que lorsque le génie de Dirac, cherchant à établir une théorie relativiste de l'électron et de l'électromagnétisme le redécouvrit quelques quinze ans plus tard qu'il acquit un nom et une reconnaissance privilégiée dans le domaine de la Science. Ainsi le concept de spineur né deux fois, d'abord entre les mains d'un mathématicien, puis dans celles d'un physicien. S'il a été depuis longtemps un instrument fondamental de la théorie quantique des champs, il est devenu au cours des vingt dernières années un outil nouveau et puissant en géométrie différentielle globale, principalement en liaison avec l'oeuvre remarquable d'Atiyah et Singer et le théorème de l'indice. A partir de lui on a pu mettre en évidence de nouveaux invariants topologiques pour une variété compacte.

Jusqu'en 1960, les champs physiques étaient tous décrits comme champs essentiellement tensoriels ou spinoriels, en accord avec la parité de leur spin, à une seule exception près, le champ gravitationnel décrit par une connexion riemannienne, dont la courbure, quelle que soit son importance, ne semblait jouer physiquement qu'un rôle subordonné. Pourtant les interactions qu'il convenait d'introduire en électromagnétisme conduisaient à prendre au sérieux le potentiel-vecteur lui-même, défini à une transformation de jauge près, et non pas seulement la 2-forme champ électromagnétique.

Il y a vingt-cinq ans, Yang et Mills, dans le but de créer un instrument capable de décrire les interactions fortes, cherchèrent à généraliser la théorie du champ électromagnétique sur l'espace-temps de Minkowski en introduisant 2-formes et 1-formes potentiels vecteurs à valeurs dans une algèbre de Lie semi-simple, dont le choix pouvait être discuté. Il fut remarqué très tôt (1) (Mme Kerbrat 1962), dès qu'on voulut étendre la théorie et les équations du champ de Yang-Mills à un espace-temps courbe et à différents algèbres de Lie, qu'en fait la 1-forme potentiel devait être conçue, sur un fibré principal de base l'espace-temps, comme une 1-forme de connexion dont la courbure donnait le champ proprement dit. Il en était ainsi pour le champ électromagnétique lui-même, la 1-forme potentiel-vecteur étant une 1-forme de connexion sur un fibré principal de groupe S^1 sur l'espace-temps. On fut ainsi conduit à donner de la charge électrique une interprétation en termes de 1-classe de cohomologie et à généraliser la situation en ce qui concerne les symétries internes, en comprenant mieux ce qu'il y avait à la base de la vieille théorie de Kaluza-Klein.

Tout cela diffusa lentement chez les physiciens théoriciens, la notion générale de connexion sur un fibré principal étant encore trop peu répandue, puis explosa brusquement en une théorie plus ou moins générale des champs de jauge dont d'autres présents parleraient mieux que moi, mais qui retrouve, sous un tout autre aspect de la physique quantique, certaines des ambitions du vieux rêve d'Einstein.

On doit noter que les solutions self-duales classiques des équations de

Yang et Mills intéressent particulièrement les physiciens. Une telle étude a été menée jusqu'à son terme, dans le cas proprement euclidien, par Atiyah et Hitchin (qui se ramenèrent à un problème de géométrie algébrique), puis par une méthode complètement différente, basée partiellement sur des fonctions quaternioniennes, par Gursey. Les retombées effectives de cette étude sur la physique, qui en fait n'est concernée que par l'espace-temps à signature hyperbolique, ne me semblent toujours pas très claires.

En spins impairs, on avait été amené à étudier, sur une algèbre de Grassmann, un anticrochet de Poisson qu'il était possible, dans mon optique, de déformer. Lorsqu'on a voulu traiter sérieusement certains problèmes de symétrie des particules physiques et des champs correspondants, on a été conduit à substituer à des algèbres de Lie des algèbres de Lie graduées et, comme les physiciens le disaient à partir de 1971, à introduire à côté des variables commutantes (correspondant aux coordonnées sur une variété différentiable) des variables anticommutantes ne bénéficiant que d'un statut formel.

Tel était le cadre physique, peu satisfaisant mathématiquement, de la théorie des supersymétries. Fort heureusement, Kostant a introduit la notion de variété différentielle graduée, notion qui a été à la fois raffinée et simplifiée par Koszul ; on peut s'en servir pour une construction mathématique correcte de ce que les physiciens nomment une supervariété.

J'aimerais faire une observation : depuis quatre ou cinq ans, tous les cours d'été de physique théorique, ceux auxquels j'ai participé, comme ceux dont j'ai lu les comptes rendus, commencent par cinq ou six leçons de géométrie différentielle concernant différentielle extérieure, fibrés et connexions, géométrie riemannienne ou symplectique, des leçons qui semblent banales aux mathématiciens depuis vingt-cinq ans au moins, mais qui apparaissent désormais comme devenues partie intégrante de la physique théorique. Cette évolution importante et après tout rapide a été rendu possible par l'action persévérante de quelques physiciens peu nombreux qui ont prêché pendant quinze ans parfois dans le désert. Il convient de leur rendre hommage, il convient aussi que tous les jeunes scientifiques prennent conscience des nouveaux moyens et modes de pensée qui sont à leur disposition.

Qu'en est-il de l'avenir ? Je ne puis que me livrer à des considérations, les unes fragiles comme toute prophétie, les autres plus assurées dans la réalité scientifique présente.

Les équations de la physique sont pour nous hyperboliques par nature, les problèmes mettant en oeuvre des équations elliptiques consistant dans la recherche de solutions correspondant, en un sens ou un autre, à un caractère stationnaire et les équations de type parabolique ne pouvant être, à nos yeux, qu'approchées ou statistiques. Ce sont essentiellement des équations non linéaires qui nous importent, le cas linéaire étant soit approché, soit exceptionnel. Cela est visible aussi bien en Relativité générale que dans les théories de jauge. Les vrais problèmes posés par la physique sont des problèmes globaux pour lesquels les solutions doivent satisfaire à des conditions aux limites aussi bien qu'à des conditions initiales.

Nous sommes donc amenés à étudier l'existence, l'unicité ou la stabilité de solutions globales de problèmes hyperboliques non linéaires. Il s'agit là de recherche pour lesquelles nulle méthode mathématique vraiment générale d'approche n'est encore conrue. De premiers résultats prometteurs ont été obtenus par Mme Choquet-Bruhat en ce qui concerne les équations d'Einstein et celles du champ de Yang et Mills.

Flato et J. Simon ont d'autre part introduit une cohomologie non linéaire sur les groupes qui permet d'approcher le problème de la linéarisation possible d'équations de la Physique Mathématique invariantes ou covariantes par un groupe et, à travers une telle linéarisation, de mettre en évidence l'existence de certaines solutions globales. La cohomologie non linéaire de Flato-Simon semble en rapport avec la cohomologie cyclique, développée et appliquée dans un autre contexte par Alain Connes et qui est une extension de la cohomologie de Hochschild pour les algèbres associatives. La cohomologie semble envahir de toute part la physique.

L'avenir me paraît donc double :

1°) Approfondissement local et, lorsque cela est possible, global de problèmes précis concernant les milieux continus et les champs, classiques ou quantiques

2°) Essai d'intelligence prodonde du rôle joué par la cohomologie en physique mathématique

Les théories les plus générales de la physique de l'avenir me semblent ne pouvoir être que de nature de plus en plus cohomologique.

C'est ainsi que le vieux savant que je suis cherche à entrapercevoir l'avenir à dix ou quinze ans. Sans doute celui-ci me démentira-t-il, mais je pense qu'il ne pourra le faire que partiellement.

(1) Voir aussi Ikeda et Miyachi Prog. of Theor. Phys. 16 (1956), 537-547.

SUPERGRAVITIES

Yvonne Choquet-Bruhat
Université Paris VI
Institut de Mécanique
4, Place Jussieu, 75005, Paris, FRANCE

Introduction

The physical motivation of original (called now simple) supergravity (space time dimension $d = 4$, one gravitino), obtained by Deser-Zumino and Ferrara-Freedman-Van Nieuwenhuisen in 1976 was twofold :

1. Search for a renormalizable quantum gravity
2. Search for a theory which would unify bosons and fermions.

These two goals were in fact united by the hope that the coupling of the graviton, boson field, with a fermi field will improve the renormalizability of the quantum theory, due to the possible minus signs introduced in the action by such a field. The unification of the laws of physics into a master law has always been an aim for Science : to assemble elementary particles - which look as different as bosons and fermions into one family, whose members can exchange their roles, is a great success.[1] Simple supergravity does it for the graviton (spin 2, metric $g_{\alpha\beta}$) and gravitino (spin 3/2, spinor valued 1 form ψ_λ) in the sense that the lagrangian admits an infinitesimal invariance - called a supersymmetry transformation - which mixes their roles, namely (ε being the parameter of the transformation, a spin 3/2 field)

$$\delta_\varepsilon \, g_{\alpha\beta} = \bar{\varepsilon}(\gamma_\alpha \, \psi_\beta + \gamma_\beta \, \psi_\alpha) \quad , \quad \delta_\varepsilon \, \psi_\lambda = D_\lambda \, \varepsilon$$

The renormalizability properties of quantum supergravity are another partial success : they are better than those of gravity alone.

A mathematical motivation of simple supergravity is the difficulty raised in a curved space time by the Rarita-Schwinger equations for the spin 3/2 field : these equations do not admit solutions with the expected generality, they are subjected to "integrability conditions" (Latremolière 1971); one also says that there is "anomalous propagation" : the system is not causal in the relativistic sense. It is already the case for the spin 3/2 equations in Minkowski space time in the presence of an electromagnetic field (Fierz and Pauli 1939 noticed the problems raised in the quantification, Velo and Zwanziger 1969 studied the non quantum system and showed the anomalous propagation). The removal of this difficulty by the coupling with another field (electromagnetic) has been tried by Madore (1975), and he obtained it by choosing a particular relation between the constants involved. Simple supergravity is indeed, at the classical level, a causal theory (cf § III) though the coupling is rather peculiar, since the fields take now their values in a graded algebra. Simple supergravity is a rigid theory : no classical matter sources can be added in a trivial way to the lagrangian with preservation of the consistency(2) of the equations : the infinitesimal supersymmetry which insures this consistency is lost if a source term other than the spin 3/2 appears in the equations. This consistency can eventually be regained if together with these new (bosonic) fields one introduces other "gravitini" (i.e. spin 3/2 fields) : these are the so called extended supergravities, N = 1, ..., N = 8(3), where N is the number of gravitini. The rule of the game is to find a lagrangian which will be the sum of the usual lagrangian for these various bosonic fields, and gravitini, and to add to these lagrangians interaction terms so chosen that the equations become consistent (equivalently, the lagrangian admit N supersymmetries).

A very tempting approach, stimulated by the success of simple supergravity on the one hand and the remarkable progress of Yang-Mills theory on the other hand , towards unification of the non gravitational fundamental interactions,has led to a revival of the Kaluza-Klein scheme, extended long ago (B. DeWitt 1965, R. Kerner 1968) from the Einstein-Maxwell system to the Einstein-Yang-Mills case. The aim now is to find a supergravity theory in a space-time of $d = 4+n$ dimensions, with one spin 3/2 field ($N = 1$). The reduction to dimension 4 of the various fields should give all the observed matter and interaction fields. Unfortunately the lagrangian of simple supergravity in dimension $d > 4$ does not lead to a consistent theory (as could be foreseen by "counting of states", cf § IV): this consistency can be obtained only by the addition of other fields, and ad hoc interaction terms. A particular success is the $d = 11$ super gravity found by Cremmer and Julia. They have constructed a supersymmetry lagrangian by adding to the 11 dimensional metric and spin 3/2 field an exterior 3-form, with a Maxwell-type action, and appropriate interaction terms : these terms are uniquely determined (only higher order interactions terms could eventually be added) by the requirement that the lagrangian admits an infinitesimal supersymmetry. This theory can be proved to be causal (Y. Choquet-Bruhat 1984) at the classical level, in the same sense as simple supergravity.

I - Einstein-Cartan theory with source a spin 3/2 field, in d-dimensional space-time.

1. Definitions

In supergravity theories the usual Einstein lagrangian is replaced by the scalar curvature of a connection $\underset{\sim}{\omega}$, metric but with torsion, that is by the Einstein-Cartan lagrangian, as it is to be expected for spinor sources with a naturally non symmetric stress energy tensor. In addition to the metric $\underset{\sim}{g}$ appears, as auxiliary unknown, the orthonormal moving frame[(4)] $\underset{\sim}{e}$.

It turns out that equations for spin 3/2 fields are proved to be consistent only if the spinors are considered to be anticommuting[(5)], that is their components in a frame are not numbers, but elements of the odd subspace $\mathfrak{a}^-$ of a $\mathbb{Z}_2$ graded algebra $\mathfrak{a}$. The equations coupling the spinors to $\underset{\sim}{e}$ and $\underset{\sim}{\omega}$ impose that these other fields take also their values in $\mathfrak{a}$; it is supposed, and it will be compatible with the equations, that they take their value in the even subalgebra $\mathfrak{a}^+$.

The product of two elements of $\mathfrak{a}^-$ (called odd) is anticommutative, the product of an element of $\mathfrak{a}^+$ (called even) with any element of $\mathfrak{a}$ is commutative.

An $\mathfrak{a}$ valued tensor, connection, spinor, at a point x of the differentiable manifold V is an equivalence class of sets of elements of $\mathfrak{a}$. A representant of a class is associated with a frame at x (natural frame, lorentz frame, spin frame); by a change of frame it changes according to the classical rules for tensor, connection, spinor.

Tensor or spinor fields, or connections on V associate to each $x \in V$ a tensor, or spinor, or connection at x.

We also suppose that $\mathfrak{a}$ is endowed with a locally convex topology : we can then define the derivatives $\frac{\partial}{\partial x^\lambda} f$ of a representant f of a tensor, spinor, or connection, in a local chart of V. Such a derivation is additive and obeys the Leibnitz formula

$$\frac{\partial(fg)}{\partial x^\lambda} = \frac{\partial f}{\partial x^\lambda} g + f \frac{\partial g}{\partial x^\lambda}$$

Covariant derivatives, tensor products and contractions are defined through the usual algebraic formulas for their representants. The order of factors is now relevant, for spinorial fields.

The moving frame $\underset{\sim}{e}$ is a set of d, $\mathfrak{a}^+$ valued, vectors on the (ordinary, C^∞) manifold V, with local coordinates x^M, $M = 0,1,\ldots d-1$.

$$\underset{\sim}{e}_A = e_A{}^M \partial_M \quad , \quad \partial_M = \partial/\partial x^M \quad , \quad A, M = 0, 1, \ldots, d-1$$

The matrix e_A^M is supposed to be invertible in $\mathfrak{a}$, that is there exists e_M^A such that

$$e_A^M\, e_M^B = \delta_A^{\ B}\, \mathbb{1} \quad , \quad e_A^M\, e_N^A = \delta_N^M\, \mathbb{1}$$

The "metric" g_{MN} is the symmetric, even valued covariant 2-tensor

$$g_{MN} = e_M^A\, e_N^B\, \eta_{AB} \quad , \quad \eta_{AB} = \mathrm{diag}(1,-1,\ldots -1)$$

One considers on V a connection $\underset{\sim}{\omega}$, metric for $\underset{\sim}{g}$, but with torsion $\underset{\sim}{S}$. The difference of $\underset{\sim}{\omega}$ and the riemannian connection $\underset{\sim}{\check{\omega}}$ of $\underset{\sim}{g}$ is a tensor, called the contorsion tensor

$$\omega_M{}^A{}_B = \check{\omega}_M{}^A{}_B + C_M{}^A{}_B \qquad \mathrm{I}(1-1)$$

The connection $\underset{\sim}{\omega}$ is metric if and only if

$$\omega_M{}^{AB} = -\,\omega_M{}^{BA} \qquad \mathrm{I}(1-2)$$

that is (since $\underset{\sim}{\check{\omega}}$ is metric), if and only if the contorsion satisfies

$$C_M{}^{AB} = -\,C_M{}^{BA} \qquad \mathrm{I}(1-3)$$

The torsion of a metric connection is

$$2\, C_{[N}{}^A{}_{M]} = S_M{}^A{}_N \qquad \mathrm{I}(1-4)$$

Using I(1-3) and I(1-4) we find that the contorsion, and thus the metric connection $\underset{\sim}{\omega}$, is determined in terms of its torsion by

$$C_M{}^A{}_B = -\,\frac{1}{2}(S_M{}^A{}_B + S^A{}_{BM} + S^A{}_{MB}) \qquad \mathrm{I}(1-5)$$

We suppose that $\underset{\sim}{S}$ (and thus $\underset{\sim}{C}$) is, like $\underset{\sim}{\check{\omega}}$, $\mathfrak{a}^+$ valued.
The curvature of the connection $\underset{\sim}{\omega}$ is the 4-tensor with components (frames corresponding to names of indices)

$$R_{MN}{}^A{}_B \equiv 2\,(\partial_{[M}\, \omega_{N]}{}^A{}_B + \omega_{[M}{}^A{}_C\, \omega_{N]}{}^C{}_B) \qquad \mathrm{I}(1-6)$$

The Ricci tensor and the scalar curvature are respectively

$$R_M{}^A \equiv e^{NB}\, R_{MN}{}^A{}_B = e^N{}_B R_{MN}{}^{AB} \quad , \quad R = e_A^M\, R_M{}^A = e_A^M\, e_B^N\, R_{MN}{}^{AB} \qquad \mathrm{I}(1-7)$$

Note that if the torsion of $\underset{\sim}{\omega}$ is not zero the Ricci tensor is in general non symmetric, $R_{MP} \neq R_{PM}$.

2. Lagrangian and equations.

A natural candidate for the lagrangian of supergravity (i.e. for a spinor valued 1-form $\underset{\sim}{\psi} = (\psi_M)$ on a d-dimensional Einstein Cartan manifold) is the real lagrangian, which is not an exact derivative

$$\mathcal{L} = \mathcal{L}_{EC} + \mathcal{L}_{3/2} ,$$

with

$$\mathcal{L}_{3/2} = k \int (\bar{\psi}_M \Gamma^{MNP} D_N \psi_P + D_N \bar{\psi}_P \Gamma^{MNP} \psi_M) \tau \qquad \text{I(2-1)}$$

$$\mathcal{L}_{E.C} = \int R \tau$$

where τ is the metric volume form; $\mathcal{L}_{3/2}$ is the generalisation to d-dimensions of the classical Rarita-Schwinger lagrangian(6). In the expression I(2-1) the derivative D_N denotes the exterior derivative on 1-forms, and the covariant derivative in the Cartan connection in spin space. Equivalently, due to the antisymmetric in N and P of Γ^{MNP} we can define D_N as the riemannian covariant derivative on tensor indices, and the derivative in the Cartan connection on spinor indices. We thus obtain covariant expressions, but the riemannian connection does not enter explicitly in the lagrangian $\mathcal{L}_S$ - no extra terms in δe in its variations need be considered.

From the expression I(1-7) of R and

$$D_N \psi_P = \partial_N \psi_P - \check{\omega}_N{}^M{}_P \psi_M - \frac{1}{4} \omega_N{}^{AB} \Gamma_{AB} \psi_P \qquad \text{I(2-2)}$$

we deduce by equating to zero the coefficient of $\delta\omega_N{}^{AB}$ in $\delta\mathcal{L}$ the value of the torsion :

$$S_A{}^P{}_B = f^P{}_{AB} + f^Q_Q{}_{[A} \delta^P{}_{B]} \qquad \text{I(2-3)}$$

with

$$f^N{}_{AB} = \frac{k}{4} \bar{\psi}_M (\Gamma^{MNP} \Gamma_{AB} + \Gamma_{AB} \Gamma^{MNP}) \psi_P .$$

The Einstein Cartan equations

$$\Sigma_M{}^A \equiv G_M{}^A - T_M{}^A = 0 , \quad G_M{}^A \equiv R_M{}^A - \frac{1}{2} C_M{}^A R$$

are obtained by equating to zero the coefficient of δe^M_A.

The stress energy tensor $T_M{}^A$ is the coefficient of δe^M_A in $- \delta\mathcal{L}_{3/2}$ deduced from

$$\Gamma^{MNP} = e^M_A e^N_B e^P_C \Gamma^{ABC} \qquad \text{I(2-4)}$$

which gives

$$T_M{}^A = -\frac{1}{2}\,\delta_{e_A{}^M}\,\mathcal{L}_{3/2} = -\,k\{\bar{\psi}_M\,\Gamma^{ANP}\,D_N\,\psi_P + \bar{\psi}_N\,\Gamma^{NAP}\,D_M\,\psi_P$$

$$+\,\bar{\psi}_P\,\Gamma^{PNA}\,D_N\,\psi_M\} + k\{\bar{\psi}_M\,\Gamma^{MNP}\,D_N\,\psi_P\}$$

The equation of motion of the source ψ_P, coefficient of $\delta\bar{\psi}_M$ is :

$$\mathcal{R}^M \equiv \Gamma^{MNP}\,D_N\,\psi_P + \frac{1}{2}\,D_N\,\Gamma^{MNP}\,\psi_P = 0 \qquad \text{I(2-5)}$$

Note that $D_N\,\Gamma^{MNP}$ is non zero because the derivative is not taken in the same connection on spinor and tensor indices; we have $\check{\nabla}_N\,\Gamma^{MNP} = 0$ and

$$D_N\,\Gamma^{MNP} = \frac{1}{4}\,C_N{}^{AB}\,(\Gamma^{MNP}\,\Gamma_{AB} - \Gamma_{AB}\,\Gamma^{MNP})$$

Identities

The first variation of the lagrangian $\mathcal{L}$, when the torsion is given by I(2-3) reduces to

$$\delta\mathcal{L} = \int 2(\Sigma_M{}^A\,\delta\,e_A{}^M - k(\delta\,\bar{\psi}_M\,\mathcal{R}^M + \bar{\mathcal{R}}_M\,\delta\,\psi_M))\,\tau$$

The lagrangian $\mathcal{L}$ is lorentz invariant. From the expression of the variation $\delta\,\psi_M$ associated to the generator U_{AB} of the lorentz group

$$\delta\,\psi_M = -\frac{1}{4}\,U_{AB}\,\Gamma^{AB}\,\psi_M$$

we deduce the identity

$$\Sigma_{[AB]} - \frac{k}{4}\,(\bar{\psi}_M\,\Gamma_{AB}\,\mathcal{R}^M + \bar{\mathcal{R}}_M\,\Gamma_{AB}\,\psi_M) \equiv 0$$

From the invariance of $\mathcal{L}$ by diffeomorphisms we deduce the Bianchi identity which reads here

$$\check{\nabla}_N\,\Sigma_M{}^N = 0 \quad \text{modulo} \quad \mathcal{R}^M = 0$$

3. Rarita Schwinger gauge

The operator $\mathcal{R}^M$ on ψ_P is degenerate; it cannot be well posed : since $\mathcal{R}^0$ contains no derivative $\partial_0\,\psi_P$ the d equations $\mathcal{R}^M = 0$ cannot determine the d quantities $\partial_0\,\psi_P|_{S_0}$ when $\psi_P|_{S_0}$ is known. A standard way to remove this degeneracy is to choose a gauge, that is to impose a further condition on ψ. A classical choice is the Rarita Schwinger gauge(7)

$$\chi \equiv \Gamma^M\,\psi_M = 0 \quad .$$

It is well known that in Minkowski space time the equations $\mathcal{R}^M = 0$ for the unknown ψ_P are a causal system with constraint : they become a hyperbolic system in the Rarita Schwinger gauge and this gauge is preserved through evolution : the same is true in a flat space time of arbitrary dimensions (or an Einstein space time satisfying $R_{MN} = \Lambda\, g_{MN}$). The fact that this statement is no more true in a curved space time is one of the origins of the theory of supergravity. We analyse below where the problem comes from.

A good hyperbolic operator for ψ_P is the Dirac operator $\not{D} \equiv \Gamma^M D_M$. A straightforward computation enables us to write the Rarita Schwinger equations I(2-5) in terms of this operator and the gauge expression χ : we find the identity (in arbitrary dimension)

$$\mathcal{R}^M \equiv - \not{D}\, \psi^M - \frac{1}{2} \Gamma^M \Gamma^N \not{D}\, \psi_N + D^M \chi + \frac{1}{2} \Gamma^M \not{D}\, \chi + r^M = 0 \qquad \text{I(3-2)}$$

with $r^M(e, \delta e, \psi)$ being given by :

$$r_M = - (D^M \Gamma^P)\, \psi_P - \frac{1}{2} \Gamma^M (\not{D}\, \Gamma^P)\, \psi_P + \frac{1}{2} (D_N \Gamma^{MNP})\, \psi_P \quad . \qquad \text{I(3-3)}$$

In a flat space time, with zero torsion, $\mathcal{R}^M$ reduces to

$$\mathcal{R}^M \equiv \Gamma^{MNP}\, \partial_N \psi_P \qquad \text{I(3-4)}$$

and we have identically

$$\partial_M \mathcal{R}^M \equiv 0 \qquad \text{I(3-5)}$$

since, in this case $\partial_M \Gamma^{MNP} = 0$ and the second derivatives commute. The identity implies, in that case, if the equation $\not{D}\, \psi_M = 0$ is satisfied, the vanishing of the wave equation for χ :

$$\partial_M \partial^M \chi + \frac{1}{2} \Gamma^M \Gamma^N \partial_M \partial_N \chi \equiv \frac{1}{2} \partial_M \partial^M \chi = 0 \qquad \text{I(3-6)}$$

While the "constraint" $\mathcal{R}^o = 0$ satisfied on S_o implies, if $\not{D}\, \psi^M = 0$ that, on S_o, $D^o \chi = 0$, if also $\chi = 0$ on S_o. The equation I(3-6), together with the vanishing of χ and $\partial_o \chi$ on S_o insures at least formally[(8)] the preservation of the gauge condition $\chi = 0$ through evolution of ψ by the hyperbolic operator $\not{D}\, \psi$, and the well posedness of the system, at least in a formal sense.

To study the preservation of the gauge condition $\chi = 0$, and well-posedness of the system $\mathcal{R}^M = 0$, when coupled with the Einstein-Cartan equations, we shall study, in the general case, $D_M \mathcal{R}^M$.

We have

$$D_M \mathcal{R}^M \equiv \Gamma^{MNP} D_M D_N \psi_P + \frac{1}{2} D_M \Gamma^{MNP} D_N \psi_P + \frac{1}{2} D_M D_N \Gamma^{MNP} \psi_P \qquad \text{I(3-7)}$$

No torsion terms appear in the Ricci identity calculated with D, because this operator has no torsion on tensor indices, and for a spinor one has for any Cartan connection the identity

$$D_{[M} D_{N]} \psi = \frac{1}{2} P_{MN} \psi .$$

We have thus

$$D_{[M} D_{N]} \psi_P = \frac{1}{2} \check{R}_{MNP}{}^{Q} \psi_Q + \frac{1}{2} P_{MN} \psi_P$$

We obtain by using the riemannian Bianchi identity and the symmetries of the riemann and the riemannian Ricci tensors an identity of the form

$$D_M \mathfrak{R}^M \equiv - \frac{1}{2} G_N{}^P \Gamma^N \psi_P - \frac{1}{4} f^P(S,\nabla S,e)\psi_P + \frac{1}{2} D_M \Gamma^{MNP} D_N \psi_P \qquad \text{I(3-8)}$$

where f^P is a known polynomial in its argument, zero if the torsion $S = 0$.

Remark The linear Rarita-Schwinger system

$$\Gamma^{MNP} \check{\nabla}_N \psi_P = 0 \qquad \text{I(3-9)}$$

in a given curved space time admits, as a particular case of the above study, the integrability condition

$$\check{G}_N{}^P \Gamma^N \psi_P = 0 \qquad \text{I(3-10)}$$

It is satisfied for every ψ_P if the space time is a vacuum Einstein space time. It is equivalent to the Rarita Schwinger gauge condition if the space time is vacuum Einstein with cosmological constant.
For the coupled Einstein-Cartan, Rarita Schwinger system one writes I(3-8) under the form :

$$D_M \mathcal{R}^M \equiv - \frac{1}{2} \Sigma_N{}^P \Gamma^N \psi_P + r \qquad \text{I(3-11)}$$

If $r \equiv 0$ the system is (formally) well posed : it will only be the case without modifications in the case $d = 4$, and by restricting the spinors to be Majorana or Weyl.

II - Simple supergravity (d = 4).

The case $d = 4$ has the remarkable property that the identity

$$D_\lambda \mathfrak{R}^\lambda \equiv - \frac{1}{2} \Sigma_\lambda{}^\rho \gamma^\lambda \psi_\rho$$

is satisfied if the spinor 1-form is Majorana valued, or a Weyl spinor of given helicity. This statement is equivalent to the property of the infinitesimal invariance of the lagrangian by the supersymmetry

$$\delta \psi_\lambda = D_\lambda \varepsilon \quad , \quad \delta e^\lambda_a = \{\bar\varepsilon \gamma^\lambda \psi_a\}$$

1. Equations.

In the case of space time dimension 4, with $\mathfrak{a}$ valued fields, $\psi \in \mathfrak{a}^-$ and $\underset{\sim}{e}$, $\underset{\sim}{\psi} \in \mathfrak{a}^+$ using the formula I(2-1) the lagrangian of the sources is

$$\mathcal{L}_S \equiv k \int (\bar\psi_\mu \gamma^{\mu\nu\rho} D_\nu \psi_\rho - D_\nu \bar\psi_\mu \gamma^{\mu\nu\rho} \psi_\rho)\ \tau$$

which we can write :

$$\mathcal{L}_S \equiv k \int 2 \{\bar\psi_\lambda \xi A^\lambda\}\ \tau$$

with[9]

$$A^\lambda \equiv \eta^{\lambda\mu\nu\rho} \gamma_\mu D_\nu \psi_\rho \quad , \qquad \text{II(1-1)}$$

and { } denotes the real part, that is :

$$2 \{\bar\psi_\lambda \xi A^\lambda\} = \bar\psi_\lambda \xi A^\lambda + \bar A^\lambda \xi \psi_\lambda$$

the total lagrangian is

$$\mathcal{L} = \int (R + 2 k \{\bar\psi_\lambda \xi A^\lambda\})\ \tau \qquad \text{II(1-2)}$$

The quantity $f^\nu{}_{ab}$ can be computed directly, or by using the general formula and we find for the torsion the real antisymmetric tensor

$$S_a{}^\nu{}_b = k\, \bar\psi_{[a} \gamma^\nu \psi_{b]} \qquad \text{II(1-3)}$$

To compute the stress energy tensor source of the Einstein-Cartan equations

$$G_\lambda{}^a = T_\lambda{}^a \qquad \text{II(1-4)}$$

we use directly the lagrangian II(1-2), with $\eta^{\lambda\mu\nu\rho}\ \tau = \varepsilon^{\lambda\mu\nu\rho}_{0123}\ d^4x$ and we find :

$$T_\lambda{}^a \equiv k\, \eta^{a\beta\nu\rho} \{\bar\psi_\beta \gamma_\lambda \xi D_\nu \psi_\rho\} \qquad \text{II(1-5)}$$

The Rarita-Schwinger equation reads[10]

$$B^\lambda \equiv \eta^{\lambda\mu\nu\rho} (\gamma_\mu D_\nu \psi_\rho + \tfrac{1}{2} (D_\nu \gamma_\mu) \psi_\rho) = 0 \qquad \text{II(1-6)}$$

Identities

The identity deduced from lorentz invariance reads here

$$\Sigma_{[ab]} + k\,\{\overline{\psi}_\lambda\ \xi\ \gamma_{ab}\ B^\lambda\} \equiv 0 \qquad \text{II(1-7)}$$

and the identity deduced from the invariance by diffeomorphisms :

$$\check{\nabla}_\mu\ \Sigma_\lambda{}^\mu - 2k\{\nabla_\alpha\ \overline{\psi}_\lambda - \nabla_\lambda\ \overline{\psi}_\alpha)\ \xi\ B^\alpha + \overline{\psi}_\lambda\ \xi\ \nabla_\alpha\ B^\alpha\} \equiv 0 \qquad \text{II(1-8)}$$

2. Supersymmetry.

It is not difficult to compute directly $D_\lambda\ B^\lambda$, using the special properties of Dirac algebra when d = 4.

We have

$$D_\lambda\ B^\lambda \equiv \eta^{\lambda\mu\nu\rho}\ \gamma_\mu\ D_{[\lambda}\ D_{\nu]}\ \psi_\rho + \frac{1}{2}\,\eta^{\lambda\mu\nu\rho}\ D_{[\lambda}\ D_{\nu]}\ \gamma_\mu\ \psi_\rho$$
$$+ \frac{1}{2}\,\eta^{\lambda\mu\nu\rho}(D_\lambda\ \gamma_\nu)D_\mu\ \psi_\rho \qquad \text{II(2-1)}$$

Using the Ricci and Bianchi identities, together with some Dirac algebra we find

$$D_\lambda\ B^\lambda \equiv -\frac{1}{2}\,\Sigma_\lambda{}^\rho\ \xi\ \gamma^\lambda\ \psi_\rho + C \qquad \text{II(2-2)}$$

where C is the $\mathfrak{a}$-valued spinor field defined by

$$C = \frac{1}{4}\,\eta^{\lambda\mu\nu\rho}\ S_\lambda{}^\alpha{}_\mu\ \gamma_\alpha\ D_\nu\ \psi_\rho - \frac{1}{2}\,T_\lambda{}^\rho\ \xi\ \gamma^\lambda\ \psi_\rho \qquad \text{II(2-3)}$$

C does not vanish for general Dirac spinors, but one can show that it vanishes for Majorana or Weyl spinors.

A Majorana spinor is egal to its charge conjugate. Its physical interpretation is that the corresponding particle (gravitino) is identical with its antiparticle[(11)]. In dimension d = 4 there exists a real representation of Dirac matrices and a Majorana spinor is taken to be a real spinor, by convention (Note(6)) $\overline{\psi}$ is then also real. For Majorana, anticommuting spinors one has the symmetries :

$$\overline{\psi}\,\varphi = \overline{\varphi}\,\psi \quad , \quad \overline{\psi}\ \xi\ \varphi = \overline{\varphi}\ \xi\ \psi \qquad \text{II(2-4-a)}$$

$$\overline{\psi}\ \xi\ \gamma_\mu\varphi = \overline{\varphi}\ \xi\ \gamma_\mu\ \psi \qquad \text{II(2-4-b)}$$

and the antisymmetries

$$\overline{\psi}\ \gamma_\mu\,\varphi = -\,\overline{\varphi}\ \gamma_\mu\ \psi \quad , \quad \overline{\psi}\ \gamma_{\mu\nu}\,\varphi = -\,\overline{\varphi}\ \gamma_{\mu\nu}\ \psi \quad .$$

We deduce from the Fierz identity :

$$(\bar\psi \varphi)(\bar\chi \lambda) = \frac{1}{4} \sum_J \varepsilon_J (\bar\psi \Gamma_J \lambda)(\bar\chi \Gamma^J \varphi)$$

where Γ_J, $J = 1,\ldots,16$ are the generators I, ξ, γ_a, $\xi\gamma_a$, γ_{ab} of the Dirac algebra and $\Gamma^J = (I, \xi, \gamma^a, \xi\gamma^a, \gamma^{ab})$, $\varepsilon_1 = -1$ and $\varepsilon_J = 1$ if $J \neq 1$.

$$T_\lambda{}^\alpha \xi \gamma^\lambda \psi_\alpha = -\frac{k}{4} \eta^{\alpha\beta\nu\rho} \sum_J \varepsilon_J (\bar\psi_\beta \Gamma_J \psi_\alpha) \xi \gamma^\lambda \Gamma_J \gamma_\lambda \xi D_\nu \psi_\rho \qquad \text{II(2-5)}$$

which reduces, because of the symmetries II(2-4) and $\gamma_\lambda \gamma^{ab} \gamma^\lambda = 0$ to

$$\begin{aligned} T_\lambda{}^\alpha \xi \gamma^\lambda \psi_\alpha &= -\frac{k}{4} \eta^{\alpha\beta\nu\rho} (\bar\psi_\beta \gamma_a \psi_\alpha) \gamma^a D_\nu \psi_\rho \\ &= \frac{1}{4} \eta^{\alpha\beta\nu\rho} S_\alpha{}^\mu{}_\beta \gamma_\mu D_\nu \psi_\rho \end{aligned} \qquad \text{II(2-6)}$$

therefore

$$C = 0 \quad .$$

The identity II(2-2) for a Majorana spinor reduces to

$$D_\lambda B^\lambda \equiv -\frac{1}{2} \Sigma_\lambda{}^\rho \xi \gamma^\lambda \psi_\rho \qquad \text{II(2-7)}$$

thus $D_\lambda B^\lambda = 0$ "on shell". The identity II(2-7) is equivalent to the infinitesimal supersymmetry of the lagrangian

$$\delta \psi_\lambda = D_\lambda \varepsilon \quad , \quad \delta e_a{}^\lambda = \bar\varepsilon \gamma^\lambda \psi_a$$

A Weyl spinor is one which takes its values in a given eigenspace of the matrix ξ. For instance we impose

$$\xi \psi_\lambda = i \psi_\lambda \quad , \quad \text{thus} \quad \bar\psi_\lambda \xi = -i \bar\psi_\lambda$$

The physical interpretation is that the gravitino is, like the neutrino taking part in weak interacting a parity violating particle.
For a pair φ, ψ of Weyl spinors one has the identities

$$\bar\varphi \psi = 0 \quad , \quad \bar\psi \xi \psi = 0$$

The identity II(2-2)for a Weyl spinor reduces to

$$D_\lambda B^\lambda \equiv \frac{i}{2} \Sigma_\lambda{}^\rho \gamma^\lambda \psi_\rho \qquad \text{II(2-8)}$$

which implies

$$D_\lambda \bar B^\lambda = \frac{i}{2} \Sigma_\lambda{}^\rho \bar\psi_\rho \gamma^\lambda \qquad \text{II(2-8bis)}$$

For a Majorana, or a Weyl spinor, the supersymmetry identity, together with the identities II(1-7) and II(1-8) enables one to prove that the $\mathfrak{a}$ valued Cauchy problem of simple supergravity is formally well posed and causal. It can be shown to be indeed well posed and causal in the classical sense if $\mathfrak{a}$ is a grassmann algebra (cf [14],[4]), that is generated by elements ζ^I, I = 1,...,N (possibly infinity) such that

$$\zeta^I \zeta^J = - \zeta^J \zeta^I \quad .$$

We should however stress that this interpretation provokes a decoupling in the equations : to satisfy them in the Grassmann algebra one must equate to zero terms of each order in the generators. One then finds that the numerical part of the metric must satisfy Einstein empty space equations, the term of lowest order (1 since odd) of ψ must satisfy the usual Rarita Schwinger equation in this ordinary curved space times in fact a causal system since this space time is vacuum Einstein. The metric couples with the spin 3/2 field only through its higher order terms in the Grassmann algebra. Note that except for the Einstein equations for the numerical part of the metric the equations are at each step a linear system for the unknowns : no singularity appears in the solutions which exist globally on the globally hyperbolic domain of existence of the numerical metric.

III - Extended supergravities.

1. Cosmological constant and massive gravitino (d = 4).

It has been proved early after the discovery of simple supergravity that the theory can be extended to the case where a cosmological constant Λ on the one hand and a mass term for the gravitino on the other hand are added to the lagrangian, provided this mass m and Λ are linked by the relation $\Lambda = 3m^2$ (Freedman and Das 1977), the consistency identity is equivalent to an infinitesimal supersymmetry with an operator D_μ modified by the addition of a zero order term, $\frac{1}{2} m \gamma_\mu$: note that such a term does not modify $D_\lambda B^\lambda$ modulo $B^\lambda = 0$.

2. Einstein-Maxwell supergravity (N = 2).

It is possible to construct a consistent Einstein-Cartan theory with a Maxwell source if one introduces two spin 3/2 fields. The natural lagrangian is ($F_{\mu\nu}$ electromagnetic field 2-form)

$$\mathcal{L}_N = \int (R + \sum_{i=1}^{2} \overline{\psi}^i_\mu \gamma^{\mu\nu\rho} D_\nu \psi^i_\rho + \frac{1}{2} F_{\mu\nu} F^{\mu\nu}) \tau$$

To present an infinitesimal supersymmetry it has to be corrected by an interaction lagrangian of the type (C_1, C_2, C_3 given numbers)

$$\mathcal{L}_I = \int \bar{\psi}_\mu^{\ i} \left[C_1 (F^{\mu\nu} + \hat{F}^{\mu\nu}) + C_2 (F^{\star\mu\nu} + \hat{F}^{\star\mu\nu}) \right] \psi_\nu^{\ j} \ \varepsilon_{ij} \ \tau$$

where $F^\star$ is the dual form of F, ε_{ij} the antisymmetric knonecker index and

$$\hat{F}_{\mu\nu} = F_{\mu\nu} + C_3 \ \bar{\psi}_\mu^{\ i} \ \bar{\psi}_\nu^{\ j} \ \varepsilon_{ij} \quad .$$

The local lagrangian possesses then two infinitesimal supersymmetries, corresponding to two identities which should lead to a proof that the corresponding system of partial differential equations, for $\tilde{e}$, A (electromagnetic potential) and the two gravitini ψ is causal (the proof has not been carried out in detail).

Extended supergravities with N = 3, 4 gravitini have been constructed by the introduction of new (bosonic) fields in the lagrangian.

IV - Kaluza-Klein supergravity.

The goal of a supergravity theory in d = 4 + N dimensions is to provide an unification of all fundamental interactions, including gravity, and Fermi matter fields. The extra N dimensions are supposed to live on a compact manifold, so small with respect to the length scale of our laboratory experiments that we do not see it. The original Kaluza-Klein theory regained the usual Einstein-Maxwell equations by equating to zero the fourteen first components of the Ricci tensor of a 5-dimensional space time V_5 with a 1-parameter isometry group G isomorphic to T^1 (equivalently U(1)), which endows V_5 with a principal fiber bundle structure. The basis, quotient $V_5 \setminus G$, is interpreted as the usual space time V_4, the fiber at $x \in V_4$ is isomorphic to T^1 (it is an orbit of G in V_5). The local trivialisations of V_5

$$\phi(\pi^{-1}(U)) \to U \times G$$

(U open set of V_4, π projection $V_5 \to V_4$) allow for the definition of adapted local coordinates in the domain $\pi^{-1}(U) \subset V_5$, namely $x^M = (x^\alpha, x^5)$ in these coordinates the metric may be written, with g_{MN} independant of x^5 and g_{55} = constant

$$ds^2 = g_{MN} \ dx^M \ dx^N = g_{\alpha\beta} \ dx^\alpha \ dx^\beta + 2 \ g_{\alpha 5} \ dx^\alpha \ dx^5 + g_{55} (dx^5)^2$$

the metric of V_4 (distance between orbits) is

$$d\sigma^2 = (g_{\alpha\beta} + (g_{\alpha 5}\ g_{\beta 5})\ g_{55}^{-1})\ dx^\alpha\ dx^\beta$$

The coefficients $g_{\alpha 5}$, components of a 1-form on V_4 are interpreted as the local electromagnetic potential. By a change of local trivialisation of V_5 (preserving g_{55}) given by

$$x'^5 = f(x^\alpha) + x^5 \quad , \quad x'^\alpha = x^\alpha$$

they transform according to

$$g'_{\alpha 5} = g_{MN} \frac{\partial x'^M}{\partial x^\alpha} \frac{\partial x'^N}{\partial x^5} = g_{\alpha 5} + g_{55} \frac{\partial f}{\partial x^\alpha}$$

which is indeed the gauge transformation of the electromagnetic potential.

The Kaluza-Klein theory has been extended to a full 5-dimensional theory by Jordan and, independantly, Thiry, who allow the 15th potential g_{55} to be also dependant on the space-time (V_4) point. One introduces then a scalar field on V_4 whose physical interpretation has varied along the years : for Jordan it represented a variable gravitational constant, for Lichnerowicz and Thiry a dielectric permeability of the vacuum (Note that for a variable g_{55} the physical interpretation of $g_{\alpha 5}$ has to be revised). Pigeaud (1975) introduced its interpretation as an elementary scalar (mesonic) field. The interpretation of the numerous scalar fields introduced by the d = 4 + N so called "Kaluza-Klein" theories is a subject of active study.

These theories extend the original Kaluza-Klein scheme not only to an arbitrary Lie group G, but also to a case where V_d is a fiber bundle with basis V_4 and fiber an homogeneous space $M = G \setminus H$: the motivation is that d = 11 = 4 + 7 seems the greatest dimension for which a construction of supergravity is possible, while the results of particle physics seem to require a group G (the Yang-Mills group) which is more than 7 dimensional (candidates are $SU(3) \times SU(2) \times U(1)$ which is 8 + 3 + 1 = 12 dimensional or SU(5) which has dimension 24). Various possibilities have been proposed for M in the framework of the 11 dimensional supergravity : $M = G = T^7$ (Cremmer and Julia), $M = S^7$, $G = SO(8)$ (Duff and Pope), $G = SU(3) \times SU(2) \times U(1)$, $M = CP^2 \times S^2 \times S^1$ (Witten, who examined also other possibilities for M) - none of these models seems completely satisfactory from the physical point of view - but what we know of "elementary" particles may also be subject to revision. In any case it is quite remarkable that there exists a consistent, 11 dimensional Kaluza-Klein supergravity, which can be proved to be causal, with just the addition, as a supplementary field, of an exterior 3-form.

1. Equations

The introduction of the 3-form $\underset{\sim}{A}$ is justified on physical grounds by the "counting of states" which must be in the same number for bosons and fermions : states are counted as the significant (i.e. which cannot be removed by change of gauge, in particular coordinates independant components also called "polarizations" of the leading term of a high frequency wave solution of the equations (or of a plane wave solution of the equations linearized around flat space time V_d).

The moving frame has the same number of states as the metric, by lorentz invariance. These can be reduced (by diffeomorphisms) to states of a symmetric 2-tensor $h_{M'N'}$ on a d-1 dimensional manifold; by the equations $h_{M'N'}$ is transverse to the direction of propagation ($n^{M'} h_{M'N'} = 0$) and traceless ($g_{M'N'} h^{M'N'} = 0$) thus for $d = 11$

$$\frac{1}{2}\, 9 \times 10 - 1 = 44 \text{ polarisations for } \underset{\sim}{e} \ .$$

The spin 3/2 field ψ_M has $d \times 2^{|d/2|} = 11 \times 32$ components, which are reduced to 10×32 significant ones by diffeomorphisms, to $9 \times 32 - 32$ by the gauge condition ($\Gamma^M \psi_M = 0$) and the equations (which imply then transversality). Fermionic states count only for 1/2 of bosonic states, thus

$$\frac{1}{2}\,(9 \times 32 - 32) = 128 \text{ polarizations for } \underset{\sim}{\psi} \ .$$

The number of missing bosonic states $128 - 44 = 84$ is the number of states of a transverse exterior 3-form on V_{11}, $C_9^3 = 84$.

To be transverse, by the equations, $\underset{\sim}{A}$ should appear in the lagrangian, principally, through a Maxwell type action = F being the exterior derivative of A :

$$\mathcal{L}_A = C_1 \int F_{MNPQ}\, F^{MNPQ}\, \tau \quad , \qquad F = dA$$

Cremmer and Julia found the following lagrangian for consistent $d = 11$ supergravity, with 1 spin 3/2 field and field A, called "3-index photon"

$$\mathcal{L} = \mathcal{L}_{E.C} + \mathcal{L}_{3/2} + \mathcal{L}_A + \mathcal{L}_{Corr}$$

In this theory the fields take as before their values in a graded $\mathfrak{a}$, $\underset{\sim}{A}$ like $\underset{\sim}{e}$ takes its values in $\mathfrak{a}^+$, $\underset{\sim}{\psi}$ always in $\mathfrak{a}^-$. The lagrangians $\mathcal{L}_{E.C}$ and $\mathcal{L}_{3/2}$ are the general ones, $\mathcal{L}_{Corr}$ with no higher than quartic terms is uniquely determined by the requirement of the existence of an infinitesimal supersymmetry. They found (C_1, C_2, C_3, C_4 are given numbers)

$$\mathcal{L}_{Corr} \equiv \int \{ C_1 \bar{\psi}_M \Gamma^{MNP} (\bar{\psi}_Q \Gamma_{NAB}{}^{QR} \psi_R) \Gamma^{AB} \psi_P + \tau^{M_1 .. M_{11}} F_{M_1 .. M_4} F_{M_5 .. M_8} A_{M_9 .. M_{11}}$$
$$+ C_3 (\bar{\psi}_M \Gamma^{MNWXYZ} \psi_N + 12 \bar{\psi}^W \Gamma^{XY} \psi^Z)(2 F_{WXYZ} - 3 \bar{\psi}_{[W} \Gamma_{XY} \psi_{Z]}) \} \tau$$

The torsion is, as in the general case, obtained as a quadratic expression in ψ by equating to zero the variation of ω.
The equations obtained by varying e, ψ and A are of the type

$$\Sigma_M{}^A \equiv G_M{}^A - T_M{}^A = 0 \quad , \quad T_M{}^A(e, \partial e, F, \psi, \partial\psi)$$

$$\mathcal{R}^M \equiv \Gamma^{MNP} D_N \psi_P + r^M = 0 \quad , \quad r_M(\psi, e, F)$$

$$\mathcal{F} \equiv \nabla . F + C_4{}^{*}(F \wedge F) + \nabla . \phi \quad , \quad \phi(\psi, e) \quad (\text{with } (\nabla . F)^{MNP} = \nabla_Q F^{QMNP})$$

Using the general identities (lorentz invariance and Bianchi), the identity $\nabla . \mathcal{F} \equiv 0$ (which could have been foreseen by the invariance of the lagrangian by the "gauge" transformation $A \rightsquigarrow A + df$), and the "supersymmetry" proved by Cremmer and Julia, it is possible to prove that the system is formally causal, and truly causal if the graded algebra $\mathfrak{a}$ is a grassmann algebra (cf Y. Choquet-Bruhat 1984).

Ground state solutions

The so called ground state solutions are particular solutions of the system of partial differential equations which possess such symmetries that in the pertubative solutions around them one can recover the known fields of elementary particle physics; a ground state is also expected to be stable against perturbation - the definition of stability being, at least at first level, that the ground state is a minimum of energy - notion which itself requires some definitions.
The ground state considered by Cremmer and Julia in their first paper is, very naturally, the 4-dimensional Minkowski space time, with direct product by the flat torus T_7, the spinor field ψ_M as well as A_{MNP} being both zero. Another solution, with a larger isometry group and which in a sense gives the dimension 4 for the ordinary space time on the one hand and implies the compacity of the remaining dimensions on the other hand is the direct product of Anti de Sitter space time with the sphere S^7, found as follows (Duff and Pope 1982).
The spinor field being again taken zero the equations reduce to

$$R_{MN} - \frac{1}{2} g_{MN} R = \frac{1}{3} (F_{MPQR} F_N{}^{PQR} - \frac{1}{8} g_{MN} F^2)$$

$$\nabla_M F^{MPQR} = - \frac{1}{576} \eta^{PQRM_1 .. M_8} F_{M_1 .. M_4} F_{M_5 .. M_8}$$

One looks for solutions of the direct product form $V_d = M \times N$, where M is the space time, with metric $g_{\mu\nu}$ and N a properly riemannian manifold with metric g_{mn}. In coordinates on V_d adapted to the product structure, x^μ coordinates on M and y^m on N the metric g_{MN} on V_d is $g_{\mu\nu}$, g_{mn}, while $g_{\mu m} = 0$. A particular solution[12] of is (Freund and Rubin 1980), if M has dimension 4

$$F_{\mu\nu\rho\sigma} = 3a\ \tau_{\mu\nu\rho\sigma}\ , \quad \text{all other components of F zero}$$

with

$$R = -12\ a^2\ g$$

(Einstein space time with cosmological constant)

$$R_{mn} = 6\ a^2\ g_{mn}$$

hence N is a compact manifold (since properly riemannian and $a^2 > 0$). Duff and Pope argue in favor of Anti De Sitter space time for M and S^7 for N on the grounds that they constitute the solution with maximal unbroken supersymmetry, that is which admits the maximum number of spinors ε satisfying the equation $\mathcal{D}_M\varepsilon = 0$, where $\mathcal{D}_M$ is the operator with $\mathcal{D}$ principal part D_M appearing in the infinitesimal supersymmetry $\delta\psi_M = \mathcal{D}_M\varepsilon$. The seven sphere[13] has the interesting property to be the only compact parallelizable manifold which is not a group : it is an homogeneous space $S^7 = SO(8) / SO(7)$, which has the Lie group of isometries of greatest dimension 28 for a 7-dimensional manifold. However SO(8) does not quite satisfy elementary particle physicists.
Kaluza Klein supergravity is a fascinating subject which has not said its last word : the interpretation of the field F of the "3 index photon" as the trace of an antisymmetrised of Yang-Mills fields is one way which may offer new possibilities (Chapline and Gibbons 1984).

NOTES

(1) Search for such unifications by introduction of graded Lie algebras had begun shortly before.
(2) We mean by "consistent" that the solutions of the classical equations are not limited by further lower order equations (integrability conditions) which result from them. Causal means furthermore that given initial data (satisfying the part, called "constraints" of the equations which depend only from them) there exists a solution, whose value at a point depends only on the initial data in the relativistic past (determined by the hyperbolic metric) of that point.

(3) It seems that the number of gravitini must be limited to 8 if one does not want to have particles with spin greater than 2 - it seems also that spin greater than 2 fields cannot be consistently coupled with gravity (C. Aragone and S. Deser 1979), because the integrability conditions involve not only the Ricci, but the full Riemann tensor.

(4) It does not introduce new "states" but only more gauge freedom, lorentz rotations in the tangent space.

(5) This hypothesis on classical (non quantum) spinor field has its counterpart in the (quantum) relation between spin and statistics - but it does not seem to be imposed by it at the classical level : in general relativity there exist models with neutrino sources.

(6) $\Gamma^{MNP} = \frac{1}{3!} \varepsilon^{MNP}_{QRS} \Gamma^Q \Gamma^R \Gamma^S$ is an antisymmetrized product. $\Gamma^M = e^M_A \Gamma^A$, with Γ^A standard (numerical) given gamma matrices, $\Gamma^A\Gamma^B + \Gamma^B\Gamma^A = -2\eta^{AB}$ and $\bar{\psi} = \tilde{\psi} \Gamma^o$.

(7) This "gauge" condition has also a physical meaning : select "pure spin" states.

(8) If the unknown were numerical valued the standard theory of hyperbolic systems would give this result. Since they are $\mathfrak{a}$ valued the classical theorems cannot be applied without further study.

(9) $\gamma^{\mu\nu\rho} = \eta^{\lambda\mu\nu\rho} \xi \gamma_\lambda$, $\xi = \frac{1}{4!} \eta_{\alpha\beta\delta\gamma} \gamma^\alpha \gamma^\beta \gamma^\gamma \gamma^\delta$, when d = 4.

(10) Note that $\xi B^\lambda = -\mathfrak{R}^\lambda$ with the notation of I.

(11) Such particles do not exist in nature.

(12) Other solutions, with $F_{mnpq} \neq 0$ have been found (Englert 1982).

(13) The search for the perturbations around the ground state has led to an intensive study of the properties of the "spherical harmonics" of S^7.

REFERENCES.

[1] C. Aragone and S. Deser. Phys. Lett. 86B (1979), 161.

[2] D. Bao, J. Isenberg, P. Yasskin. The dynamics of the Einstein-Dirac system. to appear (1983).

[3] Y. Choquet-Bruhat. The Cauchy problem in extended supergravity, N = 1 , d = 11. Communications in Maths. Phys., to appear 1984.

[4] Y. Choquet-Bruhat. The Cauchy problem in Classical Supergravity. Lett. in Math. Phys. 7, (1983) 459-467.

[5] Y. Choquet-Bruhat. "The Cauchy problem" in "Gravitation, an Introduction to Current Research" L. Witten ed., J. Wiley 1962.

[6] Y. Choquet-Bruhat. Diagonalisation des systèmes quasilinéaires et hyperbolicité non stricte. J. Maths pures et appl. 45, (1966), 371-386.

[7] Y. Choquet-Bruhat, D. Christodoulou, M. Francaviglia. "Cauchy data on a manifold". Ann. I.H.P. A t. 29, n°3 (1978) p. 241.

[8] E. Cremmer, B. Julia, J. Sherk. Supergravity, theory in 11 dimensions. Phys. Lett. 76B, 4 (1978) 409-411.

[9] E. Cremmer, B. Julia. The SO(8) Supergravity. Nuclear Phys. B 159, 1979) 141-212.

[10] G.F. Chapline and G.B. Gibbons. To appear Phys. Letters B (1984).

[11] S. Deser and B. Zumino. Consistent Supergravity. Phys. Letters 62 n°3 (1976) 335-337.

[12] S. Deser and B. Zumino. Broken supersymmetry and supergravity. Phys. Rev. Letters 38 n°25 (1977) 1433-1436.

[13] B. DeWitt. Dynamical Theory of Groups and Fields. Gordon and Breach 1965.

[14] B. DeWitt, P. Van Nieuwenhuisen and P. West. Supermanifolds and supersymmetry. (à paraître).
[15] M.J. Duff. Supergravity, the seven-sphere and spontaneous symmetry breaking. Nuclear Phys. B 219, (1983) 389-411.
[16] F. Englert. CERN preprint TH 3394 (1982).
[17] D.Z. Freedman and A. Das. Nuclear Physics B 120 (1977) 221.
[18] D.Z. Freedman, P. Van Nieuwenhuisen and S. Ferrara. Progress towards a theory of Supergravity. Phys. Rev. D, 13 n°12,(1976) 3214-3218.
[19] P.G.O. Freund and M.A. Rubin. Dynamics of dimensional reduction. Phys. Lett. 97 B (1980) 233.
[20] F.W. Hehl, P. van der Heyde, G.D. Kerlick and J.M. Nester. General Relativity with spin and torsion. Foundations and prospects. Rev. Mod. Phys. 48 (1976) 395.
[21] T. Hughes, T. Kato, J. Marsden. Well posed quasi linear second order hyperbolic systems. Arch. Rat. Mech. 63, (1977) 273-294.
[22] R. Kerner. Generalization of the Kaluza-Klein theory for an arbitrary non-abelian gauge theory. Ann. I.H.P. 9 n°2, (1968) 143-152.
[23] R. Kerner. Geometrical background for the unified field theories : the Einstein-Cartan theory over a principal fibre bundle. Ann. I.H.P. 34 n°4 (1981) 437-463.
[24] A. Lichnerowicz. Champ de Dirac, champ du neutrino et transformation CPT sur un espace temps courbe. Ann. I.H.P. I, n°3 (1964) p 233-290.
[25] P. Pigeaud. Sur de nouvelles équations de champ en théorie de Jordan - Thiry. C. R. Acad. Sci. Paris, t.280 (1975) 749-752.
[26] A. Trautman. Fiber bundles, gauge fields and gravitation in "General Relativity and Gravitation" A. Held ed, Plenum 1980.
[27] P. Van Nieuwenhuisen. Supergravity. Lectures Notes, in "Relativity, groups and topology". B. DeWitt ed., les Houches 1983 (to appear, North-Holland).
[28] E. Witten. Search for a realistic Kaluza-Klein theory. Nuclear Phys. B 186 (1981) 412-428.

SOME NONEXISTENCE THEOREMS FOR MASSIVE YANG-MILLS FIELDS AND HARMONIC MAPS

Hu Hesheng (H.S.Hu)
Institute of Mathematics, Fudan University
Shanghai, China

Introduction

The Yang-Mills theory and the theory of harmonic maps between Riemannian manifolds are two important subjects of differential geometry. They have some common features, such as they are both variational theories, being very important in theoretical physics and they both rely on the theory of non-linear partial differential equations, having almost same principal parts. Moreover, the harmonic maps from 2-dimensional space are quite similar to the Yang-Mills (Y-M) fields over 4-dimensional space, since in such cases they are both conformally invariant theories.

There are quite a lot of papers considering the solution to the Y-M equations and equations of harmonic maps. An important problem is to investigate the global existence or nonexistence of certain gauge fields or certain harmonic maps.

In the present paper we will give some nonexistence theorems. We give the concepts of the Y-M fields and harmonic maps in §1 briefly. Besides, it is emphasized that a massive Y-M field is the coupling of a pure Y-M fields and a harmonic map from the space-time to the gauge group. In §2 we condider a nonexistence theorem for massive Y-M fields on the Minkowski space-time $R^{n-1,1}$ and its generalization to some curved manifolds. In §3 we consider a nonexistence theorem for harmonic maps from Euclidean space to any Riemannian manifold and its generalizations.

§1. Pure Yang-Mills Fields and Harmonic Maps

We begin with a brief sketch of the two concepts.

(a) Y-M fields (gauge field)

Let G be a Lie group, usually being compact and linear, and g its Lie algebra. A gauge field over a Riemannian (or Lorentzian) manifold M, mathematically is defined by a connection on the principal bundle $P(M,G)$.

We shall consider gauge fields on the Minkowski space-time $R^{1,n-1}$ mainly. The metric of $R^{1,n-1}$ is

$$ds^2 = \eta_{\lambda\mu} dx^\lambda dx^\mu = -(dx^0)^2 + (dx^1)^2 + \cdots + (dx^n)^2 \qquad (\lambda, \mu = 0, 1, \cdots, n-1) \tag{1}$$

A gauge field is defined by its gauge potential (connection)

$$b = b_\lambda(x)\,dx^\lambda \tag{2}$$

which is a 1-form valued in g . The field strength (curvature) is

$$F = \frac{1}{2} f_{\lambda\mu}\, dx^\lambda \wedge dx^\mu \tag{3}$$

with

$$f_{\lambda\mu} = \frac{\partial b_\lambda}{\partial x^\mu} - \frac{\partial b_\mu}{\partial x^\lambda} - [b_\lambda, b_\mu] \tag{4}$$

The Y-M functional L or the action integral of the pure Y-M theory is defined as [1]

$$L(b) = -\frac{1}{4}\int (f_{\lambda\mu}, f^{\lambda\mu})\, d^n x \tag{5}$$

Here (,) stands for the Cartan inner product.

A pure Y-M field is a critical point of the Y-M functional, i.e. its gauge potential satisfies the Euler equations of the Y-M functional (5)

$$J_\alpha = \gamma^{\lambda\mu} f_{\alpha\lambda|\mu} = \gamma^{\lambda\mu}(f_{\alpha\lambda,\mu} + [b_\mu, f_{\alpha\lambda}]) = 0 \tag{6}$$

These equations are called pure Y-M equations.

Let S be a G-valued function. The gauge potential

$$b' = (\mathrm{ad}\,S)\,b - (dS)S^{-1} \tag{7}$$

is called the gauge transformation of b . Two gauge potentials related by a gauge transformation are considered as equivalent mathematically and physically.

(b) Harmonic maps

Let M , N be Riemannian manifolds or Lorentzian manifolds and $\phi: M \rightarrow N$ a C^2- map.

The energy integral of ϕ is

$$E(\phi) = \int_M e(\phi)\, dV_M \tag{8}$$

Here

$$e(\phi) = \underset{N}{g}_{\alpha\beta}(\phi)\frac{\partial\phi^\alpha}{\partial x^i}\frac{\partial\phi^\beta}{\partial x^j}\underset{M}{g}^{ij}(x) \qquad (i,j=1,\cdots,m;\ \alpha,\beta=1,\cdots,n) \tag{9}$$

is the expression of the energy density in local coordinates.

A map ϕ is called harmonic if it is a critical point of $E(\phi)$, i.e. a solution to the Euler equations of $E(\phi)$

$$\underset{M}{g}^{ij}\left(\frac{\partial^2\phi^\alpha}{\partial x^i x^j} - \underset{M}{\Gamma}{}^k_{ij}\frac{\partial\phi^\alpha}{\partial x^k} + \underset{N}{\Gamma}{}^\alpha_{\beta\gamma}\frac{\partial\phi^\beta}{\partial x^i}\frac{\partial\phi^\gamma}{\partial x^j}\right) = 0 \tag{10}$$

Systems of PDEs (10) is elliptic or hyperbolic, respectively, if M is Riemannian or Lorentzian [2].

(c) Massive Y-M fields

The action integral of massive Y-M fields on $R^{1,n-1}$ is a coupling of the pure Y-M functional (5) and the energy integral for harmonic maps from $R^{1,n-1}$ to the gauge group G ,

$$L_m(b,U)=\int [-\tfrac{1}{4}(f_{\lambda\mu},f^{\lambda\mu})-\tfrac{m^2}{2}(b_\lambda-\omega_\lambda,b^\lambda-\omega^\lambda)]\,d^nx \qquad (\lambda=0,1,\cdots,n-1) \tag{11}$$

Here

$$\omega_\lambda = U^{-1}\frac{\partial U}{\partial x^\lambda}$$

U is a G-valued function which is a display of "gauge" and the coupling constant m is the mass of gauge particles [3].

The Euler equations of the action integral of $L_m(b,U)$ are

$$\begin{aligned} &J_\alpha - m^2(b_\alpha-\omega_\alpha)=0 \\ &\eta^{\alpha\beta}\left(\frac{\partial b_\alpha}{\partial x^\beta}-\frac{\partial\omega_\alpha}{\partial x^\beta}-[b_\beta,\omega_\alpha]\right)=0 \end{aligned} \tag{12}$$

and the gauge transformation is

$$\begin{aligned} &b'=wbw^{-1}-(dw)w^{-1} \\ &U'=Uw^{-1} \end{aligned} \tag{13}$$

where $W\in G$. The action integral (11) is a gauge invariant.

2. Nonexistence Theorems for the Static Massive Yang-Mills Fields

In the theory of Y-M fields, one problem of considerable interest is whether there exist any nontrivial static solutions to the Y-M equations with finite energy and free of singularities over the whole $R^{1,n-1}$.

The 1st nonexistence theorem [4] was discovered by S. Deser in 1976 as follows: If $n\neq 5$, the pure Y-M equations of a compact group in n-dimensional spacetime $R^{1,n-1}$ does not admit any nontrivial static solution which has (i) no singularities (ii) finite energy (iii) the field strength approaching to zero sufficiently fast at infinity.

It is noticed that

(a) For $n=4$, there does not exist such static solution on the real space-time $R^{1,3}$.

(b) If $n=5$, such static solution do exist, since the instantons on the 4-dimensional Euclidean space R^4 may be considered as regular static solutions in $R^{1,4}$.

(c) We will show later (Remark of Theorem 2), condition (iii) can be removed and condition (ii) can be weakened.

Now we turn to the same problem for massive Y-M fields.

A massive Y-M field is called static, if through a gauge transformation, (b,U) is equivalent to (b',U') which is independent of the time variable x^0.

No loss of generalities, we assume that (b,U) is independent of x^0. Taking $W=U$ in (13), U is reduced to the unit element of G and then $L_m(b,U)$ becomes

$$L_m(b) = -\int [\tfrac{1}{4}(f_{\lambda\mu},f^{\lambda\mu})+\tfrac{m^2}{2}(b_\lambda,b^\lambda)]\,d^nx \tag{14}$$

and the equations (12) are reduced to

$$J_\alpha - m^2 b_\alpha = 0$$
$$\eta^{\lambda\mu} \frac{\partial b_\lambda}{\partial x^\mu} = 0 \tag{15}$$

The second set of equations (15) mean that the potential b satisfies the Lorentz gauge condition.

The energy momentum tensor is

$$T_{\alpha\beta} = (f_{\alpha\nu}, f^\nu_\beta) - \frac{1}{4}\eta_{\alpha\beta}(f_{\mu\nu}, f^{\mu\nu}) + m^2(b_\alpha, b^\alpha) - \frac{m^2}{2}\eta_{\alpha\beta}(b_\lambda, b^\lambda) \tag{16}$$

In particular,

$$T_{oo} = \frac{1}{2}\left[(f_{oi}, f_{oi}) + \frac{1}{2}(f_{ij}, f_{ij})\right] + \frac{m^2}{2}(b_o, b_o) + \frac{m^2}{2}(b_i, b_i) \tag{17}$$
$$(i, j = 1, 2, \cdots, n-1)$$

and

$$T_{ii} = T^\alpha_\alpha - T^o_o = -\frac{1}{2}(n-3)(f_{oi}, f^{oi}) + \frac{1}{4}(5-n)(f_{ij}, f^{ij}) + \frac{m^2}{2}(3-n)(b_\lambda, b^\lambda) + m^2(b_o, b_o) \tag{18}$$

Moreover, the conservative law

$$\frac{\partial T^\beta_\alpha}{\partial x^\beta} = 0 \tag{19}$$

holds.

The total energy of the field is

$$E = \int_{R^{n-1}} T_{oo}\, d^{n-1}x \tag{20}$$

In 1979 the author obtained the following result [5].

Theorem 1. In an n-dimensional spacetime with $n \neq 4$, the massive Y-M equations with real mass do not admit any static solution which has (i) finite energy (ii) no singularity (iii) the field strength and potential approaching to zero sufficiently fast at infinity.

Comparing Theorem 1 with Deser's result we discovered that there is a "discontinuity" as $m \to 0$ in 5-dimensional spacetime, i.e. for n=5 and $m \neq 0$, no such solution, but when m=0 such solution do exist.

Afterwards, S.Deser & C.J.Isham in their paper [6] wrote that this is the first explicit example which make us recognize that there exists a classical "discontinuity". Besides, in their paper the result is extended to the gauge field with "soft" mass.

In 1982 the author found that the finite energy condition can be weakened.

Let $\psi(r)$ be a positive, continuous and unbounded function, defined on $(0, \infty)$ and satisfying

$$\int_a^\infty \frac{dr}{r\psi(r)} = \infty \qquad (a > 0) \tag{21}$$

If the energy density T_{oo} satisfies

$$\int_{R^{n-1}} T_{oo}\, d^{n-1}x = \infty$$

and

$$\int_{R^{n-1}} \frac{T_{oo}}{\psi(r)}\, d^{n-1}x < \infty \tag{22}$$

we say that the energy is slowly divergent. It means that the total energy within the sphere of radius r approaches to infinity quite slowly as $r \to \infty$.

We have[7]

Theorem 2. Let G be a compact Lie group, m real number ($m \neq 0$), $n \neq 4$. The massive Y-M equations on $R^{1,n-1}$ do not admit any non-trivial static solution which is free of singularities and has finite or "slowly divergent" energy.

Proof. From the expression for J_o and static condition $b_{\alpha,0}=0$, we have

$$(b_o, J_o) = (b_o, f_{oi})_{,i} - (f_{oi}, f_{oi}) \tag{23}$$

Consider the integral

$$0 = \int_0^\infty \omega(r)\,dr \int_{|x|\leq r} (J_o - m^2 b_o, b_o)\, d^{n-1}x$$

where $|x| = ((x^1)^2 + \cdots + (x^{n-1})^2)^{1/2}$ and $w(r)$ will be defined later. Using (23), we have

$$0 = \int_0^\infty \omega(r)\,dr \int_{|x|\leq r} K\, d^{n-1}x + \int_0^\infty \omega(r)\,dr \int_{|x|=r} (b_o, f_{oi}) \frac{x^i}{r}\, dS$$

where

$$K = -(f_{oi}, f_{oi}) - m^2 (b_o, b_o) \leq 0 \tag{24}$$

The equality in (24) holds if and only if $f_{oi} = b_o = 0$. If K does not equal zero identically, then there exists a constant $R > 0$ and a positive constant ε such that

$$\int_{|x|\leq R_1} K\, d^{n-1}x < -\varepsilon \qquad (R_1 \geq R) \tag{25}$$

Choose

$$\omega(r) = \begin{cases} 0 & r < R \\ \dfrac{1}{r\psi(r)} & R \leq r \leq R_1 \\ 0 & r > R_1 \end{cases} \tag{26}$$

where $\psi(r)$ is a positive, unbounded, continuous function of r satisfying

$$\int_R^\infty \frac{dr}{r\psi(r)} = \infty \tag{27}$$

Then, we have

$$0 < -\varepsilon \int_R^{R_1} \frac{dr}{r\psi(r)} + \int_R^{R_1} \frac{dr}{r\psi(r)} \int_{|x|=r} \{(b_o, b_o) + (f_{oi}, f_{oi})\}\, dS$$

$$< -\varepsilon \int_R^{R_1} \frac{dr}{r\psi(r)} + \frac{1}{R} \int_{|x|\leq R_1} \frac{(b_o, b_o) + (f_{oi}, f_{oi})}{\psi(r)}\, d^{n-1}x$$

Choose R_1 sufficiently large, it is easily seen that the right side should be negative. This is a contradiction. Consequently, we should have $K=0$ identically, i.e.

$$b_o = 0, \quad f_{oi} = 0 \tag{28}$$

Thus we have

$$T_{ii} = \frac{1}{4}(5-n)(f_{ij}, f_{ij}) + \frac{m^2}{2}(3-n)(b_i, b_i) \tag{29}$$

and the conservation law becomes

$$T_{ij,i} = 0 \tag{30}$$

Consider the integral

$$0=\int_0^\infty \omega(r)dr\int_{|x|\le r} x^j T_{ij,i}\, d^{n-1}x = \int_0^\infty \omega(r)dr\int_{|x|\le r}\{(x^jT_{ij})_{,i}-T_{ii}\}d^{n-1}x$$

$$=\int_0^\infty \omega(r)dr\int_{|x|=r}(x^jT_{ij})\frac{x^i}{r}dS-\int_0^\infty\omega(r)dr\int_{|x|\le r}T_{ii}\,d^{n-1}x \tag{31}$$

It is easily seen that there exists a constant A such that

$$|T_{ij}|\le A\,T_{00} \tag{32}$$

Moreover from (29) we have

(a) If $n \le 3$, then $T_{ii} \ge 0$ and the equality holds only when $b_i=0$;

(b) If $n \ge 5$, then $T_{ii} \le 0$ and the equality holds only when $b_i=0$.

In either case, if $T_{ii} \not\equiv 0$ we have $T_{ii} > 0$ (or $T_{ii} < 0$) in some region. Hence there exist two constant $R > 0$ and $\varepsilon > 0$ such that

$$\int_{|x|\le R_1} T_{ii}\, d^{n-1}x > \varepsilon \quad (\text{or } < -\varepsilon) \qquad (R_1 \ge R) \tag{33}$$

Choosing the same $w(r)$ as that in (26), for the case (a) we have

$$0<-\varepsilon\int_R^{R_1}\frac{dr}{r\psi(r)}+A\int_0^\infty\int\frac{T_{00}}{\psi(r)}\,dS\,dr \tag{34}$$

By the assumption that the energy is finite or "slowly divergent", we can choose R_1 sufficiently large, and it is easily seen that the right side of equation (34) should be negative. This gives a contradiction again. Consequently, we should have

$$T_{ii}=0$$

For the case (b) the situation is quite similar. Consequently we have $f_{ij}=0$ and $b_i=0$. In other words, when $n\neq 4$, the solution should be a trivial one. Thus Theorem 2 is proved completely.

Remark 1. The condition for the energy in Theorem 2 cannot be omitted, because for any dimensional space-time in massive and massless Y-M field we can find static regular solutions with energy diverges sufficiently fast.

Remark 2. Consider the Yang-Mills-Higgs- Kibble field (the gauge field with "soft" mass)

$$I=\int\left(-\tfrac{1}{4}f_{\mu\nu}f^{\mu\nu}-\tfrac{m^2}{2}b_\mu b^\mu-\tfrac{1}{4}\nabla_\mu\phi\,\nabla^\mu\phi-V(\phi)\right)$$

where ϕ is a scalar invariant and $V(\phi)$ is the potential. By using the same method the result of [6] can be improved and extended to the case of "slowly divergent" energy and the classical "discontinuity" hold also for $n=5$.

Remark 3. Open problem. In the case $n=4$, does there exist a static regular solution of massive Y-M equation with finite energy or "slowly divergent" energy?

Remark 4. For the massless case $m=0$, Using (16) and starting from (31), $f_{\mu\nu}=0$ can be obtained. Thus Deser's Theorem is also easily improved.

Recently we consider the massive Y-M field over curved space-time.

Let $R\times C^{n-1}$ be a curved space-time with metric

$$ds^2=-(dx^0)^2+e^{2f}\left((dx^1)^2+\cdots+(dx^{n-1})^2\right)$$

Here ρ is a function of $x^1, x^2, \cdots, x^n$ satisfying the conditions

$$\begin{aligned} &\text{(i)} \quad 1 + L(\rho) = 1 + x^i\rho_i \geq 0 \ , \quad \rho_i = \partial\rho/\partial x^i \quad (i=1,\cdots,n-1) \\ &\text{(ii)} \quad 0 < c_1 < \rho < c_2 \end{aligned} \tag{35}$$

The following theorem is obtained by the author and Y.L.Pan[8] .

Theorem 3. In an n-dimensional curved space-time $R \times C^{n-1}$ with $n \neq 4$, the compact group Y-M field with real mass does not possess any nontrivial static solution which is free of singularities and has finite and slowly divergent energy.

As is pointed out by Sealey[9] , the condition (i) has a geometric significance, i.e. the mean curvature normal of S_r is never pointing away from zero, where S_r is the level surface $(x^1)^2 + \cdots + (x^{n-1})^2 = r^2$

For the massless case, a similar result also holds. Same as the flat space-time case, the exceptional dimension is n=5 also.

As a consequence of Theorem 3, we have[8]

Thoerem 4. If M^n is a Riemannian manifold with metric $ds^2 = e^{2\rho}((dx^1)^2 + \cdots + (dx^n)^2)$ where ρ satisfies $1 + L(\rho) \geq 0$, then M^n does not possess any nontrivial massive Y-M gauge field which has finite action or slowly divergent action.

The corresponding theorem for massless case[8] improves a result obtained by Sealey[9] .

§3. Nonexistence Theorems for Harmonic Map with Finite or Slowly Divergent Energy

It is known that the harmonic maps from Euclidean space $R^n (n > 2)$ to any m-dimensional Riemannian space M_m with finite energy must be a constant map, i.e. the image of R^n is a fixed point [2] [10].

For each harmonic map ϕ, we have a stress-energy tensor which satisfy a conservative law. Using the conservative law together with the technique in §2, the author proved that[11]

Thoerem 5. Let $\phi : R^n \to M_m$ be a harmonic map of n-dimensional ($n \neq 2$) Euclidean space R^n into an m-dimensional Riemannian manifold M_m . Suppose that the energy $E(\phi)$ of ϕ is finite or slowly divergent, then ϕ is a constant map.

In theoretical physics, the Chiral field or the nonlinear σ-model on n-dimensional Minkowski space-time $R^{1,n-1}$ is just a harmonic map ϕ from $R^{1,n-1}$ to a homogeneous Riemannian manifold M_m . If the field is static, then ϕ is a harmonic map from R^{n-1} to M_m . Hence the physical significance of Theorem 5 can be expressed as[11]

Theorem 5'. In n+1 ($n > 2$) dimensional Minkowski space-time $R^{n,1}$, there does not exist any static nontrivial Chiral field with finite energy or slowly divergent energy.

Remark 1. By using sterographic projection from $S^2 \to R^2$, we will obtain nontrivial harmonic maps from $R^2 \to M$ with finite energy. So $n=2$ is actually an exceptional case.

Remark 2. The energy condition in our theorem cannot be omitted, because we can find many regular harmonic maps from R^n whose energy does not diverge so slowly.

Remark 3. Since a solution to the Ernst equations is a harmonic map from R^n to the hyperbolic plane with metric $ds^2 = \frac{1}{\Phi^2}(d\Phi^2 + d\Psi^2)$, $\Phi > 0$

in Poincare representation, we obtain another physical meaning of Theorem 5 [11], i.e. a nontrivial solution to the Ernst equation with $n > 2$ must have infinite energy. Furthermore, the energy cannot be slowly divergent. Here the energy is in the sense of harmonic maps.

On the other hand, H.C.J.Sealey in [9] proved the theorem: Let M^n $(n > 2)$ be a conformal flat space with metric form $ds^2 = f^2(x)((dx^1)^2 + (dx^2)^2 + \cdots + (dx^n)^2)$. If $L(f) = \sum_i x^i \, \partial \log f / \partial x^i \geq -1$, then any harmonic map with finite energy from M^n to any Riemannian manifold must be a constant map.

Recently, Pan and the author [12] obtain the following more general theorem.

Theorem 6. Let M^n $(n > 2)$ be a Riemannian manifold with metric form $ds^2 = f_1^2(x)(dx^1)^2 + f_2^2(x)(dx^2)^2 + \cdots + f_n^2(x)(dx^n)^2$ satisgying the following conditions

(i) $L(f_i) = \sum_j x^j \, \partial \log f_i / \partial x^j \geq -1$

(ii) There exists a positive constant K such that $\max\limits_{1 \leq i,j \leq n} \frac{f_i}{f_j} \leq K$

(iii) For any index i, and any set of $(n-2)$ indices $j_1, \cdots, j_{n-2}$ $(\neq i)$

$$\sum_{k=1}^{n-2} (1 + L(f_{j_k})) \geq 1 + L(f_i)$$

Then, any harmonic map ϕ with finite or slowly divergent energy from M^n to any Riemannian manifold must be a constant map.

Remark 1. In the case $f_1 = \ldots = f_n = f$ the conditions (ii) and (iii) are trivial. Hence Theorem 5 and the above mentioned result of Sealey are special cases of Theorem 6.

Remark 2. We point out that the condition (i) also has the geometric meaning as that in Sealey's case.

Remark 3. Theorem 6 includes essentially the case where M^n is a direct product manifold of p conformal flat manifolds $M_1 \times \ldots \times M_p$.

Because of the limitation of space, we will not give the proof of Theorem 6 here. Instead, we give the proof only for the special case. We assume M^n be a conformally flat space C^n with metric

For each harmonic map ϕ, we have a stress-energy tensor

$$S_{ij} = \frac{1}{2} e(\phi) \underset{M}{g_{ij}} - \underset{N}{g_{\alpha\beta}} \phi^\alpha_i \phi^\beta_j \tag{37}$$

which satisfies the conservative law

$$S^i_{j;i} = 0 \tag{38}$$

here ; denotes the covariant derivative with respect to the metric of C^n. Since

C^n is conformally flat,

$$S_i^j = e^{-2\rho} S_{ij} \tag{39}$$

$$\underset{M}{\Gamma^i_{jk}} = \delta^i_j \rho_k + \delta^i_k \rho_j - \rho_i \delta_{jk} \tag{40}$$

hold true. Consider the integral

$$\begin{aligned} 0 &= \int_{|x|\le r} x^i S_{i;j}^j \, dV = \int \{ (x^i S_i^j)_{;j} - x^i_{;j} S_i^j \} \, dV \\ &= \int_{|x|\le r} (x^i S_i^j)_{;j} \, dV - \int_{|x|\le r} S_i^i \, dV - \int \underset{M}{\Gamma^i_{jk}} x^k S_i^j \, dV \end{aligned} \tag{41}$$

Using (40) and Stoke's theorem, we have

$$0 = \int_{|x|=r} \frac{e^{-\rho} S_{ij} x^i x^j}{r} \, dh - \int_{|x|\le r} S_i^i (1 + L\rho) \, dV$$

where dh is the volume element of $|x| = r$ and $dV = e^{\rho} dr dh$. From $S_{ii} = $ $= e^{2\rho} \frac{n-2}{2} e(f) \geqslant 0$ and $1 + L\rho \geqslant 0$, we see that if ϕ is not a constant map, there exist positive constants R_1 and ϵ such that

$$\int_{|x|\le r} S_{ii} (1 + L\rho) \, dV > \varepsilon > 0 \qquad (r \geqslant R_1) \tag{42}$$

Hence for $r \geqslant R_1$

$$0 < \int_{|x|=r} \frac{e^{-\rho} S_{ij} x^i x^j}{r} \, dh - \varepsilon < \frac{1}{2} \int_{|x|=r} r e^{\rho} e(\phi) \, dh - \varepsilon \tag{43}$$

multiplying the above inequality by w(r) defined in (26), and integrate, we have

$$\begin{aligned} 0 &< \frac{1}{2} \int_0^R \omega(r) dr \int_{|x|=r} r e^{\rho} e(\phi) dh - \varepsilon \int_0^R \omega(r) dr \\ &= \frac{1}{2} \int_{R_1 \le |x| \le R} \frac{e(\phi)}{\psi(r)} dV - \varepsilon \int_{R_1}^R \frac{dr}{r \psi(r)} \end{aligned} \tag{44}$$

It is easily seen that if R is sufficiently large, the right side of (44) should be negative. This is a contradiction. Hence ϕ must be a constant map.

This work is partially supported by the Chinese National Foundation of Natural Sciences.

References

1 C.N.Yang & R.Mills, Isotopic spin conservation and a generalized gauge invariance, Phys. Rev. 96(1954) 191-195.

2 J.Eells & L.Lemaire,A report on harmonic maps, Bull. London Math. Soc. 10(1978) 1-68.

3 C.H.Gu(Gu Chao-hao), On classical Yang-Mills Fields, Physics Reports 80(1981) 251-337.

4 S.Deser, Absence of static solutions in source-free Yang-Mills theory, Phys. Lett. 64B (1976) 463-465.

5 H.S.Hu (Hu He-sheng), On equations of Yang-Mills gauge fields with mass, Kexue Tongbao 25(1980) 191-195.

6 S.Deser & C.J.Isham, Static solution of Yang-Mills-Higgs-Kibble system, Kexue

Tongbao 25(1980) 773-776.

7 H.S.Hu, On the static solutions of massive Yang-Mills equations, Chinese Annals of Math. 3(1982) 519-526.

8 H.S.Hu & Y.L.Pan(Pan Yang-lian), Vanishing theorems on the static solutions of massive Yang-Mills field, Preprint of Fudan Univ. (1984).

9 H.C.J.Sealey, Some conditions ensuring the vanishing of harmonic differential forms with applications to harmonic maps and Yang-Mills theory, Math. Proc. Camb. Phil. Soc. 91(1982) 441-452.

10 S.Hildebrandt, Nonlinear elliptic systems and harmonic mappings, Preceedings of 1980 Beijing Symposium on Differential Geometry and Differential Equations, Vol.1, 481-615.

11 H.S.Hu, A nonexistence theorem for harmonic maps with slowly divergent energy. Chinese Annals of Math. 5(1984).

12 H.S.Hu & Y.L.Pan, A theorem on Liouville's type on harmonic maps with finite or slowly divergent energy, Preprint of Fudan Univ. (1984).

GEOMETRICAL APPROACH TO THE PHYSICS OF RANDOM NETWORKS

Dina Maria L.F. Santos (*)
Université Paris VI
Institut de Mécanique
4, Place Jussieu, 75005, Paris, FRANCE.

1. Introduction

Glass was discovered approximately 60 centuries ago. Yet, in spite of its many applications, it is a material not explained by Theoretical Physics.

After the publication, in 1932, of the famous Zachariasen's paper "The Atomic Arrangement in Glass" [1], the amorphous solids structure has been based on a three-dimensional random network. This model accounts for many of the typical features of these solids, such as X-ray diffraction paterns, the optical isotropy and the low thermal conductivity ; but, for some time, little progress has been made using this model. On the one hand, random networks are very difficult mathematical tools to work with, while on the other hand, they reduce or even eliminate the validity of the majority of the techniques employed by crystallography. Many theories and techniques appeared as a result of an attempt to get around these mathematical difficulties. For example : some failed attempts were made to reduce the study of amorphous solids to the study of crystals, or there have been constructed in the laboratory or on computers possible atomic arrangements compatible with expected atomic interactions. Recent important improvements that have been made by some physicists such as Dzyaloshinskii and Volvik (1978), Kléman and Sadoc (1979), Rivier and Duffy (1982) and R. Kerner (1983) allow us to analyse the amorphous solids structure in a much more serious way. It is now clear that Differential Geometry and, in particular, the theory of Multiple Fiber Bundles, has an important part to play in the study of the structure of solids, although the results obtained this way are still minor. However, the analogy with several domains of Physics explained by Fiber Bundle treatment and the results that are gradually evolving give us good reasons to believe that this mathematical theory is a suitable model for our study.

In this work, we begin with the description of a phenomenological model

---

(*) Scholarship-holder from Calouste Gulbenkian's Fondation.

that will allow us to explain why, under some conditions, random networks are energetically preferable. This model will help to understand the reasonings that brought us to consider the Fiber Bundle Theory in the study of the structure of solids. Finally, we will present an outline of the fiber bundle that could describe bi-dimensional, three-coordinated (random or crystalline) networks.

2. The 2D three-coordinated phenomenological model.

We propose a model that describes the structure of a covalent amorphous solid (such as amorphous silicon) in which all the atoms are identical and in which the atomic interactions are repulsive and central (i.e., roughly speaking, the atoms tend to be as far from each other as possible). In a first approximation, we suppose the length of bonds to be constant (unitary).
We analyse here only the 2D three-coordinated model.
From the hypothesis we have made before, the 2D networks we are going to consider must satisfy the following conditions :

1. all their polygons must be equilateral (but not necessarily perfect);
2. each one of their vortices must always belong to three and only three polygons ($N_C = 3$);

we will suppose also that

3. all the polygons in the network are convex.

It is now everywhere accepted that, in order to distinguish between amorphous and crystal materials, we must take into account the atoms that are placed beyond the closest neighboors of a certain atom. In fact, X-ray and electron diffraction results show that amorphous solids present a high degree of short-range order very similar to that one we find in the correspondent crystal. For this reason, in our study of solids structure we will consider, for each atom P, a set of points that contains, besides its three closest neighboors P_i ($i = 1,2,3$), all the other vortices Q_j of the three polygons around P (fig 1). This is called the elementary cell of P. (k_1, k_2, k_3) represents the elementary cell formed by the vortices of a k_1-gon, a k_2-gon and a k_3-gon. It contains $k_1 + k_2 + k_3 - 5$ atoms.
By elementary tripod of P we mean the set of P, its three closest neighboors P_i ($i = 1,2,3$) and the three covalent bonds.

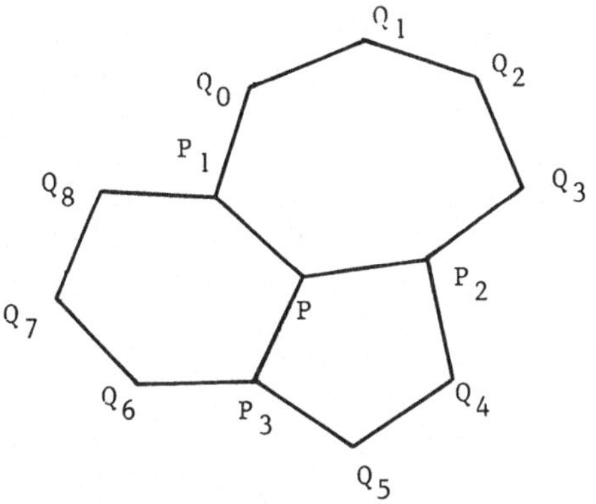

fig 1 : (5,6,7), the elementary cell of P

We can state that the information contained in the internal degrees of freedom of the elementary tripod and cell completely determines all the network.

The four perfect homogeneous networks (6,6,6), (4,8,8), (4,6,12) and (3,12,12) are the easist examples of 2D three-coordinated networks (fig. 2a)). Other examples of crystalline nets in the plane (networks produced by the repetition of one its bounded parts) are still represented on the same figure 2a). They all contain a finite number of different elementary cells. There is of course on the plane an infinity of random networks where all the polygons are equilateral (fig. 2b)). They all contain a large number (or even an infinity) of different elementary cells.

a)

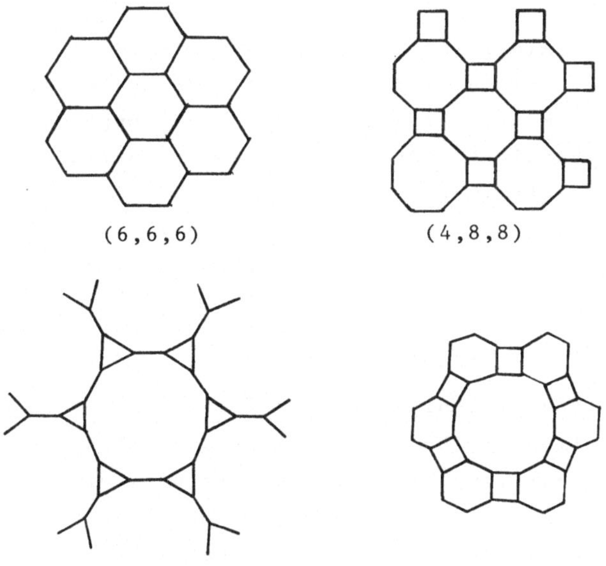

(6,6,6) (4,8,8)

(4,6,12) (3,12,12)

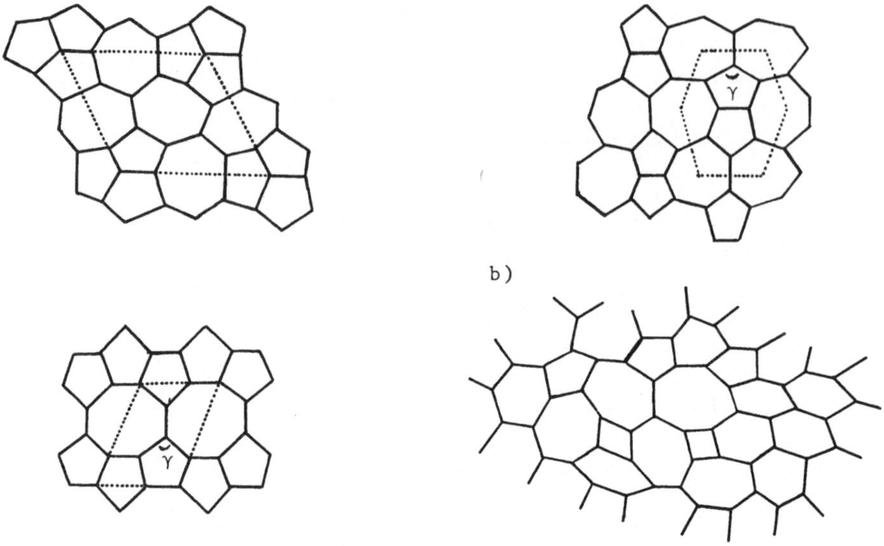

fig 2 : Some examples of 2D three-coordinated equilateral networks
a) crystalline networks b) a random network

One of the questions we want to answer is why, under some circumstances, Nature prefers random networks to the crystalline ones.
Those preferable configurations at T = 0 (T, the temperature) must correspond to the extrema of the internal energy per atom, that we are going to construct. We will suppose that this function is constituted by three terms that are the contributions of tripods, polygons and elementary cells to the energy of each atom.

FIRST TERM (TRIPOD CONTRIBUTION)
Let us call α_i $(i = 1,2,3)$ the angles between the three bonds in the tripod of P. As the atomic interactions have been supposed central and repulsive, it is reasonable to take for the contribution of the tripod to the energy of its central atom

$$U_T^{(\alpha_1,\alpha_2,\alpha_3)} = \sum_{i=1}^{3} \alpha_i^2 = \alpha_1^2 + \alpha_2^2 + (2\pi - \alpha_1 - \alpha_2)^2$$
$$= 2\,\alpha_1^2 + 2\,\alpha_2^2 + 2\,\alpha_1\,\alpha_2 - 4\pi\,(\alpha_1 + \alpha_2) + 4\,\pi^2 \quad ,$$

function that exhibits a minimum when $\alpha_i = \frac{2}{3}\pi$ $(i = 1,2,3)$, the most symmetrical configuration of the tripod. In the crystalline networks, we will take U_T, the contribution of all the tripods of the network to the total energy by atom, as the arithmetical mean value of the contributions of each tripod. It is easy to prove that

$$U_T = \frac{1}{2} \sum_{k=3}^{N} P_k \sum_{i=1}^{k} \beta_{ki}^2 \qquad (1)$$

(β_{ki} $(i = 1,\ldots,k)$ are the internal angles of the k-gon; P_k, the probability of finding a k-gon in the network; N, the maximum of the number of sides of the polygons of the network).

We will still take (1) for the case of random networks.

SECOND TERM (POLYGON CONTRIBUTION)

The contribution $U_P^{(k)}$ of a k-gon to the energy by atom will be taken proportional to its surface

$$U_P^{(k)} = \frac{1}{k} (S_k - S_k^{perf})$$

(S_k, the surface of the k-gon), because, if it would be possible to separate that equilateral k-gon from the rest of the surrounding lattice, it would take its symmetrical regular shape. We divide by k because the polygon must contribute equally to the energy of each one of its vortices. In a similar way,

$$U_P \approx \sum_{k=3}^{N} P_k U_P^{(k)}$$

THIRD TERM (ELEMENTARY CELL CONTRIBUTION)

The same reasoning on the elementary cell (k_1, k_2, k_3) allows us to take for the contribution of (k_1, k_2, k_3) to the energy,

$$U_C^{(k_1,k_2,k_3)} \approx \frac{1}{M} \sum_{i=1}^{3} S_{k_i}$$

($M = \frac{k_1+k_2+k_3}{3}$ is the number of atoms that receive the contribution of that elementary cell). The contribution of all the cells to the total energy will still be supposed proportional to the arithmetic mean value

$$U_C \approx \sum_{k_1,k_2,k_3=3}^{N} P_{(k_1,k_2,k_3)} U_C^{(k_1,k_2,k_3)}$$

where $P_{(k_1,k_2,k_3)}$ represents the probability of finding the cell (k_1,k_2, k_3) in the lattice.

In a first approximation we will neglect this third term, because particular cases have shown us that its contribution to the total energy does not modify the quantitative form of the result.

So, we have for the internal energy by atom

$$U = \frac{1}{2} \sum_{k=3}^{N} P_k \sum_{i=1}^{k} \beta_{ki}^2 + \lambda \sum_{k=3}^{N} P_k \frac{(S_k - S_k^{perf})}{k} \qquad (2)$$

(λ, the relative weight of the two contributions).

The expression of the internal energy involves then a great number of independent parameters. It is of course out of the question to study such a function, in particular for the random lattices where the number of variables is, in most cases, infinite. To get round this mathematical problem, we introduce a parametrization of 2D three-coordinated networks by a mean value angle α. It enables us to get a good approximation depending only of α to the function of internal energy (2).

We define α as :

- the geometrical mean value of the internal angles of the three polygons that constitute the unique elementary cell (k_1,k_2,k_3) of an homogenous lattice,

$$\alpha = \left[\prod_{i=1}^{3} \left(\prod_{j=1}^{k_i} \beta_{k_i j} \right) \right]^{1/m} \qquad (m = \sum_{i=1}^{3} k_i) \quad ;$$

- for a general network, we will take

$$\alpha = \left[\prod_{k=3}^{N} \left(\prod_{j=1}^{k} \beta_{kj} \right)^{p_k} \right]^{1/n} \qquad (n = \sum_{k=3}^{N} k\, p_k) \quad ;$$

this is equivalent to replacing of all the cells of the lattice by a "mean cell" consisting of p_k k-gons ($k = 3,\ldots,N$; p_k the probability of finding a k-gon among the three polygons of an elementary cell of the lattice). In order for this substitution to make sense, we must obviously suppose that the shapes of all the elementary cells of the network and of the "mean cell" are not very different and that the cells are quite homogenously distributed in the lattice.

Let us call

$$f_k = \sum_{i=1}^{k} \beta_{ki}^2$$

Some straightforward reasonings brought us to the following functions of α that are good approximations of f_k and S_k respectively :

$$F_k = 2\,\beta_k^2 + (k-2)\left[\frac{(k-2)\pi - 2\,\beta_k}{k-2} \right]^2 \qquad (3)$$

$$\mathcal{S}_k = \frac{1}{8} \operatorname{cosec}^2 \frac{\pi}{k}\, e^{A_k\alpha + B_k} \left[\sin \beta_k + (k-1) \sin \frac{(k-2)\pi - \beta_k}{k-1} \right] \qquad (4)$$

where $\beta_k = \frac{n(k-2)}{k(n-2)}\alpha$ and A_k and B_k are two parameters that can be supposed, with a good approximation, to depend only on k (independent of the lattice we are dealing with). We have obtained the following values for A_k (in rad^{-1}) and B_k

k	4	5	6	7	8
A_k	3.9718	2.5541	2.1169	1.9440	1.3525
B_k	- 8.6293	- 5.3931	- 4.4337	- 4.1147	- 2.9385

TABLE 1

We present now the errors obtained by application of the formulae (3) and (4) to some networks taken among the great number of those we have tested. We begin with the four regular 2D lattices

Lattice	(6,6,6)	(4,8,8)	(4,6,12)	(3,12,12)
Error F_k(%)	F_6 : 0	F_4 : 0.014 F_8 : 0.0048	F_4 : 0.0304 F_6 : 0.0152 F_{12}: 0.0061	F_3 : 0.2085 F_{12}: 0.0208

TABLE 2

For the family of non-perfect and non-homogenous networks we obtain varying the angle γ in the sixth lattice represented on fig. 2a)

Lattice (γ)	85°	90°	100°	107°	110°	120°	130°
Error F_5 (%)	2.34	1.43	0.28	0.003	0.02	0.58	1.89
Error F_7 (%)	1.48	1.10	0.65	0.58	0.62	1.00	1.79

TABLE 3

Considering now the function $\mathcal{S}_k$, we have obtained for the four regular nets,

Lattice	(6,6,6)	(4,8,8)	(4,6,12)	(3,12,12)
Error $\mathcal{S}_k$(%)	$\mathcal{S}_6$: 0	$\mathcal{S}_4$: 0.0060 $\mathcal{S}_8$: 0.0059	$\mathcal{S}_4$: 0.013 $\mathcal{S}_6$: 0.013 $\mathcal{S}_{12}$: 0.0097	$\mathcal{S}_3$: 0.0285 $\mathcal{S}_{12}$: 0.034

TABLE 4

and for some nets of the same family taken in table 3,

Lattice (γ)		65°	70°	80°	90°	100°	107°	110°	120°	130°
error (%)	$\mathcal{S}_5$	0.036	0.076	0.1332	0.1453	0.090	0	0.057	0.3255	0.7497
	$\mathcal{S}_7$	0.058	0	0.056	0.058	0.029	0	0.014	0.0525	0.067

TABLE 5

We have obtained a function of only one parameter α that gives a good approximation of U, the internal energy by atom. So, from now on, we will take

$$U = \frac{1}{2} \sum_{k=3}^{N} P_k F_k + \lambda \sum_{k=3}^{N} \frac{P_k}{k} (\mathcal{S}_k - S_k^{perf.})$$

We have studied the internal energy in the easier case of a lattice that is only composed by pentagons, hexagons and heptagons. From $P_5+P_6+P_7 = 1$ and Euler's theorem (in 2D three-coordinated lattices, we have always $\sum_{k=3}^{N} k P_k = 6$), we conclude that, in the particular case we are studing,

$$P_5 = P_7 = \frac{1 - P_6}{2}$$

So, U depends only of $P = P_6$, α and λ. Fig 3 a) and b) display the curves of the energy U as a function of α for different values of P and λ. They show the existence of a critical value λ_c for λ such as, when $\lambda < \lambda_c$, U displays an absolute minimum at $P = 1$ and $\alpha = \frac{2\pi}{3}$, that corresponds to the crystallization; for $\lambda > \lambda_c$, this minimum gives way to a maximum, while new minima appear for a $\alpha > \frac{2\pi}{3}$. They correspond to stable amorphous configurations. These conclusions agree with the experience. As a matter of fact, we know that in carbon structure polygons appear very easily which means that the energy of a polygon is not very high and λ is small; for silicon, on the contrary, the formation of polygons is much more difficult, and so λ is large. Then, the well known glass-forming tendency of silicon that we don't find in carbon is in agreement with the conclusions we have made before.

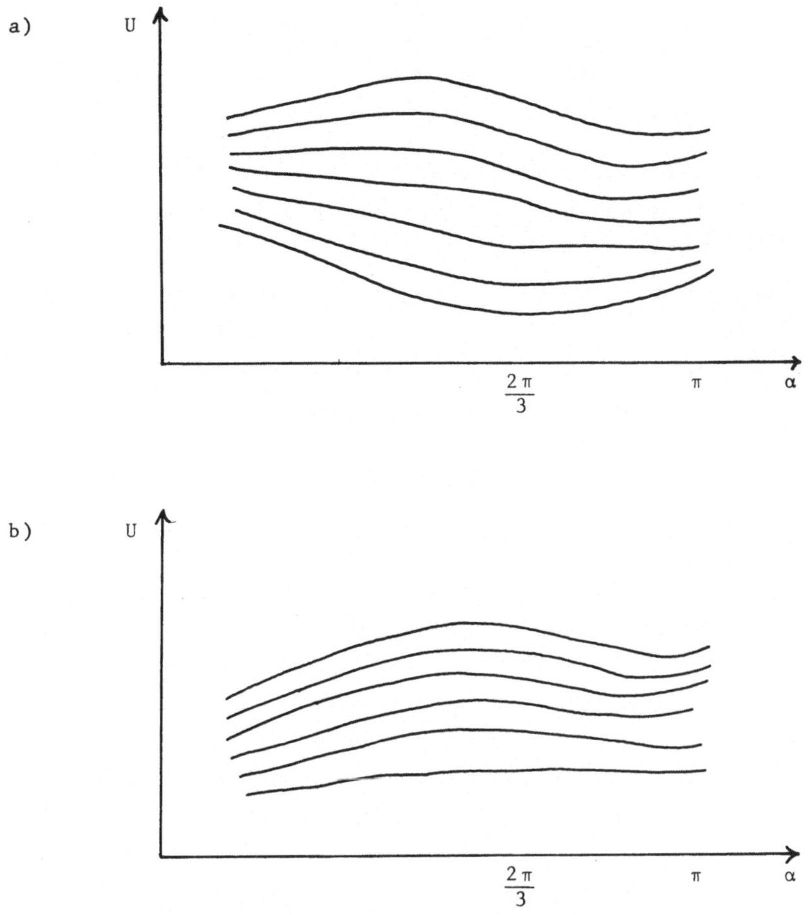

fig 3 : The curves $U_{P,\lambda} = U_{P,\lambda}(\alpha)$ for a) P=1; λ=32,34.5,36,37,39,40,42
b) P=0.7; λ=52,54,55,57,58,60

3. The description of 2D random networks in terms of Multiple Fiber Bundles.

It is well known that Classical Mechanics of a Particle can be studied by means of a Double Fiber Bundle with basis space $\mathbb{R}$, the domain of time, first fiber $\mathbb{R}^3$, the space of the positions of the particle, and second fiber $\mathbb{R}^3$, the space of velocities. The main problem in Mechanics is to determine the mouvement of a particle from a vector field X in the total space and some initial conditions. This is equivalent to determine the curve c that is the solution of the following system of differential equations

$$\begin{cases} \left.\dfrac{dx^k oc}{dt}\right|_t = X^k\,(c(t)) \\ \left.\dfrac{dv^i oc}{dt}\right|_t = X^{3+i}(c(t)) \end{cases} \qquad k,\ i = 1,2,3$$

for the initial conditions that were given.
However, the vector field X can not be given arbitrarly. The Kinematics of the particle tells us that we must always have

$$X^k(c(t)) = \left.\frac{dx^k oc}{dt}\right|_t = v^k(c(t)) \qquad k = 1,2,3$$

in order that it could correspond to the mouvement of a particle.
There is a clear analogy between what has been stated above and the description of 2D three-coordinated random networks. In fact, let us consider the fiber bundle $P(\mathbb{R}^2,E)$ with the basis space $\mathbb{R}^2$ and fiber E, the internal space of the degrees of freedom of an elementary tripod. A 2D network may be then described by a discrete section of $P(\mathbb{R}^2,E)$, assigning to each point of a discrete set of $\mathbb{R}^2$ a tripod in E. However, only some discrete sections of this bundle correspond to 2D networks, exactly those which yield closed polygons. So, we need more information and this brings us to the consideration of a second fiber F over the basis space $P(\mathbb{R}^2,E)$. The fiber of a point of $P(\mathbb{R}^2,E)$, that consists of an atom to which the shape of an elementary tripod T has been associated, is defined as the set of all the elementary cells that admit T as their central tripod.
All the 2D nets may be described by a discrete section of the Double Fiber Bundle that have been outlined. It is in the definition of these discrete sections that describe the random networks we find the analogy with Classical Mechanics. In fact, after an elementary cell has been associated to a point P of $P(\mathbb{R}^2,E)$, the tripods of the three neighboors of P are automatically determined and their elementary cells partly defined (we already know two of the three polygons of each one). This means that the elementary cells play a part that is similar to the one that velocity plays in Classical Mechanics.
Let us consider in more detail the fibers E and F.
As the tripods are supposed unoriented, the set E, of all the possible different shapes of an elementary tripod, can be parametrized by two of the three angles $\alpha_1,\alpha_2,\alpha_3$ between the three bonds. We can take, of course, $\alpha_1 \leq \alpha_2 \leq \alpha_3$ and we have always $\alpha_1+\alpha_2+\alpha_3 = 2\pi$, $\alpha_1 > \frac{\pi}{3}$ in order to make each point with only three closest neighboors and $\alpha_i \leq \pi$ $(i = 1,2,3)$ to prevent polygons from being concave. These conditions determine the quadrilateral in the plane (α_1,α_2) represented on fig. 4. To each point of

this set corresponds one and only one shape of an elementary tripod.

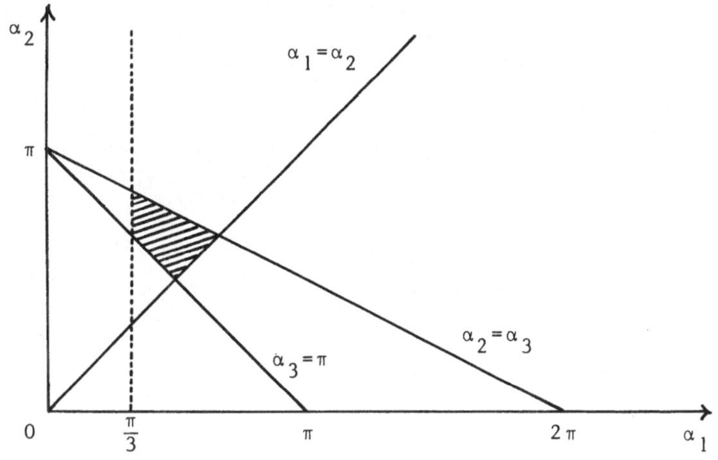

fig. 4 : The set E of all the possible different shapes of an elementary tripod.

The second fiber F over a given point (P,T) of $P(\mathbb{R}^2,E)$ has the structure of a singular foliation. Each sheaf is formed by the cells (k_1,k_2,k_3) compatible with T; its dimension is then, of course, $(k_1-3)+(k_2-3)+(k_3-3)-3 = k_1+k_2+k_3-12$. This reduces the number of possible choices of k_1,k_2,k_3 (we must have $k_1+k_2+k_3 > 12$ with, in our cases, $k_i \leqslant 12$ in order that we can consider all the regular three coordinated nets).

In the case of Classical Mechanics, the curves $\vec{x} = \vec{x}(t)$ and $\vec{v} = \vec{v}(t)$ are completely determined by minimizing the action integral

$$\int_{t_o}^{t_1} L\ (\vec{x},\vec{v})\ dt$$

All we need then is to construct a function L depending on the tripods and the polygons such that its integral is the action integral to minimize. We are not yet able to construct such a functional in terms of differential geometrical considerations, but we believe that the internal energy, we have introduced before, can be taken in a first satisfactory approximation.

ACKNOWLEDGMENT

The author expresses her gratitude to Professor Richard Kerner for his guidance and enlightening discussions.

REFERENCES.

Zachariasen , W.F. - J. Chem. Phys. Vol.54, p. 3841-3851 (1932).
Dzyaloshinskii and Volvik - J. Phys., 39, p. 693 (1978).
Kléman and Sadoc - J. Phys. Lett. 40, p. 569 (1979).
Rivier and Duffy - J. Phys. 43, p. 293 (1982).
Kerner, R. - Phil. Mag B, 47, n°2, p. 151-162 (1983).
Kerner, R. - Phys. Rev. B, 28, 4, p. 5756 (1983).
Kerner, R. and D.M. Santos - C.R. Acad. Sc. Paris, 295 (1982).
Wells, A.F. - Three-dimensional Nets and Polyhedra, J. Wiley and Sons N,Y. (1977).

THE ALGEBRA OF MULTIPLICATION OPERATORS OF STAR-PRODUCT IN $\mathbb{R}^{2n}$

J-B. Kammerer
Professeur à l'Ecole Centrale des Arts et Manufactures
92290 Chatenay-Malabry France.

M^r. Lichnerowicz has defined the star-product as a deformation of ordinary multiplication [4] and has shown , that , in $\mathbb{R}^{2n}$, there is only one formal function of the Poisson bracket (more or less an equivalence), which defines an associative product ; it is the exponential function [2] :

(1) $$f *_\nu g = f.g + \sum_{p=1}^{\infty} \frac{\nu^p}{p!} P^p(f,g) \quad ,$$

$$P^p(f,g) = \Lambda^{i_1 j_1} \dots \Lambda^{i_p j_p} \partial_{i_1 \dots j_p} f \, \partial_{j_1 \dots j_p} g \quad .$$

$\Lambda^{ij} = 1$ if $j=i+n$, -1 if $i=j+n$, 0 in the other case.

The star-product is defined here , with the twisted convolution [3] . This definition allows one to caracterize the space of multiplication operators , and to show that this star-product , under a large enough hypothesis , is equal to the star-product defined by the exponential series (1) .

I The algebras $\mathcal{S}(\mathbb{R}^{2n})$:

Suppose that $f , g \in \mathcal{S}(\mathbb{R}^{2n})$; the twisted convolution $f \hat{*}_\nu g$ is the function :

(2) $$f \hat{*}_\nu g(x) = \int_{\mathbb{R}^{2n}} f(t)\, g(x-t)\, e^{\nu\, t_\wedge x}\, dt \quad ,$$

$\nu = -\bar{\nu} \neq 0$, $x = (x_1,x_2) \in \mathbb{R}^{2n}$, $t_\wedge x = t_1 x_2 - t_2 x_1$. The star-product is the function :

(3) $$f *_\nu g = (2\pi)^{-n} F(F^* f \hat{*}_\nu F^* g) \quad .$$

F is the symplectic Fourier transform . Finally , let us introduce a "kernels composition law" :

(4) $$f_o g(x) = \int_{\mathbb{R}^n} f(x_1,t)\, g(t,x_2)\, dt \quad .$$

We have :

(5) $$f *_\nu g = (4\pi|\nu|)^{-n/2} \mathcal{F}_2^* \mathcal{C}_\nu (\mathcal{C}_\nu^* \mathcal{F}_2 f \;_o\, \mathcal{C}_\nu^* \mathcal{F}_2 g) \; ,$$

$\mathcal{F}_1$ and $\mathcal{F}_2$ are the Fourier transforms towards x_1 and x_2 ; $\mathcal{C}_\nu$ and $\mathcal{C}_\nu^*$ are two mappings , one the inverse of the other :

(6) $$\mathcal{C}_\nu f(x) = (2|\nu|)^{n/2} f(x_1 + i\nu x_2 \, , \, x_1 - i\nu x_2) \quad .$$

These three mappings are continuous of $\mathcal{S}(\mathbb{R}^{2n}) \times \mathcal{S}(\mathbb{R}^{2n})$ into $\mathcal{S}(\mathbb{R}^{2n})$ and non-degenerate ($f=0 \Leftrightarrow f_{*\nu}\bar{f} = 0$) . The mappings $\mathcal{F}_1$, $\mathcal{F}_2$, $\mathcal{T}_\nu$, $\mathcal{T}_\nu^*$ are continuous of $\mathcal{S}(\mathbb{R}^{2n})$ onto $\mathcal{S}(\mathbb{R}^{2n})$ [6] .

II The space of multiplication operators :

Suppose that $f \in \mathcal{S}'(\mathbb{R}^{2n})$, $g \in \mathcal{S}(\mathbb{R}^{2n})$, $f_{*\nu}g$ is the distribution defined in $\mathbb{R}^{2n}$:

(7) $\quad \forall \varphi \in \mathcal{S}(\mathbb{R}^{2n}) \quad , \quad \langle f_{*\nu}g , \varphi \rangle = \langle f , g_{*\nu}\varphi \rangle$

The mapping , $f \mapsto f_{*\nu}g$, is continuous of $\mathcal{S}'(\mathbb{R}^{2n})$ in $\mathcal{S}'(\mathbb{R}^{2n})$ ($f_{*\nu}g = 0 , \forall g \in \mathcal{S}(\mathbb{R}^{2n}) \Leftrightarrow f=0$) . The space of tempered distributions for which the mapping , $g \mapsto f_{*\nu}g$, is continuous of $\mathcal{S}(\mathbb{R}^{2n})$ into $\mathcal{S}(\mathbb{R}^{2n})$, is denoted by $\mathcal{O}_M^\nu$. The linear space $\mathcal{O}_M^\nu$ is a locally convex TVS by the family of seminorms :

(8) $$p_{q,B}(f) = \sup_{g \in B} \sup_{x \in \mathbb{R}^{2n}} | D^q f_{*\nu}g(x) | ,$$

$q \in \mathbb{N}^{2n}$; B is a bounded set of $\mathcal{S}(\mathbb{R}^{2n})$. The mappings :

$$f \mapsto \check{f} , \overleftrightarrow{f} , \tau_a f , x^q f , D^q f , e^{iax} f , (F\overleftrightarrow{f})_{1/\nu} ,$$

are continuous of $\mathcal{O}_M^\nu$ into itself . Let us define the product of $f \in \mathcal{S}'(\mathbb{R}^{2n})$ and $g \in \mathcal{O}_M^\nu$ by the relation (7) .

The algebra $\mathcal{O}_M^\nu$ (resp. $\mathcal{O}_M^{\bar{\nu}}$) gives to $\mathcal{S}'(\mathbb{R}^{2n})$ a structure of right module (resp. left) [1] . The mappings , $(f,g) \mapsto f_{*\nu}g$, of $\mathcal{S}'(\mathbb{R}^{2n}) \times \mathcal{O}_M^\nu$ into $\mathcal{S}'(\mathbb{R}^{2n})$ or of $\mathcal{O}_M^\nu \times \mathcal{O}_M^\nu$ into $\mathcal{O}_M^\nu$ are separatly continuous .

III Characterization of $\mathcal{O}_M^\nu$:

It is easy to shown , that the linear space $\mathcal{O}_M^\nu$ is complete and isomorphic to the space $\mathcal{L}_b(\mathcal{S}(\mathbb{R}^n))$ [7] . The mapping , $(f,g) \mapsto f_{*\nu}g$, is separatly continuous of $\mathcal{S}(\mathbb{R}^{2n}) \times \mathcal{S}'(\mathbb{R}^{2n})$ into $\mathcal{F}_2^* \mathcal{T}_\nu(\mathcal{S}(\mathbb{R}^n) \hat{\otimes} \mathcal{S}'(\mathbb{R}^n))$. Finally , let f_p $p=1,2,\ldots$ be a regularizing sequence ($f_p \in \mathcal{S}(\mathbb{R}^{2n})$) ; the sequence Ff_p converges to 1 in $\mathcal{O}_M^\nu$.

Theorem : The linear space $\mathcal{O}_M^\nu$ of multiplication operators is equal to $\mathcal{F}_2^* \mathcal{T}_\nu(\mathcal{S}(\mathbb{R}^n) \hat{\otimes} \mathcal{S}'(\mathbb{R}^n))$. The natural injections of $\mathcal{S}(\mathbb{R}^{2n})$ into $\mathcal{O}_M^\nu$ and of $\mathcal{O}_M^\nu$ into $\mathcal{S}'(\mathbb{R}^{2n})$ are continuous and their images are dense [5] .

IV Applications :

. The product of a polynom f and a tempered distribution g is given by the relation (1) .

. Let $f \in \mathcal{S}'(\mathbb{R}^{2n})$, $g \in \mathcal{O}_M(\mathbb{R}^{2n})$; we make the assumption : $\forall \nu \in]0, \nu_0]$, $g \in \mathcal{O}_M^{\nu}$. Let $q \in \mathbb{N}$; $h_\nu(f,g)$ the tempered distribution defined by the relation :

$$(9) \qquad f *_\nu g = f.g + \sum_{p=1}^{q-1} \frac{\nu^p}{p!} P^p(f,g) + \frac{\nu^q}{q!} h_\nu(f,g) .$$

If the function g has the property : $g \in \mathcal{S}(\mathbb{R}^{2n})$ or $\forall r \in \mathbb{N}^{2n}$, $\forall B$ bounded set of $\mathcal{S}(\mathbb{R}^{2n})$, $\sup_\nu p_{r,B}(g) < +\infty$; the relation (9) proves that $f *_\nu g$ has a limited development in $\mathcal{S}'(\mathbb{R}^{2n})$, when ν tends towards zero . Example :

$$f(x) *_\nu e^{iax} = f(x_1 - i\nu a_2 , x_2 + i\nu a_1) \, e^{iax} .$$

If the function f is analytic , the relation (1) exists locally ; if not , the relation (9) gives a limited development .

. Let $f \in \mathcal{O}_M(\mathbb{R}^n)$, $g \in \mathcal{O}_C'(\mathbb{R}^n)$; the distributions $1_{x_1} \otimes f(x_2)$, $\delta_{x_1} \otimes g_{x_2}$ belong to $\mathcal{O}_M^{\nu}$.

References :

1 M.A. Antonets , Letters in Math. Phys. , 2 , 1978 , p.241-245 .

2 F.Bayen , M.Flato , C.Fronsdal , A.Lichnerowicz , D.Sternheimer , Annals of Physics 111. 61-110 (1978) .

3 K.Chi Liu , J. Math. Phys. , 17 , 1976 , p.859 .

4 M.Flato , A.Lichnerowicz et D.Sternheimer , Comptes Rendus Acad. Sc. , t.283 , série A , 1976 , p.19 .

5 J-B Kammerer , Comptes Rendus Acad. Sc. , t.298 ,série I,n°4 , 1984 , p.59 .

6 C.Moreno , Produits star et analyse spectrale de certains opérateurs.

7 F.Trèves , Topological Vector Spaces , Distributions and kernels, Academic Press , New-York .

MANIFOLD b-INCOMPLETENESS STABILITY VIA A STRUCTURE OF PRINCIPAL CONNECTIONS

D.Canarutto
Istituto di Matematica Applicata
Università di Firenze

C.T.J.Dodson
Department of Mathematics
University of Lancaster

April 5,1984

Abstract

The bundle of principal connections of a principal bundle is used to study some stability problems in the theory of b-completions. It follows that b-incompleteness is stable under perturbations of the connection. This adds weight to the belief that general relativistic singularities cannot be removed by quantization.

§1. Preliminaries and notation (see [1])

By $p:E \twoheadrightarrow M$ we shall indicate a fibred manifold, and by T, T^*, V and J the tangent, cotangent, vertical and first-jet functors respectively.

A first order connection on $E \twoheadrightarrow M$ can be defined as a section $\Gamma:E \longrightarrow JE$ or, equivalently, as a certain 1-form with values in VE, $\omega_\Gamma:E \longrightarrow T^*E \underset{E}{\otimes} VE$. A structure of connections on $E \twoheadrightarrow M$ is a couple $\{C,\gamma\}$ where $\pi:C \twoheadrightarrow M$ is a fibred manifold and $\gamma:C \underset{M}{\times} E \longrightarrow JE$ is a fibred morphism over E. Any section $\tilde{\Gamma}:M \longrightarrow C$ determines a connection $\Gamma \equiv \gamma\circ(p^*\tilde{\Gamma})$. Given a structure of connections $\{C,\gamma\}$, there exists a canonical connection Λ on the fibred manifold $K \equiv C \underset{M}{\times} E \longrightarrow C$, whose connection form ω_Λ is characterized by:

$$\langle \omega_\Lambda \circ (p^*\tilde{\Gamma}), X\rangle = \langle \omega_\Gamma, T\pi_2(X)\rangle, \quad \forall \tilde{\Gamma}:M \longrightarrow C, \quad X \in TK ,$$

(we denote by $\pi_1:K \twoheadrightarrow C$ and $\pi_2:K \twoheadrightarrow E$ the canonical projections).

In the case of a principal bundle $(P,p,M;G)$ we have the structure of principal connections on $P \twoheadrightarrow M$, where $C = JP/G$. It can be seen that $JP/G \twoheadrightarrow M$ is affine: its associated vector bundle is VP/G which is (not canonically) isomorphic to $T^*M \otimes \mathcal{G}$ ($\mathcal{G}$ is the Lie algebra of G). The choice of any $p_0 \in P$ induces one such isomorphism.

§2. The principal bundle of linear frames

Henceforth $P \equiv LM$, the frame bundle of M. The canonical 1-form $\theta:TLM \longrightarrow \mathbb{R}^m$ is easily generalized to a 1-form on $K \equiv C \underset{M}{\times} LM$, and so the universal connection Λ gives rise to a degenerate symmetric bilinear form f on K, which is defined in a way similar to the Schmidt metric g_Γ induced on LM by a given connection [2]

$$f: TK \underset{K}{\times} TK \longrightarrow \mathbb{R} : (w,z) \longmapsto \theta(w)\cdot\theta(z) + \omega_\Lambda(w)\cdot\omega_\Lambda(z) .$$

For any section $\tilde{\Gamma}:M \longrightarrow C$, we shall consider the submanifold $S_\Gamma \equiv \equiv p^*\tilde{\Gamma}(LM) \subset K$ which is diffeomorphic to LM. Moreover, (LM, g_Γ) and (S_Γ, f_Γ) are isometric spaces, where $f_\Gamma \equiv f|_{S_\Gamma}$.

It will be shown [3] that the assignment of a connection $\Delta:C \longrightarrow JC$ induces a Riemannian metric g_Δ on K, which coincides with f on π_1-vertical vectors. If $\tilde{\Gamma}$ is Δ-horizontal, then the g_Γ-completion of LM can be seen as a "slice" in the g_Δ-completion of K.

§3. Stability of b-incompleteness

Our main result is the following:

Proposition. Let $\tilde{\Gamma}: M \longrightarrow C$ be a section such that M is b-incomplete with respect to Γ. Let V be an f-bounded open submanifold of K, such that the f_Γ-boundary of $V \cap S_\Gamma$ contains a point x in the f_Γ-boundary of S_Γ (this x can be identified with a point in the g_Γ-boundary of LM). Let $c:[0,1) \longrightarrow LM$ be a curve with g_Γ-endpoint x, and $\tilde{B}: M \longrightarrow C$ be a section such that $(p^*\tilde{B}) \circ c[0,1) \subset V$.
Then, c is g_B-incomplete and thus M is b-incomplete with respect to the connection B. □

For the proof, see [3], where it is also shown that a neighbourhood V of the singularity which satisfies the hypothesis of the proposition can be easily constructed. For example, it follows that b-incompleteness with respect to Γ is stable in a family of connections $\{\Gamma_\varepsilon\}$ uniformly g_Δ-converging to Γ as $\varepsilon \to 0$. Similarly, b-incompleteness is stable in a family of locally bounded conformally equivalent metric tensors on M .

Acknowledgement. C.T.J.Dodson wishes to thank the Istituto di Matematica Applicata "G.Sansone" and the "Consiglio Nazionale delle Ricerche" for hospitality and support during the course of this work.

References

[1] L.Mangiarotti and M.Modugno, *Fibered spaces, jet spaces and connections for field theories*, in Proceedings of the International Meeting on Geometry and Physics, Florence 12-15 October 1982, ed. M.Modugno, Pitagora Editrice, Bologna (1983) pp.135-165.

[2] C.T.J.Dodson, *Space-time edge geometry*, Int.J.Theor.Phys. 17,b (1978) pp.389-504.

[3] D.Canarutto and C.T.J.Dodson, *On the bundle of principal connections and the stability of b-incompleteness*, Preprint Ist.Mat.Appl."G.Sansone", Univ.Firenze (1984).

FRONT FORM PREDICTIVE RELATIVISTIC MECHANICS

NON INTERACTION THEOREM

X. Jaen, A. Molina & J. Llosa (*)
Departament de Física Teòrica, Universitat de Barcelona
Diagonal, 645, Barcelona-28 (Spain)

From a historical point of view, relativistic dynamics of directly interacting particles was initiated by Dirac[1], who proposed that a relativistic system of particles must be described in terms of a canonical realization of Poincaré algebra on a phase space of dimension 6N, that is, ten generating functions with the suitable Poisson relations. Dirac himself proposed three possible approaches to this aim, each of them consisted in assuming that as many as possible among the generators have the same simple form it would have if particles did not interact — these are called kinematical generators — while the remaining ones — called hamiltonians — would contain all the information about dynamics. These approaches are respectively called: (i) instant form, (ii) front form, and (iii) point form, depending on whether (i) the instantaneous hyperplane x^0 = constant, or (ii) the front hyperplane $x^0 + x^3$ = constant, or the hyperboloid $x^\mu . x_\mu$ = constant, are left invariant by kinematical generators.

Further progress in the instant form approach was made by Bakamjian, Thomas and Foldy[2], until the non interaction theorem revealed an inner inconsistency in the theory so far developed. The theorem states the incompatibility of world line invariance and the canonical character of position coordinates, except in case that particles do not interact. The several proofs of this theorem have been so far given in the instant form framework[3].

A later approach to the classical relativistic dynamics of particles without an intermediate field is Predictive Relativistic Mechanics[4] (PRM) which, roughly speaking, results from giving up the hamiltonian formalism in the instant form approach. Namely, points in the extended configuration space $(\vec{x}_1, \dots, \vec{x}_N, t)$ are coordinated by simultaneous positions of particles at the time t, according to a previously chosen inertial observer. Then, equations of motion are required to be like Newton (i.e.: accelerations are some given functions of positions, velocities

(*) Grup de Relativitat, Secció de Ciències (Institut Estudis Catalans)

and time). Finally, relativistic invariance is imposed at two levels: (i) changing from one inertial frame to another must not affect the equations of motion, and (ii) the world lines obtained in one inertial frame for a given set of initial data ($\vec{x}_a, \vec{v}_b, t_0$) must be also obtained in any other inertial frame for a transformed set ($\vec{x}'_a, \vec{v}'_b, t_0$). Again, when a hamiltonian formulation of PRM is attempted, requiring position coordinates to be canonical, the theory falls into non interaction [5].

Therefore, the non interaction result has been so far proved in the instant form framework only. We shall here present the proof of the same result for the front form approach and advance, without proof, that a similar result also holds in the point form framework.

In a given inertial frame $\mathcal{S}$, the equations of motion are taken as:

$$\frac{d\vec{x}_a}{d\lambda} = \vec{v}_a \qquad , \qquad \frac{d\vec{v}_a}{d\lambda} = \vec{a}_a(\vec{x}_b, \vec{v}_b, \lambda) \tag{1}$$

where $\vec{x}$ means the position of the bth particle at a time x^0_b such that:

$$x^0_b + x^3_b = \lambda \qquad , \qquad \forall b = 1, \dots, N . \tag{2}$$

Then, if $\vec{\varphi}_a(\vec{x}_b, \vec{v}_c, \lambda_0; \lambda)$ is the solution of (1), corresponding to the initial conditions:

$$\vec{\varphi}_a(\vec{x}_b, \vec{v}_c, \lambda_0; \lambda_0) = \vec{x}_a \qquad , \qquad \dot{\vec{\varphi}}_a(\vec{x}_b, \vec{v}_c, \lambda_0; \lambda_0) = \vec{v}_a \tag{3}$$

the world line of the ath particle is defined by taking $\vec{\varphi}_a$ as the spatial part and adding the time component:

$$\varphi^0_a(\vec{x}_b, \vec{v}_c, \lambda_0; \lambda) \equiv \lambda - \varphi^3_a(\vec{x}_b, \vec{v}_c, \lambda_0; \lambda) \tag{4}$$

Finally, relativistic world line invariance implies that the Poincaré transformed world lines

$$L^\mu{}_\nu \cdot \{ \varphi^\nu_a(\vec{x}_b, \vec{v}_c, \lambda_0; \lambda) - A^\nu \}$$

must be obtained in another inertial frame $\mathcal{S}'$, from a different set of initial data, that is:

$$L^\mu{}_\nu \cdot \{ \varphi^\nu_a(\vec{x}_b, \vec{v}_c, \lambda_0; \lambda) - A^\nu \} = \varphi^\mu_a(\vec{x}'_b, \vec{v}'_c, \lambda_0; \lambda'_a) \tag{5}$$

with λ'_a being given by:

$$\lambda'_a(\vec{x}_b, \vec{v}_c, \lambda_0; \lambda) \equiv L^0{}_\nu \{ \varphi^\nu_a(\vec{x}_b, \vec{v}_c, \lambda_0; \lambda) - A^\nu \} + \\ + L^3{}_\nu \cdot \{ \varphi^\nu_a(\vec{x}_b, \vec{v}_c, \lambda_0; \lambda) - A^\nu \} \tag{6}$$

The relationship between the former initial data ($\vec{x}_a, \vec{v}_b, \lambda_0$) and the new ones ($\vec{x}'_a, \vec{v}'_b, \lambda_0$) defines the so called <u>induced Poincaré transformation</u> on the extended co-phase space, the generators of which turn out to be:

$$
\left.\begin{aligned}
\vec{P}_i &= -\sum_a \frac{\partial}{\partial x_a^i} \quad , \quad i=1,2 \\
\vec{P}_- &\equiv \vec{P}_3 - \vec{P}_0 = -\sum_a \frac{\partial}{\partial x_a^3} \\
\vec{F}_i &= -\sum_a \left\{ x_a^i \cdot \frac{\partial}{\partial x_a^3} - v_a^i \cdot \frac{\partial}{\partial v_a^3} + \frac{\partial}{\partial v_a^i} \right\} \quad , \quad i=1,2 \\
\vec{J}_3 &= \sum_a \left\{ x_a^2 \cdot \frac{\partial}{\partial x_a^1} - x_a^1 \cdot \frac{\partial}{\partial x_a^2} + v_a^2 \cdot \frac{\partial}{\partial v_a^1} - v_a^1 \cdot \frac{\partial}{\partial v_a^2} \right\} \\
\vec{K}_3 &= -\sum_a \left\{ x_a^3 \cdot \frac{\partial}{\partial x_a^3} + v_a^3 \cdot \frac{\partial}{\partial v_a^3} + v_a^i \cdot \frac{\partial}{\partial v_a^i} \right\} \\
\vec{P}_+ &\equiv \frac{1}{2}(\vec{P}_0 + \vec{P}_3) = \sum_a \left\{ v_a^i \cdot \frac{\partial}{\partial x_a^i} + a_a^i \cdot \frac{\partial}{\partial v_a^i} \right\} \\
\vec{E}_r &= -\sum_a \left\{ (x_a^3 \cdot \delta_r^i + x_a^r \cdot v_a^i) \cdot \frac{\partial}{\partial x_a^i} + \right. \\
&\qquad \left. + (v_a^3 \cdot \delta_r^i + x_a^r \cdot a_a^i + v_a^r \cdot v_a^i) \cdot \frac{\partial}{\partial v_a^i} \right\} \quad , \quad r=1,2.
\end{aligned}\right\} \quad (7)
$$

where

and

$$
\left.\begin{aligned}
\vec{F}_1 &\equiv \vec{K}_1 - \vec{J}_2 \quad , \quad & \vec{F}_2 &\equiv \vec{K}_2 + \vec{J}_1 \\
\vec{E}_1 &\equiv \frac{1}{2}(\vec{K}_1 + \vec{J}_2) \quad , \quad & \vec{E}_2 &\equiv \frac{1}{2}(\vec{K}_2 - \vec{J}_1)
\end{aligned}\right\} \quad (7')
$$

Realize that, in agreement with what could be expected in the front form formalism, the first seven generators are kinematical ones and are as simple shaped as in the free particle case.

Since (5) must hold for every λ , the induced realization of Poincaré group must commute with dynamical evolution. Therefore, we have:

$$[\vec{\Lambda}_I , \vec{H}] = 0 \qquad , \quad I = 1, \ldots, 10, \tag{8}$$

where $\vec{\Lambda}_I$ stands for any generator of Poincaré and

$$\vec{H} = \frac{\partial}{\partial \lambda} + \sum_a \left\{ v_a^i \cdot \frac{\partial}{\partial x_a^i} + a_a^i \cdot \frac{\partial}{\partial v_a^i} \right\} \tag{9}$$

generates dynamical evolution on the co-phase space.

Eq. (8) provide a set of differential conditions on the accelerations, if the system is to be world line invariant. These equations read as follows:

$$\frac{\partial}{\partial \lambda} a_a^i (\vec{x}_b, \vec{v}_c, \lambda) = 0$$

$$\vec{P}_r a_a^j = 0 \quad , \quad \vec{P}_- a_a^j = 0 \quad , \quad \vec{F}_r a_a^i = - a_a^r \cdot \delta_3^i \quad , \quad r=1,2$$

$$\vec{J}_3 a_a^i = \varepsilon_{3ji} \cdot a_a^j \quad , \quad \vec{K}_3 a_a^i = 2 a_a^i + a_a^3 \cdot \delta_3^i$$

$$\vec{E}_r a_a^i + x_a^r \cdot \vec{P}_0 a_a^i = -2 v_a^r \cdot a_a^i - a_a^r \cdot v_a^i - a_a^3 \cdot \delta_r^i \quad , \quad r=1,2$$

and play the same role as the well known Currie-Hill equations in PRM.

If now, in addition, we demand that there is a Poisson structure on the co-phase space, such that:(i) the position variables are canonical coordinates related to it and, (ii) there are ten generating functions Λ_I ,I=1,...10,related to the Poincaré generators (7) acoording to:

$$\vec{\Lambda}_I = \{\Lambda_I , -\}$$

then, by using well known properties of Poisson brackets (e.g.: Jacobi identities), after a cumbersome but easy calculation, we arrive at:

$$\vec{a}_b = 0 \quad , \quad b = 1, \ldots, N,$$

which ends the proof of the non interaction theorem in the front form approach.

References:

1.- DIRAC, P.A.M., Rev.Mod.Phys. 21, 392 (1949)

2.- BAKAMJIAN, B. & THOMAS, L.H., Phys. Rev. 92, 1300 (1953)
FOLDY, L., Phys.Rev. 122, 275 (1961)

3.- CURRIE, D.G., JORDAN, T.F. & SUDARSHAN, E.C.G., Rev.Mod.Phys.35(1963)
LEUTWYLER, H., Nuovo Cimento, 37, 556 (1965)

4.- HILL, R.N., Jour. Math.Phys., 8, 201 (1967)
BEL, L., Ann. Inst. H. Poincaré 12, 307 (1970)

5.- HILL, R.N., Jour.Math.Phys. 8, 1756 (1967)

SOME NEW RESULTS ON THE VALIDITY OF HUYGENS' PRINCIPLE FOR THE SCALAR WAVE EQUATION ON A CURVED SPACE-TIME

J. Carminati and R.G. McLenaghan
Department of Applied Mathematics
University of Waterloo
Waterloo, Ontario
Canada N2L 3G1

Abstract: It is shown that every Petrov type N space-time on which the conformally invariant scalar wave equation satisfies Huygens' principle is conformally related to a special complex recurrent space-time.

We shall be discussing the question of the validity of Huygens' principle for the conformally invariant scalar wave equation (For a review of this problem up to 1980 see McLenaghan [9].)

$$g^{ab} u_{;ab} + \frac{1}{6} R u = 0, \qquad (1)$$

where g^{ab} denotes the contravariant metric tensor on a 4-dimensional space-time L_4 of signature -2, ";" denotes the covariant derivative with respect to the Levi-Civita connection and R denotes the curvature scalar. The coefficients g^{ab} and L_4 are assumed to be of class C^∞.

Cauchy's problem for Eq. (1) is the problem of determining a solution which assumes given values of u and its normal derivative on a given space-like 3-dimensional submanifold S. These given values of u are called the Cauchy data. The local existence and uniqueness of the solution of Cauchy's problem for (1) is contained in the work of Hadamard [4]. Alternate solutions have been given by other authors [1, 5, 11]. The considerations in this paper will be purely local.

The question of how the solution u of Cauchy's problem depends on the Cauchy data is of considerable importance both theoretically and in applications. Hadamard has shown that for any x_o, $u(x_o)$ depends only on the data in the interior of the intersection of the past null cone $C^-(x_o)$ with S and on the intersection itself. If for every Cauchy problem and for every point $x_o \in L_4$ the solution depends only on the data in an arbitrarily small neighbourhood of $S \cap C^-(x_o)$ we say that Eq. (1) satisfies *Huygens' principle* or is a *Huygens' differential equation*. We shall call a space-time on which (1) satisfies Huygens' principle a *Huygens' space-time* and denote it by H_4. Examples of Huygens' space-times are provided when L_4 is flat or more generally conformally flat. Hadamard posed the general problem, as yet unsolved, of determining all the

Huygens' space-times. We note that the validity of Huygens' principle for (1) is preserved under a general conformal transformation of the metric

$$\tilde{g}_{ab} = e^{2\phi} g_{ab} \ , \tag{2}$$

where ϕ is an arbitrary C^{∞} function, combined with the "gauge transformation"

$$u = e^{\phi}\ \tilde{u} \ , \tag{3}$$

which preserve the form of (1). The *only known* non-conformally flat H_4's are those that are conformally related to the *plane wave space-time* whose metric may be written as

$$ds^2 = 2dv\,[du + (Dz^2 + \bar{D}\bar{z}^2 + ez\bar{z})dv] - 2dzd\bar{z} \ , \tag{4}$$

where $D \neq 0$, and e are functions of v only. The plane wave space-time was first shown to be an H_4 by Günther [3] who used a different coordinate system.

In the continuing effort to solve Hadamard's problem a series of necessary conditions for the validity of Huygens' principle have been derived by a number of workers from Hadamard's necessary and sufficient condition. The first two non-trivial necessary conditions for Eq. (1) are

$$\text{III} \qquad S_{abk;}{}^{k} - \tfrac{1}{2} C^{k}{}_{ab}{}^{\ell} L_{k\ell} = 0, \tag{5}$$

$$\begin{aligned} \text{V} \qquad \mathrm{TS}\,(&3C_{kab\ell;m}C^{k}{}_{cd}{}^{\ell}{}_{;}{}^{m} + 8C^{k}{}_{ab}{}^{\ell}{}_{;c}S_{k\ell d} + 40\, S_{ab}{}^{k}\, S_{cdk} \\ &-8\, C^{k}{}_{ab}{}^{\ell} S_{k\ell c;d} - 24 C^{k}{}_{ab}{}^{\ell} S_{cdk;\ell} + 4\, C^{k}{}_{ab}{}^{\ell}\, C_{\ell}{}^{m}{}_{ck} L_{dm} \\ &+12\, C^{k}{}_{ab}{}^{\ell} C^{m}{}_{cd\ell} L_{km}) = 0 \ , \end{aligned} \tag{6}$$

where C_{abcd} denotes the Weyl conformal curvature tensor and

$$L_{ab} = -R_{ab} + \tfrac{1}{6} R\, g_{ab} \ , \tag{7}$$

$$S_{abc} = L_{a[b;c]} \ , \tag{8}$$

where R_{ab} denotes the Ricci tensor. The sign convertions for the Riemann and Ricci tensors are the same as those in [9]. Condition III was derived by Günther [2], while Condition V was obtained by McLenaghan [6] in the case $R_{ab} = 0$ and independently by Wünsch [12] and McLenaghan [8] in the general case. A third necessary condition which we shall call Condition VII and which is too lengthy to be given here has recently been obtained by Rinke and Wünsch [10].

The necessary conditions III, V and VII have the following consequences. Under the assumption $R_{ab} = 0$, McLenaghan [6] showed that Condition V implies that L_4 is the plane-wave space-time with metric (4); this solves Hadamard's problem in this case and in the case when L_4 is conformal to an empty space-time. The above result has been extended by Wünsch [13] to the case when $R_{ab} = \lambda g_{ab}$, who finds that Condition V implies when $\lambda \neq 0$, that L_4 is a space of constant curvature. Wünsch [13] has further extended these results to recurrent and (2x2)-decomposable H_4's, details of which may be found in his paper, obtaining only the plane wave space-time or conformally flat space-time. Rinke and Wünsch show that the only *plane fronted wave space-time with parallel rays* with metric

$$ds^2 = 2dv\,(du + m(v,z,\bar{z})dv) - 2dzd\bar{z}\,, \tag{7}$$

which satisfies Condition VII is the plane wave space-time with metric (4). This result shows that the Condition VII is required to solve Hadamard's problem since the Conditions III and V for the metric (7) do not imply that the function m has the form required for the metric (4). A possibly new Huygens' space-time presented by one of the authors [9], a solution of Conditions III and V but not conformally related to the plane wave space-time (4), is the *generalized plane wave space-time* of McLenaghan and Leroy [7] with metric

$$\begin{aligned} ds^2 = {} & 2dv\,[du + (a(z+\bar{z})u + Dz^2 + \bar{D}\bar{z}^2 + ez\bar{z} + Fz + \bar{F}\bar{z})dv] \\ & - 2(dz+az^2dv)(d\bar{z}+a\bar{z}^2dv)\,, \end{aligned} \tag{8}$$

where a, D, e and F are functions only of v. However, this space-time does not seem to satisfy Condition VII [14].

The above results suggest that every Huygens' space-time is conformally related to the plane wave space-time (4) or is conformally flat. A plan of attack for proving this conjecture is to treat separately each of the five possible Petrov types of the Weyl tensor of space-time. This is a natural approach since the Petrov type is invariant under a general conformal transformation. Petrov type N which is equivalent to the existence of a vector field ℓ satisfying

$$C_{abcd}\,\ell^d = 0 \tag{9}$$

at each point is not only the most degenerate but also the Petrov type of the plane wave space-time. This suggests that type N should be considered first. In this case we have obtained the following result:

Theorem: *For every Huygens' space-time of Petrov N there exists a coordinate system* $(u,v,z,\bar{z})$ *and a function* ϕ *such that the metric has the form*

$$ds^2 = e^{2\phi}\{2dv\,[du + (a(z+\bar{z})u + m)\,dv] - 2(dz+az^2dv)(d\bar{z}+a\bar{z}^2dv)\} \tag{10}$$

where a *is a function only of* v, *and the function* m *has the following form*

$$m(v,z,\bar{z}) = \bar{z}G(v,z) + z\bar{G}(v,z) + H(v,z) + \bar{H}(v,\bar{z}) , \qquad (11)$$

where the functions G *and* H *are given by either*

$$G(v,z) = e(v)z + f(v) , \; H(v,z) = g(v)z^2 + h(v)z , \qquad (12)$$

and e, f, g, *and* h *are arbitrary functions or* G *and* H *satisfy the differential equations*

$$\frac{\partial^2 G}{\partial z^2}(v,z) = f(v)\,[d(v)z+e(v)]^{1/d(v)} , \qquad (13)$$

$$\frac{\partial^2 H}{\partial z^2}(v,z) = [f(v)/(1+d(v))][d(v)z+e(v)]^{1/d(v)}[g(v)z+h(v)(1+d(v)) - e(v)g(v)] , \qquad (14)$$

where the d, e, f, g *and* h *satisfy certain additional algebraic constraints.*

The metric given by Eqs. (10) to (13) represent the *general solution* of Conditions III and V for Petrov type N in an appropriate conformal gauge and contains the metrics (4) and (8) as special cases. When $\phi = 0$ the metric (10) is a special case of the *complex recurrent metric* given by McLenaghan and Leroy. To complete the proof of the conjecture for Petrov type N it remains to be shown that the plane wave metric (4) is the *only* solution satisfying Condition VII.

It is probable that the conclusion of our theorem will also hold for type N space-times on which Maxwell's equations or Weyl's neutrino equation satisfy Huygens' principle, since the necessary conditions in these cases are the Condition III and the Condition V with different numerical coefficients. A detailed proof of our theorem will be published elsewhere.

One of the authors (J. Carminati) would like to thank the Natural Sciences and Engineering Research Council of Canada (NSERC) for the award of a Postdoctoral Fellowship during the tenure of which this work was completed. The work was also supported in part by an NSERC operating grant (R.G. McLenaghan).

References

[1] Y. Bruhat, Acta Math. 88, 141-225 (1952).
[2] P. Günther, Sitzungsber. Sächs. Akad. Wiss. Leipz., Math.-Naturwiss. Kl. 100, 1-43 (1952).
[3] P. Günther, Arch. Ration. Mech. Anal. 18, 103-106 (1965).
[4] J. Hadamard, Lectures on Cauchy's problem in linear partial differential equations (Yale University Press, New Haven, 1923).

[5] M. Mathisson, Math. Ann. 107, 400-419 (1932).

[6] R.G. McLenaghan, Proc. Cambridge Philos. Soc. 65, 139-155 (1969).

[7] R.G. McLenaghan and J. Leroy, Proc. R. Soc. Lond., Ser. A. 327, 229-249.

[8] R.G. McLenaghan, Ann. Inst. Henri Poincaré, Nouv. Ser., Sec. A 20, 153-188 (1974).

[9] R.G. McLenaghan, Ann. Inst. Henri Poincaré, Nouv. Ser., Sec. A 37, 211-236 (1982).

[10] B. Rinke and V. Wünsch, Beitr. Anal. 18, 43-75 (1981).

[11] S.L. Sobolev, Mat. Sb., Nov. Ser. 1, 39-70 (1936).

[12] V. Wünsch, Math. Nachr. 47, 131-154 (1970).

[13] V. Wünsch, Beitr. Anal. 13, 147-177 (1979).

[14] V. Wünsch, private communication.

ATOMIC FINE AND HYPERFINE STRUCTURE CALCULATIONS IN A SPACE OF CONSTANT CURVATURE

N. BESSIS and G. BESSIS
Laboratoire de Spectroscopie théorique
Université Claude Bernard, Lyon I
69622 Villeurbanne, France

Abstract : Space-curvature induced modifications of the electronic, fine and hyperfine structure energies and wavefunctions have been investigated when the usual Euclidean flat space is substituted by a spherical 3-space. Particularly, it has been found that the degenerate one-electron fine structure energy levels are split by an additionnal space-curvature contribution which vanishes at the traditional flat-space limit.

1. Introduction

This investigation has to be situated in the field of Atomic Spectroscopy rather than in the field of the Gravitational theories. It can be considered as a contribution to a tentative formulation of atomic physics in a curved space. The interest of calculating the energy levels of one-electron atoms in a curved space-time has been drawn recently in a series of papers /1 to 10/. In fact, the introduction of space curvature in quantum physics has been considered since a long time. Among previous works, the one of Schrödinger /11/ deserves a special mention. He, first, solved the non relativistic equation bearing his name in a space of constant curvature and put in evidence how the continuous hydrogenic spectrum is resolved into an intensely crowded line spectrum. Since the mathematical nature of the hydrogenic wave equation is not more intricate in the spherical three-space than in the flat space, it is thus possible to build up a tractable "curved-orbital" model (non relativistic or relativistic) capable of explorating, at least roughly, the space-curvature modifications of the atomic spectrum. Working in that geometrically simple space i.e. a three dimensional hypersphere of radius R imbedded in a four Euclidean space, allows us to keep a more direct parallelism between the "curved" and "flat" results and an easier extension to the many electron case. Let us recall that the space-time line element are in a space of constant positive curvature and in an Euclidean space, respectively

$$ds^2 = c^2dt^2 - R^2d\chi^2 - R^2\sin^2\chi\left(d\theta^2 + \sin^2\theta\, d\phi^2\right) \qquad (1)$$

and

$$ds^2 = c^2dt^2 - dr^2 - r^2\left(d\theta^2 + \sin^2\theta\, d\phi^2\right) \qquad (2)$$

where $0 \leqslant \chi \leqslant \pi$; $0 \leqslant r < \infty$ and in both cases $0 \leqslant \theta \leqslant \pi$; $0 \leqslant \phi \leqslant 2\pi$.

Without wanting to discuss immediatly the critical question of the order of magnitude of the curvature induced shifts and the possibilities of their detection, this model provides, within the usual framework of theoretical spectroscopy, ready to use expressions of the curvature modifications of the spectrum in situations where local curvature could be important. It can also be used as a path toward flat space calculations taking the advantage of hyperspherical parametrization. One aspect of this last point has been illustrated by recent calculations /12/. These last advantages are of interest mainly for applications in quantum chemistry and will not be further discussed hereafter. We will rather focus our attention on the determination of the atomic electronic, fine and hyperfine structure hamiltonians, wavefunctions and energy levels. We shall assume that the usual independent particle treatment of the N-electron problem including, if necessary, all the modern refinements of the technique (configuration interaction, Multiconfigurational Hartree-Fock methods) is still valid when the usual flat Euclidean space is substituted with a space of constant curvature.

2. Electronic energies and wavefunctions

In order to obtain the electronic energies and wavefunctions, the "curved" form of the many electron Schrödinger hamiltonian and of the atomic basis orbitals are required. Let us only briefly recall the main results without reproducing the details of calculation which have been given elsewhere /6/.

2.1. Many electron Schrödinger equation

The extension of the one-electron model /11/ to the many electron case, leads to the following expression of the N-electron Schrödinger equation in a space of constant positive curvature

$$\left\{ \sum_{i=1}^{N} \left(-\frac{1}{2} \nabla_i^2 + V_i \right) + \sum_{i<j} V_{ij} \right\} \Psi = E \Psi \qquad (3)$$

where

$$\nabla_i^2 = \frac{1}{R^2 \sin^2 \chi_i} \left\{ \frac{\partial}{\partial \chi_i} \left(\sin^2 \chi_i \frac{\partial}{\partial \chi_i} \right) - \ell_i^2 \right\}$$

$$V_i = -\frac{Z}{R} \cot \chi_i \quad ; \quad V_{ij} = \frac{1}{R} \cot w_{ij}$$

∇_i^2, V_i and V_{ij} are the curved form of the Laplacian, of the nucleus-electron Coulombic interaction potential $\overline{V}_i = -Z/r_i$ and of the repulsive electron-electron potential $\overline{V}_{ij} = 1/r_{ij}$, respectively ; w_{ij} is the "angular separation" between the electrons "i" and "j" on the hypersphere.

The many electron wavefunction Ψ can be determined within the usual independent particle framework as a linear combination of Slater determinants built up an atomic orbitals $\Phi_{n\ell m}(\chi, \theta, \phi)$. One has to determine the "curved" form of the $\Phi_{n\ell m}$ functions, solutions of the one electron Schrödinger equation.

2.2. *"Curved" hydrogenic orbitals*

If one sets $\Phi_{n\ell m} = \frac{1}{\sin\chi} R_{n\ell}(\chi)\, Y_\ell^m(\theta, \phi)$, where Y_ℓ^m is a spherical harmonic, it is found that the $R_{n\ell}$ function is solution of the standard Infeld and Hull (class I) factorizable equation /13, 14/

$$\left(\frac{d^2}{d\chi^2} - \frac{M(M+1)}{\sin^2\chi} + 2q\cot\chi + \lambda_S\right) R_S^M(\chi) = 0 \tag{4}$$

with $M = \ell$, $S = n - 1$, $q = ZR$ and $\lambda_S = 2\,ER^2 + 1$. Closed form expressions of λ_S and R_S^M are known /14/.

$$\lambda_S = (S+1)^2 - q^2/(S+1)^2 \tag{5}$$

$$R_S^M = N_{SM}(\sin\chi)^{S+1} \exp\left(-q\chi/(S+1)\right) P_v^{(a,a^*)}(-i\cot\chi) \tag{6}$$

where $a = -(S+1) + iq/(S+1)$; $v = S - M = \text{integer} > 0$ and $P^{(a,a^*)}$ is a real Jacobi polynomial in $\cot\chi$. From (5), one finds again the expression of the hydrogenic energies /11/ in atomic units (a.u.)

$$E_n = -\frac{Z^2}{2\,n^2} + \frac{n^2 - 1}{2\,R^2} \tag{7}$$

In a space of constant positive curvature, there are only discrete states and passing through zero of the energy levels is allowed by continuity as n increases. Let us remark that the expression (6) of the Kepler functions is no more intricate in the spherical three-space than in the "flat" space

$$\bar{R}_S^M(r) = \bar{N}_{SM}\, r^{M+1} \exp\left(-\bar{q}r/(S+1)\right) \mathcal{L}_v^{2M+1}\left(\frac{2\bar{q}r}{(S+1)}\right) \tag{8}$$

where $\bar{q} = Z$ and $\mathcal{L}_v^{2M+1}$ is a Laguerre polynomial.

In order to calculate the many electron energies and wavefunctions $\mathcal{E}$ and ψ, one has now to compute the matrix elements of V_{ij} between the "curved" functions (6).

2.3. *Multipolar expansion of the bielectronic repulsion potential*

Using the Fourier expansion of $\cos w_{ij}$ /15/, and next the hyperspherical expansion of $\sin 2k\, w_{ij}/\sin w_{ij}$ /16/, after some manipulations and introducing the traditional notation $\chi_>$ and $\chi_<$, one obtains the following expression of V_{ij} /6/

$$\frac{1}{R}\cot w_{ij} = \frac{1}{R}\cot \chi_> + \sum_{\ell=1}^{\infty} \left[C_i^{(\ell)}.C_j^{(\ell)}\right] F_\ell(\chi_>)\ G_\ell(\chi_<) \tag{9}$$

where the $C_i^{(\ell)} = C^{(\ell)}(\theta_i, \phi_i)$ are the spherical harmonic tensors.

$$F_\ell(\chi) = \frac{1}{(2\ell-1)!!}(\sin\chi)^\ell \left[\frac{d}{d(\cos\chi)}\right]^\ell \cot\chi$$

$$G_\ell(\chi) = \frac{(-)^{\ell+1}(2\ell+1)!!}{(\ell-1)!(\ell+1)!}(\sin\chi)^\ell \left[\frac{d}{d(\cos\chi)}\right]^\ell \chi\cot\chi$$

This expansion, well adapted for computing the repulsive integrals, is the "curved" homologue of the Laplace expansion

$$\frac{1}{r_{ij}} = \frac{1}{r_>} + \sum_{\ell=1}^{\infty} \left[C_i^{(\ell)}.C_j^{(\ell)}\right] r_<^\ell / r_>^{\ell+1} \tag{10}$$

At the asymptotic flat space limit ($R \to \infty$, $\chi \to 0$, $R\chi \to r$), it is easily verified that the $F_\ell(\chi)$ and $G_\ell(\chi)$ functions converge to the flat radial harmonic functions $(1/r)^{\ell+1}$ and r^ℓ, respectively.

Finally, within a non relativistic scheme, the many electron atomic "curved" energies and wavefunctions can be obtained in the spherical three-space in the same way as in the usual flat space. Some examples have been given elsewhere /17, 18/. Doubtless, a physically more consistent "curved orbital" model should be relativistic i.e. built up using the two-components "curved" Dirac orbitals.

3. Dirac orbitals and fine structure energies

3.1. Dirac equation in a spherical three-space

Starting from the generally covariant form of the Dirac equation in a Riemannian curved space-time, a convenient choice of the Dirac representation can be made which leads to the usual polar dependence (θ, ϕ) of the Dirac wavefunction. One gets the following expression of the Dirac equation for stationary states with an external electromagnetic field (V, A_χ, A_θ, A_α) /7, 9/

$$\left[\alpha_\chi\left[p_\chi + \frac{i\beta\hat{K}}{R\sin\chi}\right] + \frac{mc}{\hbar}\beta + W - \frac{1}{\hbar c}(E_T - eV)\right]\Phi(\chi,\theta,\phi) = 0 \tag{11}$$

where

$$W = -\frac{e}{\hbar c}\frac{1}{R}\left[\alpha_\chi A_\chi + \frac{\alpha_\theta}{\sin\chi}A_\theta + \frac{\alpha_\phi}{\sin\chi\sin\theta}A_\phi\right] \quad ; \quad E_T = mc^2 + E$$

$$p_\chi = -\frac{i}{R\sin\chi}\frac{\partial}{\partial\chi}\sin\chi \quad ; \quad \hat{K} = \beta(1 + \vec{\sigma}.\vec{\ell}) \quad ; \quad \beta = \begin{pmatrix} I & 0 \\ 0 & -I \end{pmatrix} \quad ; \quad \alpha_k = \begin{pmatrix} 0 & \sigma_k \\ \sigma_k & 0 \end{pmatrix}$$

$$\sigma_\chi = (\sigma^1 \cos\phi + \sigma^2 \sin\phi) \sin\theta + \sigma^3 \cos\theta$$

$$\sigma_\theta = (\sigma^1 \cos\phi + \sigma^2 \sin\phi) \cos\theta - \sigma^3 \sin\theta$$

$$\sigma_\phi = - \sigma^1 \sin\phi + \sigma^2 \cos\phi$$

I and σ^1, σ^2, σ^3 are the 2 × 2 unit and Pauli matrices.

When the external electromagnetic field in (11) reduces to the Coulombic potential, a perturbative procedure can be applied in order to obtain the "curved" Dirac orbitals and the "curved" expression of the hydrogenic fine structure energies.

3.2 "Curved" Dirac orbitals

Since, at the asymptotic flat-space limit, the function $\Phi(\chi, \theta, \phi)$ must lead to the familiar "flat" Dirac function $\overline{\Phi}(r, \theta, \phi)$, we set for Φ the following form

$$\Phi_{vkm} = \frac{1}{R \sin\chi} \begin{pmatrix} P_{vk}(\chi) & \mathcal{Y}_{\ell jm} \\ iQ_{vk}(\chi) & \mathcal{Y}_{\overline{\ell} jm} \end{pmatrix} \tag{12}$$

where $\overline{\ell} = \ell \pm 1$ for $j = \ell \pm 1/2$ and the $\mathcal{Y}_{\ell jm}$ spinor is eigenfunction of the operator $(1 + \vec{\sigma}.\vec{\ell})$ with eigenvalue $-k = j(j+1) - \ell(\ell+1) + \frac{1}{4}$; $\vec{\sigma} = (\sigma^1\ \sigma^2\ \sigma^3)$ and $\vec{\ell}$ is the orbital momentum of the electron.

It can be shown /9/ that a direct parallelism between the determination of the "flat" functions $\overline{\Phi}_{vkm}(r, \theta, \phi)$ and, within a perturbative scheme, zeroth order "curved" Dirac functions $\Phi^{(0)}(\chi, \theta, \phi)$ can be kept if we introduce the following perturbative hamiltonian

$$\mathcal{H}_1 = \frac{ic}{2R^2} \left(2R \operatorname{tg} \frac{\chi}{2}\right) \begin{pmatrix} 0 & \sigma_\chi\ \beta\ \hat{K} \\ \sigma_\chi\ \beta\ \hat{K} & 0 \end{pmatrix} \tag{13}$$

At the asymptotic flat-space limit $2R \operatorname{tg} \frac{\chi}{2} \to r$, and it is easily seen that $\mathcal{H}_1$ is of an order of magnitude $\sim 1/R^2$. Thus, in order to obtain all the $1/R^2$ contributions involved in the atomic structure calculations, it is sufficient to determine only the first order perturbed solutions of the Dirac equation (11), expanded in the basis of the unperturbed Dirac spinors $\Phi^{(0)}_{vkm}$

$$\Phi_{vkm} = \Phi^{(0)}_{vkm} - \sum_{v' \neq v} \frac{\langle \Phi^{(0)}_{v'km} \mid \mathcal{H}_1 \mid \Phi^{(0)}_{vkm} \rangle}{\overline{E}_{v'k} - \overline{E}_{vk}} \Phi^{(0)}_{vkm} \tag{14}$$

where $\overline{E}_{vk} = - \frac{1}{2} Z^2/(v + |k|)^2$ is the flat-space electronic energy and

$$\langle \Phi^{(0)}_{v'k'm'} \mid \mathcal{H}_1 \mid \Phi^{(0)}_{vkm} \rangle = - \frac{k}{2R^2} \delta_{m'm}\, \delta_{k'k}\, h_{v'v} \tag{15}$$

$$h_{v'v} = \int_0^{\pi} \left(P^{(0)}_{v'k'}\, Q^{(0)}_{vk} + Q^{(0)}_{v'k'}\, P^{(0)}_{vk}\right) \left(2R \operatorname{tg} \frac{\chi}{2}\right) d\chi$$

Hence, the "pseudo-radial" parts $P^{(0)}_{vk}$ and $Q^{(0)}_{vk}$ of the unperturbed spinor $\Phi^{(0)}_{vkm}$ can be obtained in terms of the "curved" Kepler functions $R^M_S(\chi)$. One gets

$$\begin{aligned} P^{(0)}_{vk} &= N\left\{(\gamma_2 + \gamma_1)\left(\frac{\varepsilon k}{\gamma} + 1\right)^{1/2} R_1 + (\gamma_2 - \gamma_1)\left(\frac{\varepsilon k}{\gamma} - 1\right)^{1/2} R_2\right\} \\ Q^{(0)}_{vk} &= N\left\{(\gamma_2 - \gamma_1)\left(\frac{\varepsilon k}{\gamma} + 1\right)^{1/2} R_2 + (\gamma_2 + \gamma_1)\left(\frac{\varepsilon k}{\gamma} - 1\right)^{1/2} R_2\right\} \end{aligned} \tag{16}$$

where

$$\varepsilon = E^0_T/mc^2 \quad ; \quad \gamma_1 = (k + Z\alpha)^{1/2} \quad ; \quad \gamma_2 = (k - Z\alpha)^{1/2} \quad ; \gamma = \gamma_1 \gamma_2 \quad ;$$

$$N = (\varepsilon/8\gamma)^{1/2} \left(1 + \frac{\alpha^2}{R^2} v(v + 2\gamma)\right)^{-1/2} \quad ; \quad \alpha = 1/c$$

The R_1 and R_2 functions are solutions of the equation (4) with $M = \gamma$ and $M = \gamma - 1$, respectively ; $S = v + \gamma$, $q = Z\varepsilon$ and $\lambda_S = R^2c^2(\varepsilon^2 - 1) + \gamma^2$ for both of them. As long as one is looking for the ϕ_{vkm} up to the $1/R^2$ terms, it is easily inferred that $h_{v'v}$ can be replaced by its effective value i.e. its asymptotic flat-space limit

$$\overline{h}_{v'v} = \int_0^{\infty} \left(\overline{P_{v'k}}\; \overline{Q_{vk}} + \overline{Q_{v'k}}\; \overline{P_{vk}}\right) r\, dr \tag{17}$$

where $\overline{P_{vk}}$ and $\overline{Q_{vk}}$ are related to the flat Kepler $\overline{R}^M_S$ functions (8) by the same expressions (16) as the "curved" $P^{(0)}_{vk}$ and $Q^{(0)}_{vk}$ are related to the R^M_S functions (6).

Finally, the "curved" Dirac functions are completely defined by the expression (12, 14, 16). In many electron atomic structure calculations (configuration interaction or Hartree-Fock schemes), the approximate Dirac spinors $\Phi^{(0)}_{vkm}$ could serve themselves as basis atomic orbitals.

3.3. Hydrogenic fine structure energies

Since $\lambda_S = R^2c^2(\varepsilon^2 - 1) + \gamma^2$ is given by (5) with $S = v + \gamma$, one gets the following expression of $\varepsilon = 1 - E^{(0)}_{vk}/c^2$

$$\varepsilon = \left(1 + \frac{Z^2\alpha^2}{(v + \gamma)^2}\right)^{-1/2} \left(1 + \frac{\alpha^2}{R^2} v(v + 2\gamma)\right)^{1/2} \tag{18}$$

In order to include all the $1/R^2$ curvature contributions to the fine-structure energy levels, it is sufficient to calculate the first order perturbation energy $E^{(1)}_{vk} = \langle\Phi^{(0)}_{vkm} \mid H_1 \mid \Phi^{(0)}_{vkm}\rangle$. Using (15) and (18), the Dirac energy is found to be

$$E_{vk} = \frac{1}{\alpha^2}\left\{\left(1 + \frac{Z^2\alpha^2}{(v + \gamma)^2}\right)^{-1/2} \left(1 + \frac{v(v + 2\gamma) + k^2}{2 R^2c^2}\right) - 1\right\} - \frac{k}{4 R^2} \tag{19}$$

When retaining in E_{vk} the terms up to α^2 and introducing the usual radial quantum number $n = v + |k|$, one obtains

$$E_{nk} = -\frac{Z^2}{2\,n^2} + \frac{Z^4\alpha^2}{2\,n^3}\left(\frac{3}{4n} - \frac{1}{|k|}\right) + \frac{n^2}{2\,R^2} - \frac{k}{4\,R^2} + \frac{Z^2\alpha^2}{4\,R^2}\left(1 - \frac{2n}{|k|}\right) \tag{20}$$

or, alternatively

$$E_{nk} = -\frac{Z^2}{2\,n^2}\left(1 - \frac{n^4}{Z^2R^2}\right) + \frac{3\,Z^4\alpha^2}{8\,n^4}\left(1 + \frac{2\,n^4}{3\,Z^2R^2}\right) - \frac{Z^4\alpha^2}{2\,n^3|k|}\left(1 + \frac{n^4}{Z^2R^2}\right) - \frac{k}{4\,R^2} \tag{21}$$

The two first terms in (20) are just the electronic and fine-structure hydrogenic flat-space energies and the remaining terms correspond to additional curvature contributions which vanish at the asymptotic flat-space limit. From (21), it appears that the curvature modifications of the energies increase with n as n^4/Z^2R^2. The last term $-\,k/4\,R^2$, will induce splittings of the degenerate levels of the hydrogenic spectrum which, qualitatively, in some respects, are comparable to the Lamb effect (see the Table). One notes that the Lamb shifts $\Delta E_L = \frac{Z^4\alpha^2\varkappa}{kn^3(2\ell + 1)}$ depend upon the radial quantum number n and upon the charge Z of the nucleus while the space-curvature shift $\Delta E_C = -\,k/4\,R^2$ does not.

TABLE

Hydrogenic fine-structure energies in a space of constant curvature (in a. u.)

The upper (or lower) sign corresponds to positive (or negative) curvature ;
$\varkappa = 1.159644\ \ 10^{-3}$

	Electronic energy	Flat fine structure	Curvature fine structure contributions		Lamb shift
$nd_{5/2}$	$-\frac{Z^2}{2\,n^2} \pm \frac{n^2}{2\,R^2} + \left(\frac{Z}{R}\right)^a$	$\frac{Z^4\alpha^2}{2\,n^3}\left(\frac{3}{4n} - \frac{1}{3}\right)$	$\pm\frac{Z^4\alpha^2}{4\,R^2}\left(1 - \frac{2n}{3}\right)$	$\mp\frac{3}{4\,R^2}$	$\frac{Z^4\alpha^2}{n^3}\frac{\varkappa}{15}$
$nf_{5/2}$				$\pm\frac{3}{4\,R^2}$	$-\frac{Z^4\alpha^2}{n^3}\frac{\varkappa}{21}$
$np_{3/2}$		$\frac{Z^4\alpha^2}{2\,n^3}\left(\frac{3}{4n} - \frac{1}{2}\right)$	$\pm\frac{Z^4\alpha^2}{4\,R^2}(1 - n)$	$\mp\frac{1}{2\,R^2}$	$\frac{Z^4\alpha^2}{n^3}\frac{\varkappa}{6}$
$nd_{3/2}$				$\pm\frac{1}{2\,R^2}$	$-\frac{Z^4\alpha^2}{n^3}\frac{\varkappa}{10}$
$ns_{1/2}$		$\frac{Z^4\alpha^2}{2\,n^3}\left(\frac{3}{4n} - 1\right)$	$\pm\frac{Z^4\alpha^2}{4\,R^2}(1 - 2n)$	$\mp\frac{1}{4\,R^2}$	$\frac{Z^4\alpha^2\varkappa}{n^3}$
$np_{1/2}$				$\pm\frac{1}{4\,R^2}$	$-\frac{Z^4\alpha^2}{n^3}\frac{\varkappa}{3}$

(a) the electronic energy contains the additional term (Z/R) but only for the case of negative structure

Let us mention that the calculations are also tractable when one considers, instead of the closed spherical three space, a space of constant negative curvature i.e. an open space with space-time element and Coulombic potential

$$ds^2 = c^2 dt^2 - R^2 d\chi^2 - R^2 \,\mathrm{Sh}^2 \chi \left(d\theta^2 + \sin^2 \theta \, d\phi^2\right)$$

$$V_i = - \frac{Z}{R} \left(\coth \chi_i - 1\right) \qquad (22)$$

Particularly, it can be shown /10/ how the change of sign of the space curvature amounts to make, in the above results, the changes $R \to iR$, $\chi \to i\chi$ and $\varepsilon \to \varepsilon - \frac{Z\alpha^2}{R}$. As a consequence, it is found that the sign of the curvature induced splittings and shifts in the hydrogenic fine-structure spectrum is reversed when switching from the closed spherical three-space to the open hyperbolic space (see the Table).

4. Hyperfine structure in a spherical three-space

As usually done, we shall limit ourselves to the consideration of the dipolar magnetic and quadrupolar electric hyperfine interactions between the nucleus and each electron "i". These interaction terms are respectively of the general form $W_D = \mu_N^{(1)} . T_i^{(1)}$ and $W_Q = Q_N^{(2)} . T_i^{(2)}$, where $\mu_N^{(1)}$ and $Q_N^{(2)}$ are the dipole magnetic moment and quadrupolar electric moment of the nucleus and $T^{(1)}$, $T^{(2)}$ are electronic tensors.

4.1. Hyperfine structure energy levels

Since the electronic spin and angular $(\theta_i \; \phi_i)$ dependence of the wavefunctions and interaction terms are the same in the flat and the spherical three-space, the usual expression of the hyperfine structure energies, in terms of the dipolar magnetic a(J) and quadrupolar electric b(J) hyperfine constants, still holds in the spherical three space /19/

$$E_F = \frac{1}{2} a(J) \left(F(F+1) - J(J+1) - I(I+1)\right) + b(J) \left(\frac{\frac{3}{8} K(K+1) - \frac{1}{2} IJ(I+1)(J+1)}{(2I-1)(2J-1)IJ}\right) \qquad (23)$$

where $\vec{F} = \vec{I} + \vec{J}$ is the total angular momentum of the atom

$$a(J) = \frac{1}{IJ} \mu_N \langle JJ | \sum_{i=1}^{N} (T_i)_0^{(1)} | JJ \rangle \qquad K = F(F+1) - J(J+1) - I(I+1)$$

$$b(J) = 2 \, Q_N \langle JJ | \sum_{i=1}^{N} (T_i)_0^{(2)} | JJ \rangle$$

$(T_i)_0^{(1)}$ and $(T_i)_0^{(2)}$ are the "z" components of the electronic tensors which are expected to differ when switching from the flat to the spherical three-space.

4.2. Dipolar magnetic hyperfine interaction

Infeld and Schild /20/ have already examined the solutions of Maxwell equations with a singularity at the spacial origin. Starting from their results, we have obtained the following expression of the vector potential components associated with a static magnetic dipole moment $\mu_N^{(1)}$ lying along the z axis : $A_\chi = A_\theta = 0$; $A_\phi = (\mu_N/R) \cot \chi \sin^2 \theta$. Consequently, one gets the following expression of the dipolar magnetic hyperfine interaction

$$W_D = - \frac{e}{hc} \frac{\mu_N}{R^2} \frac{\cos \chi}{\sin^2 \chi} \sin \theta \begin{pmatrix} 0 & \sigma_\phi \\ \sigma_\phi & 0 \end{pmatrix} \tag{24}$$

It is easily verified that $\sin \theta \, \sigma_\phi = - i \sqrt{2} \left\{C^{(1)} \sigma^{(1)}\right\}_0^{(1)}$ and one gets the expression of the dipolar magnetic one electron tensor

$$T^{(1)} = 2 i \sqrt{2} \, \beta_e \frac{\cos \chi}{R^2 \sin^2 \chi} \left\{C^{(1)} \sigma^{(1)}\right\}^{(1)} \begin{pmatrix} 0 & 1 \\ 1 & 0 \end{pmatrix} \tag{25}$$

$\beta_e = \dfrac{eh}{2 mc}$ is the Bohr magneton.

The only differences between "curved" and "flat" calculations will arise from the expression of the hyperfine radial parameter which is

$$f_D = \int_0^\pi \left(P_{v'k'} Q_{vk} + Q_{v'k'} P_{vk}\right) \frac{\cos \chi}{R^2 \sin^2 \chi} d\chi \tag{26}$$

and will contain additional curvature contributions.

At the asymptotic flat-sapce limit $\dfrac{\cos \chi}{R^2 \sin^2 \chi} \to \dfrac{1}{r^2}$ and one finds again the Euclidean expression /21/

$$\bar{f}_D = \int_0^\infty \left(\overline{P_{v'k'}}\, \overline{Q_{vk}} + \overline{Q_{v'k'}}\, \overline{P_{vk}}\right) \frac{1}{r^2} dr \tag{27}$$

As an illustrative example, for v' = v, one gets

$$f_D = \bar{f}_D \left(1 + \frac{(v + \gamma)^4}{Z^2 R^2 \varepsilon^2}\right) + \frac{Z \alpha \varepsilon^2 k}{2 R^2 \gamma^2} \tag{28}$$

4.3. Quadrupolar electric hyperfine constants

The expression of the quadrupolar electric interaction directly follows from the multipolar expansion of the Coulombic interaction between two particles i and j, i.e.

between each electron i and each proton "n" of the nucleus. If one introduces the "curved" expression of the quadrupole moment of the nucleus

$$Q_N^{(2)} = \sum_n e_n\, g_e\, C^{(2)}(\theta_n, \phi_n)\, G_2(\chi_n) \tag{29}$$

where e_n, g_e and $(\chi_n, \theta_n, \phi_n)$ are respectively the charge, the orbital gyromagnetic factor and the coordinates of proton "n", one gets the following expression of the quadrupolar electric hyperfine interaction

$$V_Q = -\frac{e}{R^3} F_2(\chi)\, C^{(2)} . Q_N^{(2)} \tag{30}$$

where $F_2(\chi) = \cos\chi / R^3 \sin^3\chi$

In the same way as for the dipolar magnetic structure, the only change between the "flat" and "curved" hyperfine structure calculations arise in the expression of the quadrupolar electric radial parameter

$$f_Q = \int_0^{\pi} \left(P_{v'k'} P_{vk} + Q_{v'k'} Q_{vk}\right) \frac{\cos\chi}{R^3 \sin^3\chi}\, d\chi \tag{31}$$

At the asymptotic flat-space limit $\frac{\cos\chi}{R^3 \sin^3\chi} \to \frac{1}{r^3}$ and one finds again the classical flat results /21/.

5. Conclusion

Finally, we have proposed a geometrically simple heuristic model in order to roughly investigate the space-curvature effects in atomic structure calculations. Since space-curvature concept is deeply rooted in the relativity theory, it seems theoretically more consistent to build up the model with Dirac orbitals rather than with non-relativistic orbitals. Physically, we have put in evidence the curvature modifications of the flat fine-structure hydrogenic energy levels : curvature-induced shifts of the non-degenerate levels and splitting of the degenerate levels. This curvature induced splitting should be detectable only if R is extremely small. Nevertheless, such a tractable model, putting in evidence the quantum number (n, k) dependence of the space-curvature modifications of the spectra could contribute to distinguish between the space-curvature effects and the other small perturbations of the spectra such as radiative corrections, nuclear effects ... and to extract some specific information, since it includes global effects coming from the topology of the space.

REFERENCES

/1/ P. Tourrenc and J.L. Grossiord, Nuovo Cimento B, 32, 163 (1976).
/2/ L. Parker, Phys. Rev. Lett., 44, 1559 (1980).
/3/ L. Parker, Phys. Rev. D, 22, 1922 (1980).
/4/ L. Parker, Phys. Rev. D, 24, 535 (1981).
/5/ L. Parker and L.O. Pimentel, Phys. Rev. D, 25, 3180 (1982).
/6/ N. Bessis and G. Bessis, J. Phys. A, 12, 1991 (1979).
/7/ N. Bessis, G. Bessis and R. Shamseddine, J. Phys. A, 15, 3131 (1982).
/8/ N. Bessis and G. Bessis, J. Phys. A, 16, L467 (1983).
/9/ N. Bessis, G. Bessis and R. Shamseddine, Phys. Rev. A, 29, 2375 (1984).
/10/ N. Bessis, G. Bessis and D. Roux, Phys. Rev. A (1984) (in press).
/11/ E. Schrödinger, Proc. R. Ir. Acad. Sect. A, 46, 9 (1940).
/12/ Z.J. Horak, Phys. Lett., 90 A, 31 (1982).
/13/ L. Infeld and T.E. Hull, Rev. Mod. Phys., 23, 21 (1951).
/14/ G. Hadinger, N. Bessis and G. Bessis, J. Math. Phys., 15, 716 (1974).
/15/ H.B. Dwight, Tables of integrals and other mathematical data, McMillan, New-York (1969).
/16/ V. Fock, Z. Phys., 98, 145 (1935).
/17/ N. Bessis and G. Bessis (unpublished).
/18/ R. Shamseddine, Thèse de Doctorat, Lyon (1984).
/19/ N. Bessis-Mazloum, Cahiers de Physique, 15, 345 (1962).
/20/ L. Infeld and A.E. Schild, Phys. Rev., 70, 410 (1946).
/21/ B. Judd, Cours de Troisième Cycle, Meudon (1964) (unpublished).

THEORIES OF GRAVITY AND EXPERIMENTAL TESTS IN THE POST-NEWTONIAN LIMIT

Pierre TEYSSANDIER
UA 769 "Gravitation et cosmologie relativistes"
Institut Henri Poincaré
11, rue P. et M. Curie
75231 Paris Cedex 05 (France)

The postulates upon which is founded general relativity (in particular the principle of equivalence) are in fact the underlying foundations of a large class of competing theories, the so-called metric theories of gravity. This situation has led the physicists to propose and to perform two kinds of experiments: 1) experiments which test the pertinence of the concept of metric theory; 2) experiments which permit to eliminate some theories. We show that the experiments of the first kind confirm - sometimes at a very high level of accuracy - the metric character of the gravitational interaction. To analyze the experiments of the second kind in the slow-motion, weak-field limit, the most useful tool is the parametrized post-Newtonian formalism. After a brief review of the main features of this formalism, we present the different experiments performed in order to determine the post-Newtonian parameters and we discuss the theoretical implications of the results already obtained.

I -Introduction : the status of gravity by the 1950's

In the middle of the 1950's, the theory of general relativity was standing as a prominent monument of the classical physics, on account of a) its internal consistency, b) the nature of its previsions and c) the fascinating personality of A. Einstein.

However, general relativity was then supported by a very small number of empirical confirmations. The gravitational red-shift of spectral lines from the Sun (the first test proposed by Einstein [1] in 1907) had no reliable verification because it was strongly affected by the radial motions in the solar photosphere [2]. The second classical test, the deflection of light-rays by the Sun, was infected by systematic errors [3][4] and the only certitude was that the effect was nearer to the 1.75" yielded by general relativity than to the 0.83" predicted by Soldner in 1801 on a simple Newtonian basis or by Einstein [5] in 1911 from the principle of equivalence. Moreover, the most successful test of general relativity - its ability to predict without special assumption the anomalous perihelion shift of Mercury - was not absolutely convincing since the solar quadrupole moment due to the spinning of the Sun was unknown.

Together with general relativity, there was a lot of Minkowskian theories of gravitation but none of those theories was satisfying. Indeed, the Minkowskian theories needed arbitrary coupling constants and/or specific hypotheses in order to predict the three classical effects. Moreover, they lacked in internal consistency since they did not allow to describe the electromagnetic phenomena in gravitational fields without arbitrary assumptions.

A third way consisted of the unified theories, but despite a considerable amount of researches since the first attempt by H. Weyl in 1921, this way has failed to link in a satisfying manner gravitation and electromagnetism. Moreover, the program of unification itself was questioned by the discovery of the weak and strong interactions at the subatomic scale.

Thus, briefly saying, the situation in the middle of the 1950's was a very meager experimental basis for gravity allowing a great amount of more or less arbitrary theoretical attempts.

II -The growing of experimental evidence for the relativistic theories of gravity

However, many events of the end of the 1950's opened a new era for general relativity and related topics, on both the experimental and theoretical sides :

1) The building of a rigorously covariant theory of gravitational radiation, by such workers as Pirani [6], Trautman [7] and Lichnerowicz [8] (1957-58) showed the crucial importance of the curvature tensor for the description of gravity.

2) Almost at the same time (1958), J. Weber [9] proposed to detect gravitational waves by their tidal effects on a massive bar due to the very small perturbations of the curvature tensor induced by strongly accelerated masses.

3) In September 1959, a team of scientists obtained the first recorded radar echo from a planet (Venus). The technique of radar echoes from planets and spacecrafts became rapidly a powerful tool for testing relativistic gravitational effects, thus opening the solar system as a quasi ideal laboratory for experimental physics.

4) Three events at least of the year 1960 contributed to renew the interest for experimental relativistic gravitation in terrestrial laboratories. The Pound-Rebka [10] experiment verified the principle of equivalence by a high-precision measurement of the gravitational red-shift of γ-ray photons emitted from the recoilness decay of ^{57}Co into ^{57}Fe. Hughes, Robinson, Beltran-Lopez [11] and Drever [12] set independently a relative limit of 10^{-20} on a possible anisotropy in the inertial mass of a proton. Schiff [13] proposed a new test of general relativity consisting of measuring the precession of a torque-free gyroscope induced by the so-called Lense-Thirring off-diagonal terms in the metric of the rotating Earth.

5) In 1961, Brans and Dicke [14] proposed a new pseudo-Riemannian theory of gravitation which includes general relativity as a particular case. This theory showed the need for a broader theoretical framework than general relativity alone to discuss the experimental tests of relativistic gravitation and thus introduced the general concept of metric theories of gravity. It required also higher-precision experiments in order to distinguish between the competing theories. The technological advances in such areas as nuclear physics, atomic clocks, superconducting gravimeters, radar or laser ranging, etc., have led theorists to develop a lot of mathematical tools in order to analyze the new results. Among these different methods, the parametrized

post-Newtonian formalism improved by Nordtvedt and Will [15] remains the most useful to discuss the tests of the different metric theories of gravity in the solar system.

6) After the discovery of very compact radio-sources (1960-63) with large red-shifts, many astrophysicists turned to the relativistic gravitation, in order to understand the mechanisms underlying the enormous emissions of energy involved by these quasi-stellar objects. Thus arose the natural question of the correctness of general relativity for strong gravitational fields. This question has been emphasized by the subsequent discovery of pulsars and by the discussions around the secular deceleration of the period of the binary pulsar PSR 1913 + 16 discovered by Hulse and Taylor [16] in 1974.

Thus the events about the year 1960 deeply modified the face of the relativistic gravity by the appearance of new experimental and theoretical possibilities. Henceforth, it became crucial to outline a general theoretical framework for building new theories and for interpreting the various experiments which can be performed to discriminate the theories.

III - General features of viable theories of gravity

A) The Dicke framework

The general framework for analyzing the foundamental problems of gravity has been given by Dicke in 1964 [17]. The main assumptions are :

1) The set of physical events can be represented by a four-dimensional differentiable manifold of class C^{∞} , the spacetime. No geometrical structure such as metric or affine connection on spacetime is postulated a priori.

2) The equations of gravity and of other interactions must be expressed in a form independent of the coordinate system, i.e. in a generally covariant form.

3) Like other known interaction fields, gravity has a "bosonic" character : at the classical level, it will be described by scalar, vector or tensor fields of various ranks.

4) Field equations of gravity have to be obtained from an invariant principle.

5) Nature is simple.

What could be our degree of confidence in these different assumptions ?

The assumption 1) is quite strongly supported by the present state of theoretical and experimental physics. It cannot be removed without building an entirely new physics.

The assumption 2) is a purely rational principle which must be applied to any physical theory.

The assumption 3) is strongly supported by an amount of empirical evidences (see the discussion of Dicke himself). It should be noted, however, that the theories of supergravity introduce also 3/2 - and 1/2 - spin fields to describe the gravitational interaction.

The assumption 4) is justified only by analogy with other branches of theoretical physics. This principle is not indispensable in analyzing the nature of gravity and could be removed if they were strong experimental arguments against it (e.g., the non-existence of post-Newtonian conservation laws).

We have no reason to believe that Nature is simple. The postulate 5) is in fact purely anthropomorphic : it would be more correct to say that physicists abhor too complicated models, which proves only the limitation of their intellectual abilities. Moreover, the content of this principle is not clear. Let us give an example : the principle of simplicity is often invoked to postulate that gravity is governed by partial differential equations of second order. Hence, one generally rejects for the treatment of the classical field the higher derivative equations deduced from the Hilbert action supplemented by squared curvature terms. However, it has been recently proved that the higher-order theories of gravity are equivalent to second-order scalar-tensor-tensor theories [18]. So, these theories cannot be eliminated on the basis of the preceding argument.

B) Criteria for the viability of gravitation theories

1) Universality of gravitation. The outcome of any experiment in a gravitational field has to be deduced from a small number of fundamental principles without adding "ad hoc" postulates or arbitrary constants. It is also said that a viable theory of gravity must be "complete" (Will [19]). However, this prescription should be limited because there is no theory of gravity which embodies satisfactorily the description of quantized fields.

2) Self-consistence. For any experiment, a viable theory must predict a unique result independent of the method used to get the prediction. For example, the same deviation of light by a gravitational mass must be obtained as it is computed in the eikonal approximation of the Maxwell's equations or in the zero-rest mass limit of the motion of particles.

3) Validity of special relativity in the absence of gravitational field. This criterion is tested by two kinds of experiments. First are experiments which verify that the velocity of light is independent of the velocity of the source (cf. Newman et al. [20]) or which confirm the isotropy of the propagation of light. Brillet and Hall, e.g., have tested this isotropy in 1978 to an accuracy of 2.10^{14}[21]. Second are experiments showing that the proper time of atomic and nuclear phenomena is independent of the acceleration and is given by the fundamental Minkowskian interval [20]. This second kind of experiments suggests that there exists at least one 2nd rank symmetric tensor $g_{\alpha\beta}$ on spacetime which reduces to the Minkowskian tensor $\eta_{\alpha\beta}$ when gravity is negligible.

4) Validity of Newton's law of gravity for weak masses in slow motion. Any deviation from the Newton's law can be reported to a variation of the gravitational constant G. What does experimentation say us?

a) There is no screening effect. The experiments performed with torsion balances in order to test the weak equivalence principle (see below) show that the solar attraction is not attenuated by the Earth to an accuracy of 10^{12}.

b) There is no significant variation of G with distance between attracting bodies. From the observation of the motions of planets and of spacecrafts, Mikkelsen and Newman [22] have shown that $|\Delta G/G| \leq 10^{-4}$ for distances ranging from 10^4 km to 3.10^8 km. Despite Long [23] claimed the contrary, Cavendish experiments recently performed by other experimentalists show that there is no significant deviation from the inverse-square force law over distances ranging from 5 cm to one meter. Hoskins and Newman [24], e.g., have obtained :

$$\frac{G(105\ cm)}{G(5\ cm)} - 1 = (-2 \pm 7) \times 10^{-4}$$

c) G seems to be independent of the cosmic time. Although Van Flandern [25] has claimed that $\dot{G}/G = -(6.4 \pm 2.2).10^{-11}$ /yr from lunar occultations and eclipses, a careful analysis of the radar observations from Vicking landers on Mars and other solar probes performed by Hellings et al. [26] gives :

$$\dot{G}/G = (0.2 \pm 0.4) \times 10^{-11}/yr$$

d) G seems also to be independent of the physical and chemical constitution of the attracting bodies (for experimental confirmations, see Gillies [27]).

The above criteria are sufficient to eliminate a lot of theories. The following examples are quoted from Will [19] :

Newton's theory	Is not relativistic.
Poincaré's theory and its generalizations analyzed by Whitrow and Morduch [28]; Whitehead's theory [29] in its original version	Are purely formal since they are action-at-a distance theories. Are incomplete since electromagnetic laws are set on flat spacetime without any connection with gravity. Are thus inconsistent: predict different behaviour for zero-rest mass particles and for light-rays in the geometrical optics limit.
Various vector theories as discussed by Whitrow and Morduch [28]	Describe gravity by a vector field on flat spacetime. Present the same difficulties as the Poincaré's and the Whitehead's theories.
Birkhoff's theory [30]	Has no correct Newtonian limit: speed of sound is equal to speed of light.

IV- The metric theories of gravity

Among the theories which satisfy the above criteria, those which incorporate the so-called Einstein principle of equivalence are named "metric" theories. In this section, we begin by stating the principle of equivalence and then we discuss its experimental confirmations.

A) The principle of equivalence

1) The weak principle of equivalence states that the motion of a freely falling uncharged test body in a given gravitational field depends uniquely upon the initial boundary conditions of this motion.

By a "test-body", we mean a body the size, the internal angular momentum and self-gravitating energy of which can be neglected.

These statements are impossible to precise rigorously. They are in fact heuristic propositions and their principal role is to lead to the Einstein principle of equivalence.

2) The Einstein principle of equivalence states that a) in any freely falling laboratory, so small that one can ignore the gradient of the gravitational field and in which the self-gravitating forces are negligible, the special relativity is valid and allows a correct description of nongravitational interactions; b) the fundamental nongravitational constants of physics have values independent of where and when they are measured.

The Einstein principle of equivalence plays a foundamental role in the theory of gravitation since it implies that [19]

a) there is a metric tensor g over spacetime ;

b) g defines the proper time of particles ;

c) the world lines of uncharged test bodies are geodesics of the metric g ;

d) all non gravitational fields are coupled with the metric g in the same manner (principle of universal coupling).

These properties define the class of metric theories of gravity. For such theories, the Einstein principle of equivalence allows to form immediately the non gravitational laws of physics in any gravitational field. We have simply to take the special relativistic form of the laws in terms of the Minkowskian metric $\eta_{\alpha\beta}$ and to make the replacements: $\eta_{\alpha\beta} \to g_{\alpha\beta}$, $\sqrt{-\eta}\, d^4x \to \sqrt{-g}\, d^4x$ and $\partial_\mu \to \nabla_\mu$, where ∇_μ is the covariant derivative constructed from the Christoffel symbols associated with $g_{\alpha\beta}$. It is generally agreed (without any experimental evidence) that these replacements have to be made in equations involving physically measurable fields, so that the resulting laws do not contain coupling with curvature terms, except if there are good reasons to take into account tidal phenomena (it is the case, e.g., for the spinning particles, which cannot be considered as point-like particles). This prescription is called the principle of "minimal coupling".

B) Experimental tests of the Einstein principle of equivalence

1) Tests of the weak principle of equivalence.The foundamental tests are the high-precision Eötvös-type experiments. The most precise of these tests, performed independently by Roll, Krotkov and Dicke [31] at Princeton and by Braginsky and Panov [32] at Moscow show that massive bodies of different constitution, e.g. aluminium, gold or platinum, experience the same gravitational acceleration to an accuracy of 10^{11} in the Princeton experiment and 10^{12} in the Moscow experiment.

The behaviour of individual particules such as atoms, electrons or neutrons has also been studied, but the accuracy of the results is lower than 10^3, with the exception of the experiment on neutrons performed by Koester [33].

2) Tests of the local Lorentz invariance.The special relativity (and by a consequence, the local Lorentz invariance) is verified almost every day by the phenomenological studies in elementary-particle physics. However, the measurements in this branch of physics cannot be considered as very precise tests because their interpretation needs a lot of corrections due to the complexity of weak and strong interactions, and these corrections are much greater than the influence of gravity on subatomic phenomena.

Actually, the cleanest tests of the validity of the local Lorentz invariance in freely falling laboratories are those which verify the local isotropy of mechanical and electromagnetic phenomena in a vacuum (although the Earth-bound laboratories are not truly in free fall). We have already quoted the Michelson-type experiment performed by Brillet and Hall. Another one ultrahigh precision test has been performed independently by Hughes, Beltran-Lopez and Robinson [11] and by Drever [12]. Hughes and his coworkers have used the nuclear magnetic resonance on the ^{7}Li nucleus in its ground state (nuclear spin $J = 3/2$). In an external magnetic field, this state will be split into four levels equally spaced if there is no preferred direction in space, so that a single nuclear resonance line will be observed. If any anisotropy effect is present, there will be three different intervals between the adjacent levels and a triplet nuclear resonance line will be observed. Only a single line has been found. Following the interpretation of the so-called Mach principle proposed by Cocconi and Salpeter [34], this result sets a drastic limit on the possible relative anisotropy $\Delta m/m$ of the inertial mass of the proton in the ^{7}Li nucleus: $\Delta m/m \leq 10^{-20}$. It should be noted, however, that such an experiment has also been interpreted by Dicke and Peebles [19][35] as a strong evidence of the non-existence of an extra-metric second rank tensor universaly coupled to matter.

3) Tests of the invariance with respect to the local position.This class of tests is mainly constituted by the red-shift experiments and by the verifications of the constancy of the nongravitational coupling constants.

a) Gravitational red-shifts. A red-shift experiment consists to compare the frequency of two identical standard clocks H_1 and H_2 at rest at different points in a static gravitational field. We assume the validity of the weak principle of equivalence and of the local Lorentz invariance in a small freely falling laboratory. Under these assumptions, the proper time measured by a standard clock H moving with respect to the laboratory may be written as (see, e.g., [19])

$$c^2 d\tau^2 = F_H^2(U)\left(c^2 dt_M^2 - dx_M^2 - dy_M^2 - dz_M^2\right) \qquad (1)$$

where dt_M is the proper time determined by a fundamental standard clock such as the Marzke-Wheeler clock [36][37] at rest in the freely falling laboratory and the dx_M^i are the orthogonal Cartesian coordinates defined from dt_M in order to have the velocity of light locally equal to c in all directions. U is the gravitational potential in the static field under study and F_H is a conformal factor which may depend on the type of standard clocks H used for the test. If F_H is a function of U, the rate of a clock H depends on its location in a gravitational field, so there is a violation of the invariance with respect to the local position.

Let (t,x,y,z) be a static coordinate system in a uniform gravitational field. This coordinate system is accelerated upward in the +z direction with the acceleration $\vec{g}$ $-\vec{\nabla}U$ relatively to a freely falling system of reference. Hence, in the (t,x,y,z) system, the proper time (1) may be written as (see, e.g;, [38]) :

$$c^2 d\tau^2 = F_H^2(U)\left[c^2\left(1+\frac{gz}{c^2}\right)^2 dt^2 - dx^2 - dy^2 - dz^2\right]$$

This formula implies that the proper time measured by a clock H placed at rest at the height z (dx = dy = dz = 0) is given by

$$\Delta\tau = F_H(U)\left(1+\frac{gz}{c^2}\right)\Delta t \tag{2}$$

Suppose now that the static clock H_1 emits toward the static clock H_2 electromagnetic signals separated by the interval Δt_1. Since the frame (t,x,y,z) is static, the interval Δt_2 between the reception of two successive signals at H_2 will be equal to Δt_1. So, taking into account the expression (2), the shift between the ticks of H_1 and H_2 defined by

$$Z = (\Delta\tau_1 - \Delta\tau_2)/\Delta\tau_1$$

will be approximatively given by

$$Z \simeq (1+\alpha)\Delta U/c^2 \tag{3}$$

where $\alpha \simeq -c^2 F_H'(U_1)/F_H(U_1)$ and $\Delta U = g(z_1 - z_2)$. The formula (3) holds also for the gravitational shift of spectral lines defined by $Z = (\nu_2 - \nu_1)/\nu_2$, where $\nu = 1/\Delta\tau$.

If the local invariance with respect to position is not violated, we have $\alpha = 0$ and

$$Z \simeq \Delta U/c^2 \tag{4}$$

The first entirely successful measurements of the frequency shift of the photons were the experiments performed by Pound, Rebka and Snider [10][39] with γ-rays ascending and descending a height of 22.5 m. The limit thus yielded on α was $|\alpha| \leq 10^{-2}$.

Since these pioneering experiments, the advance in the technology of frequency standards of ultrahigh stability has permitted to compare directly standard clocks placed at different points. An example is the direct comparison of two cesium beam atomic clocks situated respectively at $z_1 = 250$ m and $z_2 = 3500$ m above the sea level, performed by Briattore and Leschiutta [40].After two series of experiments, it has been found for the gain of the standard clock at mountain altitude $\Delta\tau/\tau = (33.8 \pm 6.8)$ ns/day and $\Delta\tau/\tau = (36.5 \pm 5.8)$ ns/day. These results are compatible with the standard prediction (4) ($\Delta\tau/\tau$ = 30.6 ns/day). More recently, Vessot, Levine et al. [41] have obtained $|\alpha| \leq 2.10^{-4}$ by comparing the signals generated by an hydrogen-maser clock located in a rocket to a similar clock on the Earth.

In a somewhat different way, the equivalence between a static gravitational field and an uniformly accelerated frame has also been verified in the quantum limit by using a neutron interferometer (Colella, Overhauser and Werner [42]).

b) Constancy of the non gravitational coupling constants. The most significant limits on possible cosmological variations in the nongravitational coupling constants have been obtained by measuring the ratio of two isotopes of samarium ($^{147}Sm/^{149}Sm$) found in the Oklo natural reactor, in Gabon. The value of this ratio shows that the neutron capture cross section for ^{149}Sm has not been significantly modified over a period of 2.10^9 years. This constancy implies the following limits (Shlyakhter [43]):

Fine structure constant α : $\dot{\alpha}/\alpha < 5.10^{-18}$/year

Weak interaction constant β: $\dot{\beta}/\beta < 10^{-12}$/year

Strong interaction constant g_s^2: $\dot{g}_s/g_s < 4.10^{-19}$/year

The amount of empirical evidence in favor of the Einstein principle of equivalence has led almost all the theorists to consider the metric theories of gravity as the only acceptable theories. Let us recall that in metric theories, matter and other nongravitational interactions are coupled to the gravitational field by the metric g only. However, other fields than the metric (scalars, vectors, etc) can also enter in the description of the gravitational interaction. These additional fields do not interact directly with the matter, but they contribute to the metric by the field equations.

V - The parametrized post-Newtonian formalism

The comparison of the predictions of the different metric theories is relatively simple when the massive bodies move slowly and generate weak gravitational fields, as it is the case in the solar system. The most useful framework to discriminate the competing theories is the parametrized post-Newtonian (PPN) formalism, elaborated under its actual form mainly by Nordtvedt and Will [15][19].

A) Assumptions underlying the PPN formalism

1) An oversimplified model of universe is adopted. The Galaxy (and possibly the Local Cluster) excepted, all the bodies of the universe are supposed to be very far from the

post-Newtonian system under examination, so that the local matter can be considered as isolated in an extended vacuous region. The density distribution of the distant matter is assumed to be homogeneous and isotropic (cosmological principle).

2) A freely falling frame of reference in which the universe appears isotropic is chosen (universe rest frame). One assumes that this frame can be mapped by quasi-Cartesian coordinate systems (X^0, X^i) such that the metric components $g_{\alpha\beta}$ differ from the Minkowski tensor by very small quantities $h_{\alpha\beta}$ (weak-field limit).

3) One supposes that $h_{\alpha\beta} \to 0$ as the distance $|\vec{X}-\vec{X}'|$ between the field point $\vec{X}$ and a point $\vec{X}'$ inside the matter of the post-Newtonian system becomes very large.

At this point, a deep difficulty arises in the PPN formalism. Since the universe is considered as homogeneous and isotropic at a large scale, the metric must tend asymptotically to a cosmological ds^2. Hence, the $g_{\alpha\beta}$ cannot tend to $\eta_{\alpha\beta}$. Let us emphasize that this problem is not satisfactorily solved in the PPN formalism. Usually, one assumes a set of cosmological boundary conditions for the dynamical variables of the metric theories (see [19]). For example, one imposes upon a scalar field Φ or upon a vector field K^μ the following asymptotic behavior compatible with the symmetry of the universe

$$\Phi(x^0, x^i) \to \Phi_0(x^0) \; ; \qquad K^\mu(x^0, x^i) \to K(x^0)\,\delta^\mu_0$$

Upon the metric, one has to impose

$$g_{\alpha\beta}(x^0, x^i) \to \{c_0(x^0), -c_1(x^0), -c_1(x^0), -c_1(x^0)\}$$

Then, in order to put the asymptotic values of $g_{\alpha\beta}$ in the standard post-Newtonian form, one makes the transformation

$$dx^{\bar{0}} = \sqrt{c_0(x^0)}\, dx^0 \,, \qquad dx^{\bar{i}} = \sqrt{c_1(x^0)}\, dx^i$$

However, this infinitesimal transformation is not integrable in a dynamical universe ($c_1(X^0) \neq$ cst). So it does not define a coordinate transformation.

Let us now abandon this problem and see the general PPN metric.

B) <u>General form of the PPN metric</u>

1) It can be easily verified that the post-Newtonian approximation involves terms of order v^4/c^4 in g_{00}, v^3/c^3 in g_{0i} and v^2/c^2 in g_{ij}, where $\vec{v}$ is the 3-velocity of a typical element of the fluid matter constituting the post-Newtonian system.

2) Using the postulate of isotropy of the universe and the arbitrariness of the quasi-Cartesian coordinate systems x^α compatible with the preceding assumptions, one retains in

the most usual version of the PPN formalism only ten post-Newtonian functionals to develop the metric of any gravitational theory :

$$U = \int \frac{\rho'}{|\vec{x} - \vec{x}'|} d^3x'$$

$$(\vec{U})_{ij} = \int \frac{\rho'(x - x')^i (x - x')^j}{|\vec{x} - \vec{x}'|^3} d^3x'$$

$$\Phi_W = \int \frac{\rho'\rho''(\vec{x} - \vec{x}')}{|\vec{x} - \vec{x}'|^3} \cdot \left(\frac{\vec{x}' - \vec{x}''}{|\vec{x} - \vec{x}''|} - \frac{\vec{x} - \vec{x}''}{|\vec{x}' - \vec{x}''|} \right) d^3x' d^3x''$$

$$\mathcal{A} = \int \frac{\rho'[\vec{v}' \cdot (\vec{x} - \vec{x}')]^2}{|\vec{x} - \vec{x}'|^3} d^3x'$$

$$\Phi_{(\alpha)} = \int \frac{\rho' A'_{(\alpha)}}{|\vec{x} - \vec{x}'|} d^3x' , \qquad A'_{(\alpha)} = \left(v'^2, U', \Pi', p'/\rho' \right)$$

$$\alpha = (1, 2, 3, 4)$$

$$\vec{V} = \int \frac{\rho' \vec{v}'}{|\vec{x} - \vec{x}'|} d^3x'$$

$$\vec{W} = \int \frac{\rho'[\vec{v}' \cdot (\vec{x} - \vec{x}')](\vec{x} - \vec{x}')}{|\vec{x} - \vec{x}'|^3} d^3x'$$

where ρ' is the rest mass of matter at the point $\vec{x}'$, p' the pressure at $\vec{x}'$, $\vec{v}' = d\vec{x}'/dt$, Π' the internal energy par unit rest mass. Units are chosen so that G(today) = c = 1.

Many other metric functionals could be formed but only the above terms appear in the known theories when one assumes the universe to be isotropic and the matter of the post-Newtonian system to be a perfect fluid. (The case of a slightly anisotropic universe, e.g., has been studied by Nordtvedt [44]).

Since ten functionals are introduced to take into account the post-Newtonian gravitational effects, the standard post-Newtonian metric contains ten dimensionless PPN parameters denoted γ , β , ξ (or ξ_W), α_1, α_2, α_3, ζ_1, ζ_2, ζ_3, ζ_4. Each metric theory of gravity predicts specific values of these PPN parameters. In general relativity $\gamma = \beta = 1$ and other parameters ξ , α_i , ζ_j vanish.

3) Until now, we have used a universe rest frame. But slightly anisotropies in the cosmic background radiation reveal that the solar system is in motion relative to this rest frame with a velocity $\vec{w}$ ($w \simeq 390$ km.s^{-1}) [45]. For that reason, the PPN metric has to be expressed in a moving quasi-Cartesian coordinate system. One takes an approximate form of the Lorentz transformation compatible with the post-Newtonian limit (Will [19]). It should be noted however that, despite what is claimed by Will, such a transformation differs from the post-Galilean transformations found by Chandrasekhar and Contopoulos [46] by the absence of any gravitational term. Because of this absence, the physical meaning of the quasi-Lorentzian transformation used in the PPN formalism is questionable. A justification should be found on a different basis, which requires further investigations. There is another difficulty : it is supposed in making a quasi-Lorentzian transformation that wt is of the same order as x, which is true for short durations only. In fact, the characteristic length of the solar system is $L \simeq 3.10^8$ km , so that the duration $L/w \simeq 10^6$ s (i.e. 10 days). Hence, it seems difficult to justify the computation of the secular relativistic effects in the usual PPN formalism. It should be noted also that some

problems of coordinate transformations are encountered in Geodesy, as Boucher and Lestrade point it out in their lecture.

These questions remaining open, we shall henceforth admit the validity of the post-Galilean transformations. Then the PPN metric in a frame moving with the velocity $\vec{w}$ with respect to the universe rest frame is given by

$$g_{00} = 1 - 2U + 2\beta U^2 + 2\xi \Phi_W - (2\gamma + 2 + \alpha_3 + \zeta_1 - 2\xi)\Phi_1$$
$$- 2(3\gamma - 2\beta + 1 + \zeta_2 + \xi)\Phi_2 - 2(1 + \zeta_3)\Phi_3 - 2(3\gamma + 3\zeta_4 - 2\xi)\Phi_4$$
$$+ (\zeta_1 - 2\xi)\mathcal{A} + (\alpha_1 - \alpha_2 - \alpha_3) w^2 U + \alpha_2 w^i w^j U_{ij} - (2\alpha_3 - \alpha_1) w^i V_i$$

$$g_{0i} = \frac{1}{2}(4\gamma + 3 + \alpha_1 - \alpha_2 + \zeta_1 - 2\xi) V_i + \frac{1}{2}(1 + \alpha_2 - \zeta_1 + 2\xi) W_i$$
$$+ \frac{1}{2}(\alpha_1 - 2\alpha_2) w^i U + \alpha_2 w^j U_{ij}$$

$$g_{ij} = -(1 + 2\gamma U)\,\delta_{ij}.$$

The Tables 1 and 2 give the significance of the PPN parameters and their values in some metric theories of gravity.

Table 1. Significance of the PPN parameters

$\xi \neq 0$	$\Rightarrow$	They are local gravitational effects due to the position and/or the orientation of the experimental apparatus relative to the surrounding matter (preferred-location effects).
$\alpha_i \neq 0$	$\Rightarrow$	They are local gravitational effects due to the motion of the system relative to the rest frame of universe (preferred-frame effects).
$\alpha_3 = \zeta_1 = \zeta_2 = \zeta_3 = \zeta_4 = 0$	$\Leftrightarrow$	There is an energy-momentum vector P^μ conserved in the post-Newtonian limit. The corresponding theories are called "semiconservative" theories.
$\alpha_1 = \alpha_2 = \alpha_3 = 0$, $\zeta_1 = \zeta_2 = \zeta_3 = \zeta_4 = 0$	$\Leftrightarrow$	There are an energy-momentum vector P^μ and an angular momentum tensor $J^{\mu\nu}$ conserved in the post-Newtonian limit. The corresponding theories are called "fully conservative" theories.

Table 2. Values of the PPN parameters in some metric theories

Theories and their gravitational fields	Arbitrary functions or constants	PPN parameters γ	β	ξ	α_1	α_2	α_3 ζ_j
General relativity $g_{\alpha\beta}$	none	1	1	0	0	0	0
Brans-Dicke [14] $g_{\alpha\beta}$, Φ	ω	$\frac{1+\omega}{2+\omega}$	1	0	0	0	0
Will-Nordtvedt [15] $g_{\alpha\beta}$, K^μ; $K^\mu \to K\delta^\mu_0$	none	1	1	0	0	$\frac{K^2}{1+\frac{1}{2}K^2}$	0
Rosen [47] $g_{\alpha\beta}$, $\eta_{\alpha\beta}$; $\eta_{\alpha\beta} = (1,-1,-1,-1)$; $g_{\alpha\beta} \to (c_0,-c_1,-c_1,-c_1)$	none	1	1	0	0	$\frac{c_0}{c_1} - 1$	0
Ni [48] $g_{\alpha\beta}, \eta_{\alpha\beta}, t, \Phi, \psi_\mu$; $\eta_{\alpha\beta} = (1,-1,-1,-1)$; $g_{\alpha\beta} \to \eta_{\alpha\beta}$, $\Phi, \psi_\mu \to 0$	$f_1(\Phi), f_2(\Phi), f_3(\Phi)$, e	γ	β	0	α_1	α_2	0
	f_1, f_2, f_3 are three arbitrary functions of the scalar field Φ. $\gamma, \beta, \alpha_1, \alpha_2$ depend on the asymptotic behaviour of the $f_i(\Phi)$.						
Whitehead [29] $g_{\alpha\beta}$, $\eta_{\alpha\beta}$; $\eta_{\alpha\beta} = (1,-1,-1,-1)$	none	1	1	1	0	0	$\alpha_3 = 0$; $\zeta_i \neq 0$
	Action-at-a-distance theory reinterpreted as a metric theory						

VI - Determination of γ, β and α_1

The most recent determination of γ, β and α_1 is derived from the analysis of solar system astronomical data, particularly the ranging data provided by the Vicking landers on Mars. This analysis, like the previous ones, requires the knowledge of the PPN equations of motion.

A) PPN equations of motion for photons

In any metric theory, it can be deduced from the Maxwell's equations that the photon trajectories are null geodesics. Hence, the PPN equations for the motion of photons are

$$\frac{d^2\vec{x}}{dt^2} = (1+\gamma)\left[\vec{\nabla}U - 2\vec{n}(\vec{n}.\vec{\nabla}U)\right]$$
$$\vec{n}.\frac{d\vec{x}}{dt} = 1-(1+\gamma)U \qquad (5)$$

where $\vec{n}$ is the unit 3-vector defining the initial direction of the photon.

B) PPN equations of motion for massive bodies

We consider that the post-Newtonian system under study is a set of N massive bodies labeled by a = 1,2,..., N. The equations of motion of the center of mass of each body are deduced from the conservation of the energy-momentum tensor $T^{\mu\nu}$: $T^{\mu\nu}{}_{;\nu} = 0$ (for the definitions and the detailed calculations, see [19]). In the limit of the PPN approximation, these equations may be written as

$$\vec{a}_a = \frac{d^2\vec{x}_a}{dt^2} = \vec{a}_{a(self)} + \vec{a}_{a(Newt.)} + \vec{a}_{a\,(N\,bodies)} \qquad (6)$$

1) $\vec{a}_{a(self)}$ is a term of self-acceleration which can be decomposed into two parts :

$$\vec{a}_{a(self)} = \vec{a}^{\,I}_{a(self)} + \vec{a}^{\,II}_{a(self)} \qquad (7)$$

$\vec{a}^{\,I}_{a(self)}$ involves the internal structure of the body. This term is null if $\alpha_3 = 0$ and $\zeta_1 = \zeta_2 = \zeta_3 = \zeta_4 = 0$ (semiconservative theories). Moreover, whatever the values of α_3 and ζ_j are, this self-acceleration is zero for spherically symmetric bodies. So, it is negligible in the solar system.

$\vec{a}^{\,II}_{a(self)}$ is a term proportional to α_3, which involves the motion of each part of the body relative to the universe rest frame. If $\alpha_3 \neq 0$, this self-acceleration is proportional to the rotation rate and to the self-gravitational energy of the body. Thus it could produce observable effects in the solar system.

2) $\vec{a}_{a(Newt.)}$ is the quasi-Newtonian acceleration of the body, defined by the tensorial equations

$$(m_I)^{jk}_a\, a^k_{a\,(Newt.)} = (m_P)^{\ell m}_a\, \mathfrak{U}^{\ell m}_{,j}$$

where $\mathfrak{U}^{\ell m}$ is a gravitational tensor defined by

$$\mathfrak{U}^{\ell m} = \sum_{b\neq a} (m_A)^{ms}_b \frac{n^\ell_{ab}\, n^s_{ab}}{|\vec{x}_b - \vec{x}_a|} \qquad \vec{n}_{ab} = \frac{\vec{x}_b - \vec{x}_a}{|\vec{x}_b - \vec{x}_a|}$$

$(m_I)^{jk}_a$ and $(m_P)^{\ell m}_a$ are respectively the inertial mass and the passive gravitational mass tensors of the body a ; $(m_A)^{ms}_b$ is the active gravitational mass tensor of the body b. These tensors are constructed from a scalar mass m and internal energies of the body, including the self-gravitational energy. In some metric theories, $(m_I)^{jk} \neq (m_P)^{jk} \neq (m_A)^{jk}$. The difference between the inertial mass and the passive gravitational mass tensors of a massive body (Nordtvedt effect) is too small to be measured in a laboratory by an Eötvös-type experiment , but could be eventually detected in the motion of the Moon around the Earth (see below). On the other hand, the difference between the passive and active gravitational mass tensors can be tested in laboratory (see Kreuzer experiment).

Let us note that $(m_I)^{jk}_a = (m_P)^{jk}_a = (m_A)^{jk}_a = m_a \delta^{jk}$ in general relativity.

3) $\vec{a}_{a(\text{N bodies})}$ is a purely post-Newtonian acceleration which involves the masses, the velocities and the relative positions of the N bodies, and the motion of the system relative to the universe rest frame. This acceleration produces several effects. In particular, it is responsible for the well known anomalous perihelion shift of Mercury.

C) Determination of γ, β and α_1

The most accurate data for determining the parameters of the solar system actually come from the ranging to the Viking landers on Mars (Hellings [49]). By a numerical integration of the PPN equations of motion (5) and (6), the time-of-flight τ_i of a photon emitted at the time t_i from a tracking station on Earth and reflected by a spacecraft on the surface of Mars can be expressed as a function of t_i and of constant parameters such as

$\vec{x}_{0k}$, $\dot{\vec{x}}_{0k}$: the initial heliocentric positions and velocities of the planets and of the Moon.

GM_k : the masses of the planets and of Ceres, Pallas and Vesta.

The mass distribution of 200 asteroids.

The length of the astronomical unit in meters.

$J_2^{\odot}$: the quadrupole moment of the Sun.

γ , β , α_1 : the PPN parameters that characterize the gravitational field of the solar system (the other PPN parameters are assumed to be zero).

Other parameters characterizing the planets and the tracking stations, etc...

So, we can write

$$\tau_i = F(t_i, \vec{x}_{0k}, \dot{\vec{x}}_{0k}, GM_k, J_2^{\odot}, \gamma, \beta, \alpha_1, \ldots)$$

The best values of the solar system parameters are determined by the method of least squares estimation. The results given by Hellings are

$$\gamma - 1 = (-0.7 \pm 1.7) \times 10^{-3} \tag{8}$$

$$\beta - 1 = (-2.9 \pm 3.1) \times 10^{-3} \tag{9}$$

$$\alpha_1 = (2.1 \pm 1.9) \times 10^{-4} \tag{10}$$

$$J_2^{\odot} = (-1.4 \pm 1.5) \times 10^{-6} \tag{11}$$

VII - Constraints on ξ, α_i and ζ_i

A) The Lunar Eötvös experiment

We have seen in Chap. V that many metric theories predict a violation of the weak principle of equivalence for massive bodies (Nordtvedt effect [50]). For the sake of simplicity, let us suppose that the post-Newtonian system is a set of N spherically symmetric bodies. Then the quasi-Newtonian acceleration of the body a is given by

$$\vec{a}_{a\,(Newt.)} = \left(1 + \eta \frac{\Omega_a}{m_a}\right) \vec{\nabla} U \tag{12}$$

$$U = \sum_{b \neq a} \frac{(m_A)_b}{|\vec{x}_b - \vec{x}_a|}$$

where m_a is the scalar mass of the body a, Ω_a its self-gravitational energy

$$\Omega_a = -\frac{1}{2} \iint_a \frac{\rho' \rho''}{|\vec{x}' - \vec{x}''|} d^3x' \, d^3x'' \tag{13}$$

and η is a linear combination of PPN parameters

$$\eta = 4\beta - \gamma - 3 - \frac{10}{3}\xi - \alpha_1 + \frac{2}{3}\alpha_2 - \frac{2}{3}\zeta_1 - \frac{1}{3}\zeta_2$$

It results from Eqs. (12) and (13) that the Earth and the Moon fall toward the Sun with slightly different accelerations if $\eta \neq 0$, since $(\frac{\Omega}{m})_{\oplus} = -4.6\ 10^{-10}$ and $(\frac{\Omega}{m})_{☾} = -0.2.10^{-10}$. This Nordtvedt effect produces a deformation of the orbit of the Moon in the direction of the Sun [51].

The best analysis of the data obtained during six years from the Lunar-Laser-Ranging Experiment gives (Shapiro et al. [52])

$$\eta = 0.001 \pm 0.015$$

So, the Eötvös experiment using the Earth-Moon system shows that there is no breakdown of the weak equivalence principle for massive self-gravitating objects to an accuracy of 10^{11}. This result may be compared with the laboratory experiments performed by Roll-Krotkov-Dicke and Braginski-Panov (see Chap.IV).

B) Geophysical tests

Several metric theories predict preferred-location effects ($\xi \neq 0$) and/or preferred-frame effects ($\alpha_i \neq 0$). These effects imply that the local value G_L of the gravitational constant which appears in the Newton's law, as measured by a Cavendish experiment, should depend on the orientation of the apparatus relative to the external universe and/or on the velocity of the laboratory relative to the universe rest frame.

Because of the Earth's rotation and orbital motion around the Sun, the preferred-location and/or preferred-frame effects involve different types of post-Newtonian variations for the locally measured gravitational constant G_L :

1) Diurnal and semidiurnal variations, analogous to the 12 and 24 hours solid-Earth tides. Using an ultrahigh precision superconducting gravimeter, Goodkind and Warburton [53] have searched these anomalous PPN tides. They have obtained

$$|\alpha_2| < 2\times 10^{-3}\,, \qquad |\xi| < 10^{-3} \tag{14}$$

2) Annual variations, which produce a purely spherical deformation of the Earth. This deformation causes an annual variation of the diurnal rotation of the Earth which can be compared with the observations. Taking into account the variations due to the atmospheric winds, one obtains [19]

$$|\alpha_3 + \tfrac{2}{3}\alpha_2 - \alpha_1| < 0.02$$

C) Orbital tests

Preferred-location and preferred-frame effects can also influence the orbits of celestial bodies. In particular, the presence of α_3 in the self-acceleration term $\vec{a}^{\,\mathrm{II}}_{(\mathrm{self})}$ and of α_i, ξ in the term $\vec{a}_{(\mathrm{N\ bodies})}$ (see Eqs. (6) and (7)) produces a secular perihelion shift of the planets, which is not included in the classical relativistic shift due to γ and β. The comparison of the theoretical contribution with the observed perihelion positions for Mercury and Earth leads to the limit [19]

$$|49\,\alpha_1 - \alpha_2 - 6.3\times 10^5\,\alpha_3 - 2.2\,\xi| < 0.1$$

if one assumes $J_2^{\odot} < 5.10^{-6}$. If one takes now into account the limits (10) and (14), this result shows that

$$|\alpha_3| < 2\times 10^{-7}$$

It should be noted that the tighter limit $|\alpha_3| < 2.10^{-10}$ is obtained from the observations of pulsars, although the behaviour of these compact objects cannot be rigourously described in the post-Newtonian approximation [19].

D) Equality of the passive and active gravitational masses

In some metric theories of gravity, the active gravitational mass tensor differs from the passive gravitational mass tensor. For the massive bodies on which we can experiment in a laboratory, the two tensors are isotropic and their only measurable difference is due to the nuclear electrostatic energy E_e. The ratio of the active mass to the passive mass is then given by (Will, [54]) :

$$m_A/m_P = 1 + \frac{1}{2}\zeta_3 (E_e/m_P)$$

For a nucleus of atomic number Z and of mass number A, a semiempirical formula gives

$$m_A/m_P = 1 + 3.8 \times 10^{-4} \zeta_3 Z(Z-1) A^{-4/3}$$

As a consequence, we have for two bodies of different nuclear constitution

$$\frac{(m_A/m_P)_1 - (m_A/m_P)_2}{(m_A/m_P)_1} = 3.8 \times 10^{-4} \zeta_3 \left[Z_1 (Z_1 - 1) A_1^{-4/3} - Z_2 (Z_2 - 1) A_2^{-4/3} \right] \tag{15}$$

Kreuzer [55] has compared the active and passive gravitational masses of fluorine (Z = 9, A = 19) and bromine (Z = 35, A = 80) by using a Cavendish balance. The conclusion was

$$\frac{|(m_A/m_P)_{Fl} - (m_A/m_P)_{Br}|}{(m_A/m_P)_{Fl}} < 5 \times 10^{-5}$$

If Eq. (15) is taken into account, the Kreuzer experiment yields

$$|\zeta_3| < 0.06$$

VIII - Conclusions

We can now summarize the results obtained in the solar system :

$$\left.\begin{aligned} \gamma - 1 &= (-0.7 \pm 1.7) \times 10^{-3} \\ \beta - 1 &= (-2.9 \pm 3.1) \times 10^{-3} \\ \alpha_1 &= (2.1 \pm 1.9) \times 10^{-4} \end{aligned}\right\} \text{(Ranging data from the Vicking landers on Mars)}$$

$$4\beta - \gamma - 3 - \frac{10}{3}\xi - \alpha_1 + \frac{2}{3}\alpha_2 - \frac{2}{3}\zeta_1 - \frac{1}{3}\zeta_2 = 0.001 \pm 0.015$$

(Lunar laser ranging experiment)

$|\alpha_2| < 2 \times 10^{-3}, \quad |\xi| < 10^{-3}$ (Post-Newtonian Earth tides)

$|\alpha_3 + \frac{2}{3}\alpha_2 - \alpha_1| < 0.02$ (Rotation rate of the Earth)

$|\zeta_3| < 0.06$ (Equality of active and passive gravitational masses)

$|\alpha_3| < 2.10^{-7}$ (Perihelion shifts)

At this point, several observations can be made :

1) The PPN parameters γ, β, α_i and ξ are actually known to an accuracy about 10^3. At the contrary, the parameters ζ_j are totally unknown, except ζ_3. It will be the merit of future experimentations to set limits on them.

2) General relativity agrees with all experiments.

3) Many other metric theories can be made compatible with all the tests performed in the solar system if their adjustable constants and their arbitrary cosmological boundary conditions are constrained in a convenient way. For example, the Brans-Dicke theory is consistent with the values of the PPN parameters if $\omega > 500$.

This situation forces the physicists to perform entirely new tests. New tests can be post-Newtonian experiments in laboratory like those proposed by Braginsky, Caves and Thorne [56] or post-post-Newtonian effects, such as the gravitational radiation, the behaviour of the binary pulsar PSR 1913 + 16 or the cosmological tests. These last areas remain open domains which can provide the "ultimate tests" of gravitational theories - as Will says about PSR 1913 + 16 - or reveal entirely new properties of gravity.

References

[1] A. Einstein, Jahrb.Radioact.Elekt. 4, 411 (1907).
[2] For a detailed story of the solar red-shift problem, see E.G. Forbes, Annals of Science (London 17, 129 (1961).
[3] J. Chazy, La théorie de la relativité et la mécanique céleste, Vol.1, pp. 252-256 (Gauthier-Villars, Paris, 1928).
[4] B. Bertotti, D. Brill and R. Krotkov, in Gravitation : An Introduction to Current Research, ed. L. Witten, pp. 1-48 (Wiley, New York, 1962).
[5] A. Einstein, Annalen der Physik 35, 898 (1911). On the computation of Soldner, see P. Lenard, Annalen der Physik 65, 593 (1921).
[6] F.A.E. Pirani, Phys.Rev. 105, 1089 (1957).
[7] A. Trautman, Bull.Acad.Polon.Sc.Cl.III, 5, 273 (1957).
[8] A. Lichnerowicz, C.R.Ac.Sc.Paris 246, 893 (1958).
[9] J. Weber, Gravity Research Foundation Prizes Essays, April 1958 and April 1959.
[10] R.V. Pound and G.A. Rebka Jr, Phys.Rev.Lett. 4, 337 (1960).

[11] V.W. Hughes, H.G. Robinson and V. Beltran-Lopez, Phys.Rev.Lett. 4, 342 (1960).
[12] R.W.P. Drever, Phil.Mag. 6, 683 (1961).
[13] L.I. Schiff, Proc.Nat.Acad.Sc. (U.S.A.) 46, 871 (1960); Phys.Rev.Lett. 4, 215 (1960).
[14] C. Brans and R.H. Dicke, Phys.Rev. 124, 925 (1961).
[15] C.M. Will and K. Nordtvedt Jr, Astrophys.J. 177, 757 (1972) ; C.M. Will, Astrophys.J. 185, 31 (1973).
[16] R.A. Hulse and J.H. Taylor, Astrophys.J. 195, L 51 (1975).
[17] R.H. Dicke, Experimental Relativity. In Relativity, Groups and Topology, ed. C. De Witt and B. De Witt, pp. 165-313 (Gordon and Breach, New York, 1964).
[18] P. Teyssandier and Ph. Tourrenc, J.Math.Phys., 24, 2793 (1983) ; J.P. Berthias, P. Teyssandier and Ph. Tourrenc, to be published in the Proceedings of the 13th International Colloquium on Group Theoretical Methods in Physics (University of Maryland, 1984).
[19] C.M. Will, Theory and Experiment in Gravitational Physics (Cambridge University Press, Cambridge, 1981).
[20] D. Newman, G.W. Ford, A. Rich and E. Sweetman, Phys.Rev.Lett. 40, 1355 (1978).
[21] A. Brillet and J.L. Hall, Phys.Rev.Lett. 42, 549 (1979).
[22] D.R. Mikkelsen and M.J. Newman, Phys.Rev.D 16, 919 (1977).
[23] D.R. Long, Nature 260, 417 (1976) ; Nuovo Cim. 62B, 130 (1981).
[24] J.K. Hoskin and R.D. Newman, in 10th International Conference on General Relativity and Gravitation : Contributed papers, Vol.2. p. 984 (1983).
[25] T.C. van Flandern, Astrophys.J. 248, 813 (1981).
[26] R.W. Hellings, P.J. Adams, J.D. Anderson, M.S. Keesey, E.L. Lau, E.M. Standish, V.M. Canuto and I. Goldman, preprint.
[27] G.T. Gillies, The Newtonian Gravitational Constant : An Index of Measurements, Bureau International des Poids et Mesures (France, 1983).
[28] G.J. Whitrow and G.E. Morduch, in Vistas in Astronomy, Vol.6, ed. A. Beer, pp. 1-67 (Pergamon Press, Oxford, 1965).
[29] A.N. Whitehead, The Principle of Relativity (Cambridge University Press, Cambridge, 1922).
[30] G.D. Birkhoff, Proc.Nat.Acad.Sci. (U.S.A.) 29, 231 (1943).
[31] P.G. Roll, R. Krotkov and R.H. Dicke, Ann.Phys. (N.Y.) 26, 442 (1964).
[32] V.B. Braginsky and V.I. Panov, Sov.Phys.JETP 34, 463 (1972).
[33] L. Koester, Phys.Rev.D 14, 907 (1976).
[34] G. Cocconi and E. Salpeter, Nuovo Cim. 10, 646 (1958).
[35] P.J. Peebles and R.H. Dicke, Phys.Rev. 127, 629 (1962).
[36] R.F. Marzke and J.A. Wheeler, in Gravitation and Relativity, ed. H.Y. Chiu and W.F. Hoffmann, pp. 40-64 (Benjamin, New York, 1964).
[37] See B. Bertotti, in Evidence for Gravitational Theories : Proceedings of Course 20 of the International School of Physics "Enrico Fermi", ed. C. Møller, pp. 174-201 (Academic Press, New York, 1962).
[38] C.W. Misner, K.S. Thorne and J.A. Wheeler, Gravitation (Freeman, San Francisco, 1973).
[39] R.V. Pound and J.L. Snider, Phys.Rev. 140, B 788 (1965).
[40] L. Briattore and S. Leschiutta, Nuovo Cimento 37B, 219 (1977).
[41] R.F.C. Vessot, M.W. Levine, E.M. Mattison, E.L. Blomberg, T.E. Hoffman, G.U. Nystrom, B.F. Farrel, R. Decher, P.B. Eby, C.R. Baugher, J.M. Watts, D.L. Teuber and F.D. Wills, Phys.Rev.Lett. 45, 2081 (1980).
[42] R. Colella, A.W. Overhauser and S.A. Werner, Phys.Rev.Lett. 34, 1472 (1975).
[43] A.I. Shlyakhter, Nature 264, 340 (1976).
[44] K. Nordtvedt Jr, Phys.Rev.D 14, 1511 (1976).
[45] G.F. Smoot, M.V. Gorenstein and R.A. Muller, Phys.Rev.Lett. 39, 898 (1977).
[46] S. Chandrasekhar and G. Contopoulos, Proc.Roy.Soc. (London) 298A, 123 (1967).
[47] N. Rosen, Gen.Rel. and Grav. 4, 435 (1973) ; Ann.Phys. (N.Y.) 84, 455 (1974).
[48] W.T. Ni, Phys.Rev.D 7, 2880 (1973).
[49] R.W. Hellings, preprint.
[50] K. Nordtvedt Jr, Phys.Rev. 169, 1014 (1968) ; Phys.Rev. 169, 1017 (1968).
[51] K. Nordtvedt Jr, Phys.Rev. 170, 1186 (1968).
[52] J.J. Shapiro, C.C. Counselman III and R.W. King, Phys.Rev.Lett. 36, 555 (1976).
[53] R.J. Warburton and J.M. Goodkind, Astrophys.J. 208, 881 (1976).
[54] C.M. Will, Astrophys.J. 204, 224 (1976).
[55] L.B. Kreuzer, Phys.Rev. 169, 1007 (1968).
[56] V.B. Braginsky, C.M. Caves and K.S. Thorne, Phys.Rev.D 15, 2047 (1977).

SURVEY OF RELATIVISTIC EFFECTS

IN GEODESY AND FUNDAMENTAL ASTRONOMY

C. BOUCHER
Institut Géographique National

J.F. LESTRADE
Bureau des Longitudes

1) Introduction.

This survey has been established by starting from geodetic and astrometric problems. Consequently it is necessary to understand that the adopted approach is neither theoretical nor even physical, but metrological : the precision of measurements which is presently achievable implies that the complete relevant model must include relativistic effects.

The list of concepts and techniques which is given in this paper is certainly unexhaustive and will be even more in the future when improved precision will give access to smaller effects.

A few efforts has been done up to now in the geodetic or astrometric community to review theese problems. One can mention for instance Boucher 1979, Moritz 1980, Soffel, Ruder and Schneider 1983 and the existence of a Working Group on theese topics within the International Association of Geodesy.

Much more information can be found in the field of experimental gravitation i.e. the investigation of any experimental effect which can add about our knowledge of the validity of the various theories of Gravitation (Will 1981, Tourrenc and Teyssandier 1984).

As it can be seen later, many of theese experiences belong to Geodesy or Astrometry.

2) Theoretical framework.

2-1 Reference frames and time scales.

Several reference systems are current used in Geodesy and Astrometry :

- terrestrial systems are linked to the Earth in its diurnal rotation so that points located at its surface are almost fixed in such systems. Among them, one can distinguish topocentric systems used for laboratory experiements or measurements performed at the surface of the Earth, from geocentric systems used by geodesy for any global survey of the Earth. Such systems have an equatorial orientation.

- celestial systems are such that directions of remote objects such as stars or quasars are almost fixed. On distinguish between geocentric and barycentric (barycenter of the solar system) systems.

Such systems must be defined and realized with enough accuracy to express geocentric coordinates of points on the Earth at the mm level and barycentric directions at the milliarcsecond level over several decennies.

Time scales occur in three major ways :

- the fundamental time scale, now TAI as defined and determined by BIH.
- time argument in ephemerides and Earth rotation
- clock readings.

Proper definitions and models has to be choosen in order to satisfy present and forthcoming relative accuracy of 10E-14 to 10E-16.

The relativistic modelling of spatial and temporal reference systems is the quasi-cartesian coordinate systems (x^α) where the metric can be expressed as

$$g_{\alpha\beta} = \eta_{\alpha\beta} + h_{\alpha\beta}$$

where η is the Mikowski metric and h a small quantity (10^{-5} in the Solar System).

For any such coordinate system one can associate :

- a coordinate time t by

$$x^o = ct$$

- a 3 D vectar $\bar{x} \in \mathbb{R}^3$ by

$$\bar{x} = (x^i)_{1 \leqslant i \leqslant 3}$$

A particularly interesting class of such systems are PPN coordinates (Will 1974, Misner, Thorne, Wheeler 1973 or Will 1981). This system can be used satisfactorly in the Solar System for any viable metric theory of Gravitation, the distinction between theese theories being done by the numerical values of the PPN parameters (β, γ..) The transformation between two PPN systems is achieved by a combination of spatial rotations, post-galilean transformation and gauge transformation.

Nevertheless, it is our opinion that several questions have to be clarified :

- in the PPN frame, the velocity in of the origin of the system with regard to the mean rest have of the Universe is hold as a constant. The validity of such an assumption must be investigated especially for non-barycentric PPN systems. For instance a geocentric PPN system can be defined only at some epoch.

- a complete analysis of motion of test particles from sources outsid the solar system such as stars or quasars cannot be done in PPN system. The possible influence of galactif matter and the rest of the Universe must be taken into account, especially for long term stability (see section 3.6). The adoption of a PPN system for a barycentric space-time frame is presently the nominal model. For instance conversion between a barycentric coordinate time and proper time as given by clock in the vicinity of the Earth has been extensively studied (e.g. Moyer 1981, Guinot 1980).

The relativistic modelling of a geocentric celestial or terrestrial (i.e. non rotating or rotating) frame is still subject to research and controversies. The general technique could be applied in aur opinion :

a) to define the motion of the center of the Earth $P(\tau)$ in the barycentric frame.

b) to define at some epoch τ_0 an orthonormal tetrahedron e_α :

$$e_\alpha \cdot e_\beta = \eta_{\alpha\beta}$$

such that $e_0 = u(\tau_0)$ is the 4- velocity of P at τ_0.

c) to choose the evolution of e_α as a mobile frame at $P(\tau)$ such that :

$$e_0 = u(\tau) \text{ at any } \tau$$

$$\nabla_u e_\alpha = -\Omega\, e_\alpha$$

where
$$\Omega = \Omega_{FW} + \Omega_{SR}$$

Here Ω_{FW} generates a Fermi - Walker transport and Ω_{SR} a spatial rotation.

d) to define finally local normal coordinates from e_α.

With this construction, a geocentric celestial fram can be defined with $\Omega_{SR}=0$ and a terrestrial with Ω_{SR} modelling the diurnal Earth rotation.

if $a = \nabla_u u = 0$ ie if P follow a geodesic, the celestial coordinates become the so-called Fermi normal coordinates.

This type of approach can be found in Ashby and Allan 1979 for the time coordinate, Ashby 1980, Ashby and Bertotti 1983 for a more complete study. One must also mention the introduction of topocentric systems in Tourrenc 1981 and a systematic use of Clifford algebra for rotating frames in Hamilton 1981.

We feel that further clarifications and explanations are necessay, both for spatial systems and time scales.

On this last point, recommandations by I AU and CCIR (CCIR 1978) about TDB, TDT and TAI time scales has to be reformulated into such a study.

2-2 Motion of the Planets and of the Moon

The high accuracy acheived in astrometry with the new techniques of observation (radar, laser, VLBI) makes the relativistic effects quite critical in the physical models used to adjust to these measurements.

In the 60's, planetary distances observed from time-delays measured with a powerful Radar and compared to predictions of the orbital motion of planets were thought to be a fourth test of the Theory of General Relativity (Shapiro, 1964). At the end of the 60's, the Earth-Moon distance measured from Lunar Laser Ranging and the possible violation of the Principle of Equivalence Between inertial and gravitational mass leading to a large perturbation in the orbital motion of the

Moon (Nordtvedt, 1968) generated a lot of interest for testing a fundamental hypothesis.

The PPN (Parametrized-Post-Newtonian) formalism (Will, 1974) is the general framework in which one usually wishes to develop the theory of the motions of the bodies of the Solar System. The expansion of the metric is limited to $1/c^2$-terms and the motions of the Solar System bodies are derived by successively assuming that each one is massless and follows a geodesic in the metric curved by the other masses, actually mainly by the Sun. The resulting differential equations of motion are lengthy and can only be numerically integrated e.g. the Jet Propulsion Laboratory ephemeridis DE102 (Newhall, 1983).

The equations of motion for the more realistic case of rotating finite bodies in the extended PPN formalism of Will and Nordtvedt are presented in Dallas 1977 but not integrated.

A further sophistication of the equation of motion could be to account for the gravitational radiation of the Solar System and its effects on orbital motions.

The possible variation of the gravitational constant $\dot{G}/G$ (Canuto and Goldman, 1982) and its consequence for the motion of the planets have been derived and analysed with a large set of data at JPL (including optical observations, Radar and Laser ranges and the ultra precise radio metric data of the Viking mission on Mars) and led to the result $\dot{G}/G = (0.2 \pm 0.4)\, 10^{-11}$year (Hellings et al, 1983).

Besides these numerical integrations in a very general formalism, analytical solutions of the equations of motion have been given for a much more simple model. The Schwarzschild problem generalized to three systems of coordinates and including the paramaters γ et β can be represented by the metric (Brumberg, 1981) :

$$g_{oo} = (1-2m/r)+2\,[\beta-\alpha]\,(\frac{m}{r})^2$$

$$g_{or} = 0$$

$$g_{rr} = -1+2\,[\gamma-\alpha]\,(\frac{m}{r})+2\,[\alpha-r\,\alpha']\,(\frac{m}{r})$$

where α is a parameter selcting the systems of coordinates, and $m = GM_\odot/c^2$.

Solutions for the Moon can be found in Brumberg and Ivanova 1981 and Lestrade and Chapront-Touze 1982 ; amplitudes and periods of the relativistic perturbations in the Lunar orbital motion are given. The largest of these terms in the Earth-Moon distance is 1 meter with a period of half of the Lunar orbital period.

Using the above parametrized Schwarzschild metric, similarly, analytical expressions for the relativistic perturbations in motions of planets valid for three systems of coordinates can be found in Lestrade and Bretagnon 1982 . Tables show there are perturbations of a few kilometers with orbital period of the planet in the semi-major axis of Mercury and Mars.

In order to obtain complete solutions, these relativistic corrections can be added to high-precision analytical expressions for the Newtonian perturbations which dominate the orbital motion of any body of the Solar System. Modern theories of motion in the Newtonian framework are

VSOP82 (Bretagnon, 1982) for the Sun and all the planets, ELP2000-82 (Chapront-Touze and Chapront, 1983) FOR THE Moon, TOP82 (Simon, 1983) for Jupiter, Saturn, Uranus and Neptune.

2-3 Motion of artificial satellites and space probes.

In the same way that for the natural bodies of the Solar System, the problem is to obtain numerical information sufficient to estimate the position (and velocity) of the object (center of mars of the satellite or probe) at a given epoch and in a well defined reference system. This estimation should of course be consistent with the achievable precision of tracking measures, which is now of about up to 1-10cm for an Earth satellite (equivalent range between a ground station and the object) and 1m for space probes.

Of course the complete dynamical model has up to now various uncertainities which exceed the inner precision of individual measurements : gravity field, non gravitational forces (drag, solar pressure...) and propagation effects on electromagnetic links (optical or radio)...

Presently, the best accuracy reached in the a posteriori determination of satellite orbits can be illustrated as :

- 50 cm for a short arc of the SEASAT-1 satellite
- 1m for a long arc of the LAGEOS satellite (7 years)

A present challenge is to reach below 10 cm in order to be able to make a fruitful analysis of satellite radar altimetric data over oceans.

The relativistic approach of this problem is similar to the one discussed for planets.

Concerning the reference systems, satellite ephemerides must be expressed in a geocentric systems, while probe ephemerides must be expressed in a barycentric system.

The analytical expression of the major relativistic effects from a geocentric Schwartz-Schild metric has been derived for an Earth satellite by several anthors : Singer 1956 for the secular shift of the perigee, Rubincam 1977 for secular and periodic terms.

The main secular effect can reach 15" per year for a low satellite.

The introduction of a consistent relativistic model into numerical dynamical softwares such as GEODYN (NASA), UTOPIA (University of Texas) or GIN (CNES) is still under consideration (see e.g. Melbourne 198).

2-4 Propagation of electromagnetic signals.

The propagation of electromagnetic signals is modelled by a null geodesic, within the optical geometric limit of Maxwell equations. (Misner, Thorne, Wheeler 1973, Will 1981).

The problem is the solution of the equation of a null geodesic

$$\frac{d^2x^\alpha}{d\lambda^2} + \Gamma^\alpha_{\beta\gamma}\frac{dx^\beta}{d\lambda}\frac{dx^\gamma}{d\lambda} = 0$$

$$g_{\alpha\beta}\frac{dx^\alpha}{d\lambda}\frac{dx^\beta}{d\lambda} = 0$$

with one of the following baundary conditions

a) $x^i(t_1) = x^i_1$ $\qquad x^i(t_2) = x^i_2$

b) $x^i(t_o) = x^i_o$ $\qquad \frac{dx^i}{dt}(t_o) = \left|\frac{d\bar{x}}{dt}\right|_o n^i_o$

The problem has be widely studied for $U \simeq \frac{\mu}{|\bar{x}|}$. See Moyer 1971, Boucher 1978, Murray 1981, Fanselow 1984, Will 1981.

Two expressions are currently derived, one givin the time delay :

$$|\bar{x}_2 - \bar{x}_1| = c(t_2 - t_1) - (1+\gamma)\frac{\mu}{c^2}\,\mathrm{Log}\,\frac{|\bar{x}_2| + \bar{n}_1.\bar{x}_2}{|\bar{x}_1| + \bar{n}_1.\bar{x}_1}$$

the other are giving the bending $\delta\theta$:

$$\delta\theta = (1+\gamma)\frac{\mu}{c^2 d}\left(\frac{\bar{n}_1.\bar{x}_2}{|\bar{x}_2|} - \frac{\bar{n}_1.\bar{x}_1}{|\bar{x}_1|}\right)$$

$$\bar{d} = \bar{n}_1 \mathrm{X} (\bar{x}_1 \mathrm{X} \bar{n}_1)$$

Special expressions can be derived with $\bar{x}_1$ at the infinity.

The general model in the Solar System is currently obtained by adding the contribution of each body, especially Sun and Earth (Fanselow, 1984).

High order expressions have also been published, e.g. Enstein, Shapiro 1980.

3) Measurement techniques.

3-1 Satellite laser ranging.

The principle is the measurement of the propagation delay of a laser pulse transmitted by a ground laser, reflected by a satellite and received by a telescope colocated with the laser. Precision of a few cm is now reached. A network of a few tens of such equipments is presently operational around the world. Due to its dynamical qualities, the LAGEOS satellite is a quite convenient target for high accuracy modelling.

Relativistic effect are therefore carefully investigated for it (Martin, Torrence 1981, Martin, Torrence, Misner 1983, Ashby, Bertotti 1983).

The proper treatment of this problem needs still to be published. We can identily the following steps :

a) definition of a terrestrial frame $(\bar{x}, t)$ (see section 2-1)

b) expression of the motion of the ground station $\bar{x}(t)$.

c) expression of the motion of the satellite $\bar{x}_s(t)$ using a relativistic dynamical model (see section 2.3).

d) conversion of clock readings at tranmission τ_e and reception τ_r from proper time to coordinate times t_e and t_r.

e) determination of the reflection time t_s such that null geodesics are convecting $\bar{x}(t_e)$ to $\bar{x}_s(t_s)$ and $\bar{x}_s(t_s)$ to $\bar{x}(t_r)$ respectively (see section 2.4).

3-2 Lunar Laser Ranging and Radar Ranging in the Solar System :

Lunar Laser Ranging developped at the end of the 60's provides measurements of the Earth-Moon distance with subdecimeter precision.

Studies of the dynamics of the Earth-Moon system and of the Earth rotation are enhanced with this new data set. Pulses of photons are sent by a Laser-telescope, bounced on reflectors installed on the Moon back to the Earth bound telescope and time-delays between departures and returns of these pulses are recorded with an ultra-precise terrestrial clock.

The Lunar Laser Ranging modeling involves several relativistic corrections :

a) dynamical corrections arising from the relativistic perturbations in the orbital motion of the Moon (see paragr. 2-2).

b) photon propagation corrections arising from the speed of light between the Earth and the Moon resulting from the metric mainly determined by the relative positions of the Sun, Earth and Moon. The correction is given by Brumberg 1981 , see also section 2.4

c) Clock corrections arising from the discrepancy between proper time of a terrestrial clock and coordinate-time of the metric. A terrestrial clock is used to register the time-delays of the pulse travelling from the telescope on the Earth to the reflector on the Moon and back. The relationship between proper time and coordinate time is derived in Moyer 1981 using a metric accounting for the masses of the Sun, Earth, Moon, Jupiter and Saturn. Corrections a), b), c) are often provided by different authors and some care must be taken to check that they are all calculated from a metric expressed with the same coordinate system. Although, in General Relativity, the measurable quantities, angles and time-delays, should not be coordinate-dependent, each of these corrections, individually considered, is coordinate-dependent . This is their combination when modeling measured quantities which is independent of the system of coordinates.

At the beginning of the 60's, powerful Radar were pointed at Venus for detection of echos. This type of observation was later conducted on Mercury, Mars and Galilean Satellites. The unknown topography of these bodies limits the accuracy of such ranging to more than a kilometer ; this however was a very significant progress in Astrometry and especially for the determination of the Astronomical Unit (Shapiro, 1965).
Later, the precision of Radar Ranging was improved in the Viking Mission in 1976 by pointing at the orbiters or landers of Mars which were equipped with transponders (a device to re-emit a received radio signal). The precision in Ranging to Mars became 6 meters.

Radar Ranging are similar to Laser Lunar Ranging as far as relativistic corrections are concerned. The three types of corrections mentioned above exit and are treated the same way.

The earliest result in testing the validity of General Relativity with Radar ranges to Mercury showed that the relativistic advance of its perihelion predicted by Einstein was accurate and yielded the values unity for the two PPN-formalism parameters γ and β Shapiro et al. (1972) with 10 per cent and 20 per cent error bars, respectively. Later, the precise Viking ranges set a very strong limit on the parameter γ found to be unity within 0.1 per cent error bar(Reasenberg et al. 1979).

3-3 Satellite radio tracking.

Satellite radio tracking systems are widely used for navigation, positioning and time transfer. Several techniques are used : radar, Doppler, ranging, interferometry.

Each technique involve the propagation of the phase of the radio carrier or the modulation (phase or group propagation). Two operational systems are particularly used : Transit (5 satellites) and Global Positioning System-GPS (18 satellites when fully operational).

Relativitic model is to be developped according to a technique similar to section 3-1.

The one way Doppler effect (satellite to Earth) has been studied by several authors, e.g. Boucher 1978, Harkins 1979, Jenkins 1969, Gaposchkin Wright 1968, and the effect on positioning accuracy due to neglected relativistic Doppler correction is presented e.g. in Malyevac Tanenbaum 1981.

Relativistic effects an time transfer via radio links are also widely investigated (e.g. Ashby Allan 1979, CCIR 1978).

3-4 Radio Interferometry : Very long Baseline Interferometry :

The VLBI (Very Long Baseline Interferometry) technique consists in observing the same celestial radio-source with at least two radio-telescopes a few thousands kilometers apart to measure time-delays taken for a radio wave front to propagate between these two telescopes. This technique is rapidly becoming a powerful tool in Astrometry and Geodesy. The present accuracy is approaching the milliarcsecond level for positions of celestial radio-sources and the contimeter level for locations of radiotelescopes on intercontinental baselines.

Several relativistic effects have to be accounted for :

a) the differential retardation of the radio wave front when crossing the Solar System along the slightly different lines of sight of the two radiotelescopes. This can amount to almost one nanosecond which is large by VLBI standard. Practically, in the physical model used to analyse VLBI data, the expression given in section 2.4 is differentiated for a source at infinity.

b) the orbital motion of the Earth induced what is classically called aberration. The current approach in the existing physical model is the framework of Special Relativity (Fanselow, 1984). However, the increasing accuracy of the recent VLBI observations makes such transformation a questionable approximation (Fujmoto et al., 1982).
The Earth can be taken as a body "falling" in the gravitational field of the Solar System and the radio wave front from the celestial radio-source taken as propagating in a coordinate system at rest with respect to the Universe. Then, a general coordinate transformation is necessary to transform the time-delay which is traditionally computed first in Solar System barycentric coordinates and then transformed into geocentric, even topocentric, coordinates in which time measurements are actually conducted.

The principle of arbitrary choice of the coordinate system in General Relativity should enhance elaborate studies to coherently choose the most practical one in the future.

3-5 Ground and Space Astrometry

The goal of astrometric observations of stars and planets is the construction of a practical celestial reference system in order to carry out stellar and planetary dynamics studies. Although this may seem contradictory, the method traditionally used up to now is to select reference stars and the motion of the Sun to establish a fondamental celestial reference system with respect to which, then, a larger class of these objects is studied.
The precision of these traditional astrometric observations witn meridians is limited at a few hundedths arcsecond level.

But, recently, new techniques are already or potentially reaching the millarcsecond level. Very Long Baseline Interferometry (VLBI) provides positions of 130 extragalactic radio sources with a few milliarcsecond precision (Fanselow et al., 1984), and the future ESA astrometric satellite, HIPPARCOS, should yield positions, proper motions and trigonometric parallaxes of 100 000 stars with the same precision (Kovalevsky, 1980). These two fondamental celestial reference systems, extragalactic and stellars, should be tied in order that the most practical system, the HIPPARCOS one (which includes many more objects than the VLBI one), be rendered as stable, in the inertial sense, as is the VLBI system constructed with much fewer sources but that are at cosmological distances. This tie could be made with radio-stars which can be observed by both techniques, VLBI and HIPPARCOS (Lestrade et al., 1984).

The relevant relativistic corrections are :

a) The deflexion of the light the gravitational field of the Solar System, and mainly by the Sun. The linear dependence of this deflexion on the distance between the center of the Sun and the point of closest approach of the light ray makes the amount of deflexion still 5 milliarcseconds for a direction at 90 degrees from the Sun.

b) The metric that is used to depict extragalactic radio source positions (VLBI reference system). The Robertson-Walker metric is dominated by the mass and energy at the scale of the Universe. The metric used to depict star positions (HIPPARCOS reference system) is dominated by the mass and energy at the scale of the Galaxy. A third fondamental celestial reference frame that is currently constructed by astronomers is the planetary reference system which basically is the coordinate system used to write the differential equations of motion of the Solar System bodies and to fit their solution to modern Radar and Laser observations. The metric used in this latter case is dominated by the mass and energy at the scale of the Solar System.
Identification of the coordinates between these three celestial reference systems which are indepedently constructed is a major problem for the astrometrists. Which coordinate transformation must be used to transform from one of these systems to another requires elaborate studies for these coming years.

3-6 Inertial technology and gravimetry

The definition and evolution of spin for rotating bodies such as gyroscopes has been studied in several publications (Misner, Thorne and Wheeler 1973, Will 1981, Boucher 1981, Eby 1979, Barker and O'Connel 1979, Teyssandier 1972...).

The spin $\bar{S}$ is precessied by $\bar{\Omega}$:

$$\frac{d\bar{S}}{d\tau} = \bar{\Omega} \times \bar{S}$$

where $\bar{\Omega} = \bar{\Omega}_T + \bar{\Omega}_G + \bar{\Omega}_{LT} + \bar{\Omega}_{PF}$ accumulates four effects :

- Thomas precession

$$\bar{\Omega}_T = -\frac{1}{2}\,\bar{v} \times \bar{a}$$

where $\bar{v}$ is the velocity and $\bar{a}$ the spatial part of the 4 - acceleration of the center of mass of the spinning body.

- geodetic precession

$$\bar{\Omega}_G = (\gamma + \frac{1}{2})\,\bar{v} \times \bar{\nabla} U$$

where U is the gravitationnal potential.

- Lense - Thirring precession due to spinning attracting bodies
- preferred frame precession

$$\bar{\Omega}_{PF} = -\frac{1}{2}\alpha_1 \; \bar{w} \times \bar{\nabla} U$$

where $\bar{w}$ is the velocity of the (PPN) system (see sect. 2.1)

Such a situation can be applied to various cases of interest for geodesy or astrometry :
- attitude of an Earth satellite
- spaceborne gyroscope
- fixed gyroscope an the Earth (gyrotheodolite)
- moving gyroscope or the surface of the Earth (inertial surveying and navigation).

Altnrough theese effects are presently below the level of current instrumental errors in the case of ground measurements, they are considered in the case of high precision space experiments (HIPPARCOS, Everitt experiment...)

In the cas of gravity measurements, the variations of the locally measured gravitational constant G_L give variations of the observed gravity. Theese effects, nul for General Relativity, are discussed in Will 1981. See also Warburton and Goodkind 1976 , who derive the following results on PPN parameters from supraconducting gravimeters :

$$|\alpha_2| < 4\text{x}10^{-4} \qquad |\xi| < 10^{-3}$$

4 - Conclusion

We hope that this review has shown the works already achieved but also the investigations which have to be done, in a closer cooperation between theoreticians, geodesists and astronomes. It is sur that new effects are to be added soon at the survey done in this paper.

Références

ASHBY (N.), ALLAN (D.W.) - (1979) Radio Science 14.4,649-669

ASHBY (N.) - (1980) Relativistic Kepler problem and construction of a local inertial frame. Univ. of Colorado. Dept of Physics Boulder, CO.

ASHBY (N.), BERTOTTI (B.) - (1983) Relativistic perturbations of an Eartn Satellite. Preprint.

BARKER (B.M.), O'CONNEL (R.F.) - (1979) Gen. Rela. Grav. 11 149-175.

BOUCHER (C.) - (1978 Relativistic correction to Satellite Doppler Observation. Proceed. IAG-SSG 4.45 meeting 'The Mathematical Structure of the Gravity Field' Lagonissi, Greece.

BOUCHER (C.) - (1981) Relativistic effects in inertial technology for geodesy. Proceedings 2nd International Symposium on Inertial technology for Surveying and Geodesy. Banff, Canada.

BRETAGNON, P., 1982, Astron. Astroph., 114, 278.

BRUMBERG, V.A., 1981, in Reference Coordinate Systems for Earth Dynamics, Ed. Gaposchkin and B. Kolaczek, Reidel, 283.

BRUMBERG, V.A., and Ivanova, T.V., 1981, Proceedings of VIERMA colloquium in Dubrovnik.

CANUTO, V.M., Goldman, I., 1982, Astroph. and Space Scien., 86, 225.

CCIR - (1978) Relativistic effects in a terrestrial coordinate time system. Report 439.3 14th Plein. Meeting CCIR Kyoto. Vol. VII.

CHAPRONT-TOUZE,M., Chapront J., 1983, Astron. Astroph., 124, 50.

DALLAS, S.S, 1977, Cel. Mech., 15, 111.

EBY (P.) - (1979) Gen. Rel. Grav. 11 111-117.

EPSTEIN (R.), SHAPIRO (I.I.) - (1980) Phys. Rev. D 22.12 2947-2949.

FANSELOW, J.L., 1984, Observation Model and Parameter Partials for the JPL VLBI Parameter Estimation Software MASTERFIT, JPL publication 83-39.

FANSELOW, J.L., Sovers, O.J., Thomas, J.B., Purcell, G.H., Cohen, E.J., Rogstad, D.H., Skjerve, L.J., Spitzmesser, D.J., 1984, Radio Interferometric Determination of Source Positions Utilizing the Deep Space Network Antennas - 1971 to 1980, Astron. J., in press.

Fujmoto M.K., Aoki, S., Nakajima, K, Fukushima, T., Matsuzaka, S., 1982, in Proceedings of Symposium 5 : Geodetic Application of Radiointerferometry, Tokyo, May 7-8, 1982, p. 26.

GASPOSCHKIN (E.M.), WRIGHT (J.P.) - (1968) SAO Spec. Rep. 283.

GUINOT (B.) - (1980) Temps atomique et Relativité. Séminaire de Mécanique Céleste. PARIS.

HAMILTON (J.D.) - (1981) Can.J.Phys. 59 213 224.

HARKINS (M.D.) - (1979) The relativistic Doppler shift in satellite tracking. Radio Science 14-4 pp. 671-675.

HELLINGS et al., 1983, Ap. J.,

JENKINS (1969) Astron. J. 74.7 960-963

KOVALEVSKY, J., 1980, Cel. Mech., 22, 153.

LESTRADE, J-F, Chapront-Touze , M., 1982, Astr. Astroph., 116, 75.

LESTRADE, J-F, Bretagnon, P., 1982, Astr. Astroph., 105, 42.

LESTRADE, J-F, Preston R.A., Niell, A.E., Mutel, R.L., Phillips, R.B., 1984, IAU Symposium 109, Astrometrie Techniques, Gainesville, Florida, Jan. 9-12, 84, Reidel (in press).

MALYEVAC (C.A.), TANENBAUM (M.) - (1981) Relativistic effects on Doppler point positioning results. AGU Spring Meeting Baltimore.

MARTIN (C.F.), TORRENCE (M.H.) - (1981) Effects of General Relativity on LAGEOS semi-major axis. AGU Spring Meeting, Baltimore.

MARTIN (C.F.), TORRENCE (M.H.) MISNER (C.W.) - (1983) Progress on General Relativistic effects affecting LAGEOS data réduction.

MELBOURNE (1983) Project MERIT Standards. USNO Circ. 167.

MISNER (C.W.), THORNE (K.S.), WHEELER (J.) - (1973) Gravitation. Freeman San Francisco.

MORITZ (H.) - (1980) Relativistic effects in Reference Frames. IAU coll. 'Reference coordinate systems for Dynamics' Warsaw.

MOYER (T.D.) - (1971) JPL Tech. Rep. 32-1527.

MOYER (T.D.) - (1981) Cel. Mech., 23, 57.

MURRAY (C.A.) - (1981) Mon.Not. R.Astr. Soc. 195,639-648.

NEWHALL, XX, Standish E.M., Williams, J.G., 1983, Astron. Astroph., 125, 150.

NORDTVEDT, K, 1968, Phys. Rev., 170, 1186.

REASENBERG, R.D., et al., 1979, Ap. J., 234, L219.

RUBINCAM (D.P.) - (1977) Celestial Mech. 15 21-33.

SHAPIRO, I.I., 1964, Phys. Rev. Let., 13,789.

SHAPIRO, I.I., 1965, in Proceedings IAU Symposium 21, Paris, Bulletin Astronomique, 25, 177.

SHAPIRO, I.I., Pettengill, G.H., Ash, M.E., Ingalls, R.P., Campbell, D.B., Dyce, R.B., 1972, Phys. Rev. Let., 28, 24, p. 1594.

SIMON, J.L., 1983, Astron. Astroph., 120, 197.

SINGER - (1956) Phys. Rev. 104.1 11-14

SOFFEL (M.), RUDER (H.), SCHNEIDER (M.) - (1983) Veroff. Bayer. Kommis. fur die Intern. Vermess. Astron. Geod. Arbeiten 43,173-187.

TEYSSANDIER (P.) - (1972) Let. Nuovo Cimento 5-16 1038-1043

TOURRENC (P.) - (1981) J. de Phys.42. 12 suppl. 441-449.

TOURRENC (P.), TEYSSANDIER (P.) - (1984) La Recherche.

WARBURTON (R.J.), GOODKIND (J.M.) - (1976) Ap. J. 208 881-886.

WILL, C.M., 1974, in "Experimental Gravitation", ed. by Bertotti, Academic Press, New-York and London.

WILL (C.M.) - (1981) Theory and Experiment in Gravitational Physics. Cambridge Univ. Press.

RELATIVISTIC EFFECTS IN HEAVY IONS

J.P. Briand
Université Pierre et Marie Curie and Institut Curie
11, rue Pierre et Marie Curie
75231 Paris Cedex 05

The atomic structure theory has been mainly dealing, up to now, with very light atoms or valence electrons in heavy atoms, i.e. in cases where the relativistic corrections only play a minor role. The relativistic corrections, which are known to scale, in atoms, as the fourth power of the atomic number Z, can then be mainly studied in the inner shells of heavy atoms, i.e. in cases where a very large number of electrons are simultaneously present. A pertinent study of these effects, getting rid of the expected and uncontrolled cancellations of the various terms in a many-electron atom, should only be carried out in heavy one- or two-electron systems, which are obviously some very unstable and unusual objects. In the last decade, heavy atoms having only a very small number of electrons (one or two) have been produced and studied. In the next few years one can expect a great improvement in the production of very heavy elements like Uranium or super heavy ones (Z=184) as well as in the precision of the measurements. This opportunity opens up a new area in atomic physics for the study of relativistic effects or quantum electrodynamics corrections. We shall then briefly discuss :

. the present experimental situation in the study of few electron-heavy ions and the predictible improvements in the next few years,

. the study of relativistic and QED corrections in hydrogenlike heavy ions (the Dirac equation),

. and the present approaches in the relativistic many-body problems in atoms (mainly heliumlike ions).

I. Production and study of few electron-heavy ions

Heavy highly stripped ions are mainly produced in hot plasmas.

. Solar corona plasmas

. Terrestrial plasmas (Tokamaks or laser produced plasmas).

The solar corona is made of helium and metallic ions up to Z=26 (Iron) which is the end of the nucleosynthesis chain. The contents in Iron atoms in the corona is very large and of the order of parts per thousand. During solar flares the temperature of the corona, in special locations, can reach few keV (i.e. 10, 20, 30 millions of degrees). At such temperatures the Iron ions can be fully stripped. The heaviest highly stripped ions which can be observed in the solar corona are then the hydrogenlike and heliumlike Iron ions. These ions are yet very difficult to observe because x rays are absorbed by the earth atmosphere and the spectrometers must be embarked in orbital containers. Another

difficulty is that the spectrometer must be pointed on a certain active area and one has to wait for flares.
A satellite named SSM (Solar Maximum Mission) was launched in 1980 to study these ions. Unfortunately it began to spin and very soon stopped giving information. After being fixed the satellite is now operating and provides new data. Similar ions can also be produced either in laboratories using hot plasmas produced in Tokamaks or in solid targets irradiated by powerful lasers. The highest temperatures and densities in both cases are similar to those observed in the solar corona, and the same kind of ions (Iron contaminants...) are produced. These sources of x rays have been used in the past two years (Alcator C, MIT) to study ions up to hydrogenlike Argon (Z=18){1}. Another method to produce heavy highly stripped ions is the so-called Beam Foil Excitation Technique.
In this method the ions are ultimately stripped by passing through thin Carbon foils the heavy ion beams delivered by the most powerful accelerators of nuclear physics. With such a technique, the ultimate degree of ionisation of an atomic system (i.e. bare nucleus) is obtained when the velocity of the ion entering the Carbon foil is that of the orbital velocity of the (1s) electron in the projectile. Owing to the large mass of the ions compared to that of the electrons, this corresponds to huge ion energies ranging from some MeV up to few hundreds of GeV ! Until recently the most powerful accelerators could not reach energies larger than 10 MeV/amu, allowing only the production of bare ions up to the Iron element... again. Since 1982, new accelerators have been completed (GANIL in France, BEVALAC in Berkeley...) providing enough energy to produce now any kind of ions up to bare Uranium, an opportunity which will open up a new area in atomic physics.
At the same time, new types of ion sources have been developed which now produce, but at a very low velocity compared to that of the ions in the beam foil technique, few electron heavy ions up to Z=56. This technique, which allows to carry out Doppler free spectroscopy and then very precise energy measurements, will certainly be extended to the heaviest elements in the next two years.
So far, the precision of the energy measurements of the characteristic lines (e.g. Lyman α lines), whose wavelengths are in the x-ray range, was of the order of 20-50ppm. Very recently (unpublished results) a precision of the order of few ppm has been reached for hydrogenlike Argon (Z=18) and an increase of precision by at least one order of magnitude is expected in the next two years. In conclusion, we are now entering a new domain of atomic physics in which any kind of ions can be studied with a precision comparable to that obtained in optics for very light elements.

II. Study of the one-electron ions and check on the validity of the fundamental theories

The most fundamental theories in physics like the Schrödinger theory or quantum electrodynamics, have always been checked, with a great confidence, on simple atomic systems like the hydrogen atom. The relativistic or quantum electrodynamics effects in atoms can only be accurately studied in two-body systems for which exact

calculations are possible. These effects scaling very fastly with Z ($\sim Z^3$, Z^4) must then be studied in the heaviest hydrogenlike ions. The energy levels of these ions are calculated as the sum of the Dirac eigenvalues plus the QED corrections. So, in principle, only a global check of both theories seems to be possible. Fortunately, in a certain number of cases, owing to the different orders of magnitude of both effects or of their scaling laws with Z, separate checks can be carried out at least with a certain degree of precision.
In hydrogen atoms the relativistic corrections or the QED shifts are small but precisely measurable owing to the extraordinary precision which can be reached in the spectroscopic measurements in the visible or in the radio frequency range.

Let us first study the measurement of the relativistic effects in hydrogen atoms and in heavy hydrogenlike ions.
The fully covariant theory, which is believed to be perfectly exact, is given by the Dirac equation. An approximate value of the energy levels can be given using the eigenvalues of the Schrödinger equation which can be corrected for the most important relativistic corrections (first order terms), mass variation, spin-orbit interaction, and Darwin correction. The question is now: Have the energy values, predicted by the Dirac equation, been carefully compared to the experimental ones ? or, in other words, Is there any measurable difference between the "exact" value and the one predicted from a simple corrected Schrödinger eigenvalue ? In the case of the hydrogen atom, the energy difference between these two calculations, which represents the relativistic corrections whose order is greater than one, is only of $\sim$0.27ppm. The present precision in the measurement of the best known constant in physics -the Rydberg constant- being of the order of 2ppm, it is not presently possible, in the case of the hydrogen atom, to check these relativistic corrections (an improvement in the measurement of the Rydberg constant is however soon expected {2}). When considering heavy hydrogenlike ions, the situation is much better because the corrections increase with the atomic number faster (Z^4) than the energy of the level (Z^2). The heaviest element for which an absolute energy measurement of the Lyman α lines has been made (i.e. Iron), the theoretical energy difference is equal to 1.57 eV while the absolute experimental error is only equal to 0.6 eV {3}.
Another illustration of the usefulness of very heavy hydrogenlike ions to study the relativistic effects in the atoms is the measurement of the 2p level (2p3/2-2p1/2) fine structure interval. The physical origin of the 2p level splitting is the spin-orbit interaction. In the first theory of the atomic structure this splitting was calculated in the framework of the perturbation method. The idea was to compute, following the formulae of the classical magnetostatics, the interaction energy W of the magnetic moment of the electron $\vec{\mu}$ in the magnetic field $\vec{B}$ created by its rotation around the nucleus ($W=\vec{\mu}.\vec{B}$). This lead to the well known formula

$$\Delta E_{so} = \frac{1}{48}\,\alpha^4 m_o c^2 .$$

In this kind of calculation, the existence of a magnetic moment for the electron is postulated and the diamagnetic effect in the spin-orbit interaction is neglected (this is the so-called magnetostatic approximation). The Dirac equation, on the other hand, directly leads to the existence of the magnetic moment of the electron and provides, we believe, the exact interaction energy without any approximation. The difference between the first theory and the Dirac theory can then be considered as mainly equal to the diamagnetic effect in the spin-orbit interaction, i.e. to the "reaction" interaction between the magnetic moment of the electron and the magnetic field (effect of $\vec{\mu}$ on $\vec{B}$). To check the Dirac equation and the exactness of this reaction effect, which is one of the main questions in relativity, one must be able to be sensitive, in the experiments, to the energy difference between the two kinds of calculation. In the case of the hydrogen atom, again, the experimental energy separation between the 2p levels -known, as usual in optics, with a very high precision ($\sim$10 ppm) {4}- cannot conclusively discriminates between the two calculations ($\Delta E_{Dirac} - \Delta E_{Schrödinger} << 1$ ppm).
In heavy hydrogenlike ions, these effects are much larger and, even with a lower experimental precision, it has been possible very recently to be sensitive to such a difference, as shown in Table I.

	Schrödinger + magnetostatic correction	Dirac	difference δ	experimental error	Ref.
Z=26 Fe	20.69 eV	21.17 eV	0.48 eV	0.20 eV	{5}
Z=36 Kr	75.6 eV	79.55 eV	3.95 eV	1.4 eV	{5}

Table I. Comparison of the diamagnetic effects {5} in the spin-orbit splitting with the present error bars of the measurements.

These two examples show how sensitive the measurements on heavy one-electron systems can be, and how it is now possible to check some fundamental problems in relativity like these reaction effects. These experiments also allow to check quantum electrodynamics which for very heavy element play an important role. Recently it has been shown {6} that transient super heavy elements (Z=184) can be formed in U→U collisions {6} and there is no doubt that, in the next two years, new experiments in atomic physics will re-initialize discussions on some basic theoretical problems in relativity.

III. Relativistic many-body problem in atoms

The most fundamental theories in physics supposed to be well understood, the quantum many-body problem yet remains to be solved in an approximate way. Basically it is the main problem in atomic physics. For light elements, i.e. in the non relativistic case, this problem seems to have been definitively solved (this means that the precision of the calculations is now more accurate than the effect of relativity). For heavy elements where the relativistic effects are expected to be significant, there are many theoretical methods that have been developed. Unfortunately the experiments on heavy atoms comprising a great number of electrons cannot discriminate between those different methods. In this section we shall then be mainly dealing with the three-body systems, i.e. the heliumlike heavy ions which are now available for experiments, as described in section I, and can provide relevant tests of the theories.

The total energy of a heliumlike ion can be calculated neglecting the three-body relativistic effects, i.e. only adding all binary interactions, as the sum of the interaction energies (h_i) of each electron in the field of the nucleus and a term representing the repulsion energy between the two electrons

$$E = h_i + h_2 + \langle \frac{e^2}{r_{1,2}} \rangle .$$

In the non relativistic case, this last term is simply the expectation value, calculated by various methods, of the electrostatic interaction energy and the h_i energies are directly given from the Schrödinger equation.

In the relativistic case, the h_i terms can be calculated from the Dirac equation but the electron-electron interaction term must include, in addition, all the magnetic effects like spin-spin magnetic interaction, retardation,... It is the so-called Breit term.

In both cases, the h_i energies are calculated in the framework of the self-consistent field and the electron-electron interaction term through a perturbation method.

In the non relativistic case, at the first order of the perturbation theory, the overall electric field in which a given electron is moving, is considered as static and averaged in time. At the second order, by various successful methods, the instantaneous correlation effects between both electrons can be now calculated with a perfect precision.

In the relativistic case, at first order only, a similar method can be used to get the expectation value of the Breit term.

This kind of calculation, performed some time ago, had never been checked due to the lack of experimental methods to produce or study heavy heliumlike ions, and the need of extra studies on higher order terms in the evaluation of the Breit interaction never appeared. The rapid development of the experimental methods, in the past few years, has recently enhanced the interest in this field and lead theoreticians to be "on duty" towards these problems.

The calculation of the Breit term through a perturbative method and the self-consistent field model seemed to be well understood even if it had never been directly checked in heavy heliumlike ions. What seems now more difficult, yet fundamental, is to study the relativistic (or QED) instantaneous effects in the correlation between the two electrons, i.e. the time dependent effect in the relative "instantaneous" movement of the two electrons via, for instance, the magnetic spin-spin interaction or the retardation effects.

Very recently, the transition energy of some heavy heliumlike ions has been measured {7} {5} and compared to various calculations. In Fig.1 is presented, for instance, the x-ray spectrum emitted by the heaviest heliumlike ion prepared for the moment $\{(1s)(2p)\,^{3,1}P_{1,2} \rightarrow (1s)^2\,^1S_0$ transitions in heliumlike Krypton$\}$.
At the same time and at the same degree of precision as the experiments, some calculations including all the terms have been carried out by J.P. Desclaux to help understand these experimental data. These calculations provide data which do not differ far from other kinds of calculations (partial and sometimes complete ones).

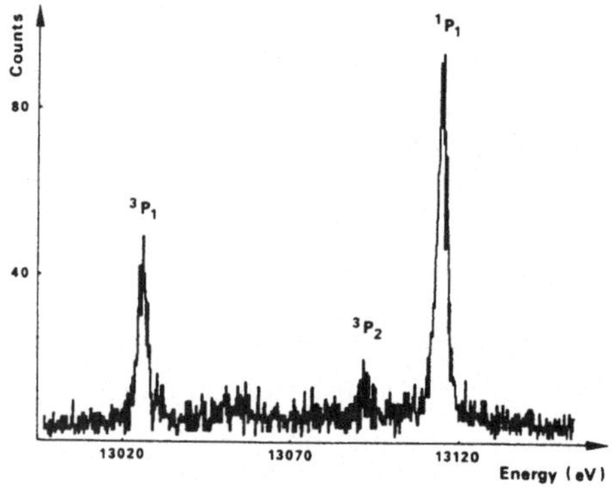

Fig. 1. X-ray spectrum of the $(1s)(2p)\ ^1P_1\ ^3P_{1,2} \rightarrow (1s)^2\ ^1S_0$ transitions in heliumlike Krypton.

In Table II are presented the various contributions to the transition energy in the heliumlike Krypton and the experimental values.

Contribution (eV)	$^1P_1\rightarrow{}^1S_0$	$^3P_1\rightarrow{}^1S_0$	$^3P_2\rightarrow{}^1S_0$
Dirac-Fock	13,139.97	13,051.88	13,116.88
Breit interaction	-15.97	-16.02	-16.46
Electrostatic correlations	1.10	1.10	1.09
Magnetic correlations	0.44	0.44	0.44
Hydrogenic Lamb shift	-11.49	-11.64	-11.49
Screening of the Lamb shift	0.87	0.82	0.80
Total	13,114.92	13,026.58	13,091.28
Experiment	13,114.32 ± 0.75	13,026.61 ± 0.75	13,091.2 ± 1.5

Table II. Various contributions to the transition energy (in eV).

The good agreement between theory and experiments, also observed in Fe and Ar heliumlike ions {7}, seems to prove that the relativistic part of the electron-electron interactions will be soon well understood at the present degree of accuracy. While it is not surprising that the static (time averaged) part of the spin-spin (mainly) interaction, which is the largest part of the Breit term, agrees with experiment, the second order terms (instantaneous correlations) seem also to be at the right order of magnitude.
Two of these second order terms whose importance is of the order of the error bar, or slightly larger, have however to be considered in prospect of a very close increase of the experimental precision : the magnetic correlation effects and the three-body QED effects.
The magnetic correlation effect, which is the relativistic counterpart of the electrostatic usual correlation effect, is expected to play a quite important role in the energy of heavy elements. (It is expected to scale like Z^3 while the electrostatic correlation effect is roughly constant over Z). This effect, supposed to be quite small in very light atoms (i.e. a very small part of the total spin-spin interaction with an order of $\sim 10^{-9}$ of the total energy) has recently been calculated by Desclaux {5}{7}. The mathematical method of calculation, similar to that used for the electrostatic correlation effect, consists in introducing in the Breit term some extra configurations of the same parity. The result, in the case of Krypton ions, is presented in Table III. This result on the magnetic correlation effect is of the order of half an eV, i.e. very close to the experimental error bars.
The second term is relevant to quantum electrodynamics. The Lamb shift of the heliumlike $(1s)^2$ ground state cannot be simply calculated as twice the (1s) Lamb shift of an hydrogenlike ion, the electric field felt by a given electron being slightly different from that of the nucleus. There is no straightforward way, for the moment, to calculate the three-body QED effect and a first attempt has been made, neglecting exchange between both electrons, to use a simple screening model. This leads, for instance in the case of Krypton, to a (1s) Lamb shift in heliumlike Iron which is 0.87 eV smaller than in the hydrogenlike, and then larger than the experimental error bar. This is, for the moment, the state of art and it is not surprising that this kind of rough approximation still agrees with experiment. There is however no doubt that these relativistic effects (or the QED effects from which, at the second order, they cannot be separated) will be re-examined as soon as the experimental precision is increased. This is what can be expected in the next two or three years which are supposed to be critical in this field.
In order to conclude and go far beyond the expected evolution, one must underline that the still unobserved three-body problem in relativity (the non additivity of binary interactions) has recently been re-investigated and seems to be accesible, in the next decade, to experiments in heavy or super heavy lithiumlike ions.

Configu-ration	Total energy*	ΔE**	Breit term	Weight⁺
$1s^2$	35,270.75	0	-16.81	0.99998
$+2s^2$	35,271.14	0.39	-16.70	0.730(-5)
$+2p^2$	35,271.76	1.01	-16.47	0.115(-6)
$+3s^2$	35,271.78	1.03	-16.45	0.164(-6)
$+3p^2$	35,271.84	1.09	-16.41	0.407(-6)
$+3d^2$	35,271.91	1.16	-16.37	0.581(-6)

* Without Breit and relative corrections

** Contribution of the correlation energy due to electrostatic interactions

\+ Coefficient squares of the configuration state function as determined by the self-consistent Dirac-Fock process. Numbers in parentheses are powers of ten.

Table III. Contribution of various configurations to the value of the Breit term (in eV).

References

(1) E. Källne, J. Källne, P. Richard, M. Stockli, J. Phys. B 17 115 (1984).

(2) T. Hansch, private communication.

(3) J.P. Briand, M. Tavernier, P. Indelicato, R. Marrus, and H. Gould, Phys. Rev. Lett. 50 832 (1983).

(4) J.C. Baird, J. Brandenberger, Gondaira Ken-Ichiro, and H. Metcalf, Phys. Rev. A 5 564 (1972).

(5) J.P. Briand, P. Indelicato, M. Tavernier, O. Gorceix, D. Liesen, H.F. Beyer, B. Liu, A. Warczak, and J.P. Desclaux, Zeit. für Phys. 318 1 (1984).

(6) See, for instance, B. Muller, Invited talk at the 9th International Conference on Atomic Physics, Seattle (1984).

(7) J.P. Briand, M. Tavernier, R. Marrus, and J.P. Desclaux, Phys. Rev. A 29 3143 (1984).

THE INTERFEROMETRIC DETECTION OF GRAVITATIONAL WAVES

A. BRILLET
Laboratoire de l'Horloge Atomique
C.N.R.S. E.R. 132
Bât. 221 - Université Paris XI
91405 Orsay Cedex, France

I. Introduction

The experimental research on the detection of gravitational waves (G.W.) started in the 1960's with J. Weber's pioneering work [1]. The first "Weber bars", consisting of aluminium cylinders whose deformations were monitored by piezoelectric transducers, have now been reproduced and improved in different places. Up to now, their evolution included cooling to less than 4°K, the use of new materials such as Al 5056, Niobium, or Sapphire in order to increase the bar mechanical Q factor, the development of ultra low noise transducers and amplifiers (parametric devices, SQIDs), the achievement of very efficient acoustic and seismic isolation filters, new developments in data analysis and a lot of other technical progress. When J. Weber, in 1969 [2], interpreted his first results as being apparently positive, this triggered some activity in the theoretician's world (astrophysicists and relativists). Since then, much effort has been made concerning the modelization of collapsing stars and the evaluation of their gravitational radiation emission, as they are the best candidates for the production of frequent and intense gravitational pulses.

Much progress is still needed, and expected, in this domain, resulting from the elaboration of realistic models including a good hydrodynamic study and the use of numerical relativity techniques, but since 1976 [3], it is already clear that none of the G.W. sources under study is able to produce frequent signals (1 per week) with an amplitude h larger than 10^{-21}. This limit is 5 orders of magnitude smaller than Weber's signals, 3 orders smaller than the sensitivity of today's cryogenic Weber bars, and about 3 times smaller than the quantum limit of a supposedly perfect bar operating around 1KHz. From this point, one had to choose between two possibilities :

- the first one was to keep improving the Weber bars down to the quantum limit, while trying to develop new ideas and new techniques in order to beat the quantum limit. This resulted in some creative rethinking about the quantum theory of measurement, with the concept of "quantum non demolition" (Q.N.D.) [4,5] which should allow one to reach sensitivities below the quantum limit, although the experimental implementations of this idea are not easy.

- the second one was to reconsider the possibility of using other techniques, particularly the interferometric one, which is more complex than the Weber bar, but has the main advantages of not suffering from this quantum limit and of being a broad-band detection technique. This was certainly a difficult decision, for those researchers who had already given much effort in building Weber bars, to start from zero again and move to a whole new technology. But the groups in Glasgow (R. Drever, J. Hough) and in Münich (H. Billing) did it, convinced that interferometers were much more promising than bars.

The pioneer of this technique was R. Forward, who built a small size, interferometer in the late 60's and improved it until 1978 [6]. He was able to obtain a photon-noise limited sensitivity, but with a low power, so that the sensitivity was lower than Weber bars and it did not attract much attention. Simultaneously, R. Weiss was starting a very detailed theoretical analysis of the ultimate sensitivity and of the noise sources of an interferometer [7]. Today, prototype interferometer are being studied in Glasgow, Münich, Caltech (R. Drever), M.I.T. (R. Weiss), and we are starting a project in Orsay. The goal of these prototypes is to reach a photon noise limited sensitivity for light powers of 100 Watt or more. This would guarantee that a large (kilometer-size) interferometer should be able to detect events in which $h \simeq 10^{-21}$. Meanwhile, projects for kilometer size antennas are already being prepared (R. Weiss, J. Hough).

In the following of this paper, which is mainly intended for people who know about G.W. but have not followed the development of the interferometric technique, we will recall the principle, optical design, and noise analysis of an interferometer and end up with some ideas about what will be done in Orsay.

II. Principle of an interferometric detector of G.W.

An interferometric detector is a two arms interferometer, with which one wants to detect the anisotropy of space resulting from the action of a G.W. The measured quantity is just the difference in the times taken by the light to make a round trip in the two arms (which are typically orthogonal). This small time difference ($\lesssim 10^{-24}$s) produces a small phase shift (10^{-10} fringe) of an interference pattern. This effect has been calculated by R. Weiss [8] for any orientation and polarization of a plane G.W. The resulting formulas are tedious so we will consider the simplest case of a plane wave propagating along Oz perpendicularly to the plane of the interferometer, with its polarization axis oriented along the interferometer arms (xOy).

Then, this weak G.W. produces a perturbation of the metric tensor which can be written:

(1) $h_{xx} = - h_{yy} = h(w)e^{i(kz - wt)}$ all other components hij being null.

Then the transfer function of the antenna is

(2) $\frac{\Delta t(w)}{h(w)} = t_s \, \mathrm{sinc}\left(\frac{w t_s}{2}\right) . e^{-i \frac{w t_s}{2}}$, where ts is the time the light wave spends in each antenna arm, and $\mathrm{sinc}\, x = \sin x / x$. If the light makes b roundrips in each arms of length L, then ts = $\frac{2bL}{c}$. If w.ts $\ll$ 1, then $\Delta t(w) = h(w) t_s$ The transfer function is maxima when w.ts = π . The corresponding phase shift is

(3) $$\Delta \phi_{Max}(w) = \frac{2\pi c}{\lambda} h(w) . \frac{\pi}{w}$$

III. Interferometer design

Some variations are possible in the design of the interferometers and different groups have made different choices. The common points to all the projects are still many :

- the length of the interferometer arms will be the largest practical one, since the sensitivity increases with the length, up to b.L = 150 km (wts = 0.5 , $w = 2 \pi \times 10^3$ Hz). Increasing the length also decreases the effect of most noise sources, as we will see later. Practically, the length of the antennas will lie between 1 and 10 km, the beam being folded 10 to 100 times in order to reach w.ts $\sim \pi$.

- the beam pathes will be evacuated to a high vacuum (10^{-6} Torr or less) in order to suppress acoustical noise and index fluctuations. Fiber optics interferometers for instance have to be excluded, although they would be very handy, because the index fluctuations of any material would be much too large.

- in order to approximate free falling masses, and to avoid seismic noise, the mirrors and beamsplitter of the interferometer cannot be rigidly connected together, nor to the earth. They will be suspended, in a pendular way, to a vibration isolated stand, and some servo system will be needed to prevent oscillations of the pendulum and to maintain the alignment of the interferometer.

The simplest implementation of a G.W. interferometer is the multipass 2 - waves interferometer, that Michelson himself was using 100 years ago. If the length difference between the 2 arms is less than half a wavelength of light, it can be used with broadband illumination, but it may be difficult to realize this condition in a kilometric system. The M.I.T. group is studying this solution, with a broadband Argon laser. The Münich group prefers to rely on a frequency stabilized Argon laser, which will allow a path difference of 1 m, and reduce the problems of scattered light. The groups in Glasgow and Caltech are studying more sophisticated interferometers in which the multipass is replaced by a Fabry-Perot resonator. The finesse of the Fabry-Perot is then the equivalent of the number of passes (2b). The advantages of the Fabry-Perot are : more versatility, a smaller transverse size of the beam which reduces the transverse size of the vacuum tubing, and, may be, a lower sensitivity to scattered light. The drawbacks are : more servo loops, since each arm must be kept resonant with the laser light ; probably some loss of contrast ; the need for a better vacuum and for a better laser frequency stabilisation. Otherwise, the ultimate sensitivity is the same for all designs, and the final choice will be a compromise between performance, simplicity and cost.

IV. Noise analysis

Let us first remember the problem precisely. The largest signals expected correspond to a pulse of gravitational radiation, with $h \sim 10^{-21}$ for 1 ms. We will rather use the linear measure of the spectral density of signal

$$(4) \qquad h(f) \sim 3.10^{-23} \ Hz^{-1/2} \qquad \text{for} \quad 0 < f < 10^3 \quad ,$$

as a goal for the interferometer sensitivity.

A) Internal noise sources

We call so all the noise sources which are present even the apparatus (light source, beamsplitter, mirrors) is perfectly symmetrical and isolated from the rest of the world.

1) Poisson noise (photon shot noise)

This is the most fundamental one, the result of the quantification of light. In a Michelson interferometer, the number N of output photons per unit time is

$$(5) \qquad N = \frac{P_0}{2h\nu} (1 - \cos \Phi) \quad ,$$

where P_0 is the light input power, h the Planck constant, ν the light frequency and Φ the phase difference between the 2 arms. A phase shift $\Delta\Phi$ produces a change in the number of photons

$$(6) \qquad N = \frac{P_0}{2h\nu} \Delta\Phi \ \sin \Phi$$

If the light source is coherent (laser), the fluctuations of the number of photons obey Poisson statistics. The mean square fluctuation is

$$(7) \qquad \delta N^2 = \frac{N}{\tau}$$

for a measurement time τ , so that a phase shift $\Delta\Phi$ is observable only if

$\Delta N^2 > \delta N^2$:

(8) $$P_0 = \frac{h\nu}{\tau\,\Delta\phi^2 \cos^2\phi/2}$$

This condition is least stringent if ϕ = 0 (dark fringe). For $h \sim 10^{-21}$, $\tau = 10^{-3}$s, eq. (3 and 7) give $P_0 > 80$ Watt. This is a serious problem, since no single frequency visible laser is able to deliver more than 10 Watt.

2) Thermal noise of the mirrors

The masses constituting the mirrors will see their internal vibration modes thermally excited. This will result in displacements of their surface along the optical axis. For a mirror of mass m, resonant frequency ω_0 (1 mode analysis), surtension coefficient Q, the Nyquist force has a spectral density

(9) $$F^2\,(f) = \frac{4KT.\omega_0}{Q}\ \mathrm{N/Hz}$$

and the response of the oscillator to this force is

(10) $$x^2\,(f) = \frac{4KT.\,w_0}{mQ\,[(w_0^2 - w^2)^2 + (\frac{w_0 w}{Q})^2]}\ \mathrm{m^2/Hz}$$

In practice one will try to make the masses so that their resonant frequency is well above the G.W. frequency band, so that $x^2\,(f) = \frac{4KT}{mQw_0^3}\,\mathrm{m^2/Hz}$, which corresponds to

a strain sensitivity $h(f) = \left[\frac{x^2(f)}{L^2}\right]^{1/2} \simeq \frac{7.10^{-21}}{L}\ \mathrm{Hz^{-1/2}}$,

for m = 100 kg , $Q = 10^5$, $w_0 / 2\pi$ = 5KHz and T = 300 K .

This is one reason why a room temperature antenna cannot be shorter than a few hundred meters. This description of thermal noise is obviously very incomplete since we considered only one vibration mode. But a more detailed analysis [8] shows that these corrections are small in a well-designed system.

3) Index fluctuations

In a cylindrical light beam of length L and diameter D, the mean number of atoms is

(11) $\langle N\rangle = \rho L\,(\pi D^2/4)$ where ρ is the average density of atoms.

The r.m.s. fluctuation of N is $\langle\delta N^2\rangle$ = N , so the fluctuation of the index of refraction, $n = 1 + \alpha\rho$, is :

(12) $$\langle\delta n^2\rangle = \frac{4\alpha^2\rho}{L\pi D^2}$$

α being the atomic polarizability. The characteristic time for this fluctuation is $\tau = D/v$ where v is the average velocity of the atoms across the beam, so that

(13) $$\langle\delta n^2(f)\rangle = \frac{4\alpha^2\rho}{L\pi D v}$$

In a multipass Michelson system, there are b beams with uncorrelated fluctuations and

$$h_M(f) \sim \left[\frac{4\alpha^2\rho}{Lb\pi DV}\right]^{1/2} \quad Hz^{-1/2} ,$$

whereas in a Fabry-Perot system $h_{P.F.}(f) = \left[\frac{4\alpha^2\rho}{L\pi DV}\right]^{1/2} = h_M(f) \times \sqrt{b}$

At a pressure of 10^{-6} Torr, for nitrogen gas, one gets

$$h_M(f) = 10^{-24}\ Hz^{-1/2} \quad \text{and} \quad h_{P.F.}(f) \sim 10^{-23}\ Hz^{-1/2}$$

in a kilometric system (b = 100) with a beam diameter of 5cm. This effect can be decreased by increasing L and D and by decreasing the pressure. It requires a high or ultrahigh vacuum, unless one considers unreasonably large beams.

B) External perturbations

a) Frequency fluctuations of the light source

They are unimportant in a perfectly symmetrical interferometer, but the slightest asymmetry restricts the tolerable fluctuations (δf). If the length of the 2 arms differs by ΔL for instance, the condition is $\frac{\delta F(f)}{F} < \frac{L}{\Delta L} \times h(f)$, so that one needs to realize $\delta F(f) < 1.5\ 10^{-3}\ Hz/Hz^{1/2}$ if ΔL = 1cm in a kilometric interferometer. This is already 6 orders of magnitude smaller than the fluctuations of a commercial laser, but only a factor of 10 better than our present stabilized laser. Other sources of asymmetry in the mirror reflectivities, or the light scattering contribute to limit δF. In a Fabry-Perot system, there is the additional condition that the total frequency fluctuation of the laser has to be much smaller than the width of the resonance of each arm (1 KHz). This is one more constraint which requires precise and ultrafast servo loops in order to reduce both the slow drifts and the fast fluctuations of the laser. To reach the necessary stability is very difficult but possible with known technologies

b) Power fluctuations of the light source

Their direct effect on the interferometric signal can be suppressed by proper design (symmetry + use of a dark fringe). Since these conditions cannot be fulfilled perfectly, we also use a standard phase modulation technique, which gives a frequency transposition of the signal at a few Megahertz, where the laser does not have excess noise.

But there are still some ways by which the power fluctuations around the G.W. frequency couple into the signal : these are radiation pressure effect, the radiometer effect, and also the thermal expansion of the mirror coatings. The noise spectral density associated to these effects is larger at low frequencies. We expect them to be troublesome only below 100 Hz.

c) Fluctuations of the beam geometry

They also couple in through asymmetries and misalignement of the apparatus. They have to be filtered out, as close as possible to the interferometer input, in order to avoid additional fluctuations of acoustical or thermal origin. This can be done either actively with moving mirrors, or passively, by filtering the laser beam through a pinhole, a resonator, or a fiber. The passive techniques are more efficient against beam geometry fluctuations, but they introduce more losses, and they may not stand very high powers (heating, nonlinear effects in the fiber ...).

d) Seismic noise

In the frequency range between 10 Hz and 1 KHz, the ground motions are mostly generated by local disturbances such as wind, breaking ocean waves and human made noise. Their amplitude is then quite variable. In a quiet site these motions can be represented by [9] :

$$(14) \qquad \delta x\,(f) \sim \frac{10^{-8}}{f^2} \quad m \times Hz^{-1/2}$$

These ground motions will be quite uncorrelated in a long baseline interferometer. At 1 KHz, for L = 1 km, the corresponding strain sensitivity would be :

$$h \sim 10^{-17}\ Hz^{-1/2}$$

if the mirrors were rigidly connected to the ground. If the mirror is a high Q pendulum of frequency f_0, then the sensitivity limitation by seismic noise is

$$h\,(f) \sim \frac{10^{-11}}{f^2} \left(\frac{f_0}{f}\right)^2 \quad Hz^{-1/2}$$

A 1 Hz pendulum should be sufficient to get $h\,(1000) \leqslant 10^{-23}\ Hz^{-1/2}$, but the difficulty increases rapidly if one wants to make an antenna sensitive at lower frequencies. In practice, it will be necessary to further isolate the suspension point of the pendulum. Furthermore such a passive system does not bring any isolation at low frequencies, so that an active servo loop is needed there to prevent large but slow displacements of the mirrors.

With today's technology, it will be possible to decrease the seismic noise level below the Poisson noise for frequencies higher than 100 Hz.

e) Thermal noise of the suspension

At frequencies well above the pendulum resonance, the thermal noise (CF(10)) takes the form

$$\delta x^2\,(f) = \frac{4KT\,W_0}{W^4\,Qm} \quad m^2/Hz\,, \qquad W_0 = 2\pi f_0\,, \qquad W = 2\pi f$$

or the equivalent strain

$$h\,(f) = \frac{1}{L(\pi f)^2} \left[\frac{KT\,W_0}{mQ}\right]^{1/2}$$

For an ideal quartz fiber pendulum of length $l = 1$ m, $W_0 = \pi$, $Q \sim 10^7$, $m = 100$ kg and $L = 1$ km, one gets

$$h\,(f) \sim \frac{3.10^{-19}}{f^2}\ \mathrm{Hz}^{-1/2} \quad \text{at room temperature.}$$

This is larger than our goal 3.10^{-23} $\mathrm{Hz}^{-1/2}$ below 100 Hz. Further improvements will include cooling and/or the use of other kinds of suspensions (electrostatic, magnetic ...).

f) Other noise sources

R. Weiss [8] has also considered the noise due to cosmic rays, and the influence of varying electric and magnetic fields. He concludes that these effects are negligible if obvious precautions are taken.

This summary of the possible noise sources shows that the sensitivity we are aiming at is very difficult to achieve, but it also shows that this is possible in the range of 100 Hz to a few Kilohertz, and it allows us to point out the main difficulties.

V. The Orsay project

Our goal is, within 5 or 6 years, to operate a prototype ($L = 5$ to 10 m) with a phase sensitivity better than 3.10^{-11} rd $\mathrm{Hz}^{-1/2}$, in order to demonstrate the feasibility of a kilometric antenna. We wish to put the emphasis on the aspects of the problem which have not yet been studied by the other groups in order to be complementary with them so as to encourage a real collaboration, mainly with the european groups. We will first study the problems of the laser stabilization, which we are already close to master, and the problem of getting enough power in the interferometer. There are 2 solutions to this problem : the first one is to add coherently the beams from a few lasers. We have shown that this technique is viable [10] and we think it could provide about 50 Watts with 4 or 5 lasers. The second one is what Drever calls "recycling the light" [11] ; it consists in placing the interferometer inside a resonator in order to multiply the power. It may be possible to gain a factor of 100 this way, but this has not been tried yet. Furthermore it has some interesting theoretical aspects, which are being studied at the Institut Henri Poincaré by P. Tourrenc and his group. Another important point we will study soon is how to symmetrize an interferometer in order to reduce the effect of many noise sources.

VI. Conclusions

We hope we have convinced the reader that large interferometers do have the potential sensitivity to detect bursts of gravitational radiation at the level where they are predicted today. Whether this will be realized within 5 years or 15 years, it is still too early to say, but there are no fundamental obstacles. The work to be done is not only a huge experimental effort ; we also need more theoretical work concerning the prediction of the amplitude and the frequency spectrum of these events, and also how these signals could be used as tests of the gravitation theories.

References

[1] J. Weber, Phys. Rev., 117, 306 (1960).
[2] J. Weber, Phys. Rev. Lett., 22, 1320 (1969).
[3] L. Smarr, Editor, Sources of Gravitational Radiation (Proceedings of the Batelle-Seattle Workshop) Cambridge, 1978).
[4] V.B. Braginsky and Yu. I. Vorontsov, Usp. Fiz. Nank., 114, 41 (1974).
[5] C. Caves and al., Rev. Mod. Phys., 52, 341 (1980).
[6] R. Forward, Phys. Rev., D17, 379 (1978).
[7] R. Weiss, Quaterly Prog. Rep. Research Lab. Electronics, M.I.T. 105 (1972).
[8] R. Weiss, N.S.F. report (grant PHY-8109581).
[9] N. Robertson, Ph. D. Thesis, University of Glasgow (1981).
[10] C. Man et A. Brillet, same issue.
[11] R. Drever, p. 321, in Gravitational Radiation, N. Deruelle et T. Piran editors, North-Holland (1983).

THE DEVELOPMENT OF LONG BASELINE GRAVITATIONAL RADIATION DETECTORS AT GLASGOW UNIVERSITY

J. Hough, S. Hoggan, G.A. Kerr, J.B. Mangan, B.J. Meers, G.P. Newton, N.A. Robertson and H. Ward

Department of Natural Philosophy, University of Glasgow, Glasgow G12 8QQ, Scotland

R.W.P. Drever

California Institute of Technology, Pasadena, California 91125, USA and Department of Natural Philosophy, University of Glasgow, Glasgow G12 8QQ, Scotland

Introduction

Our work at Glasgow is devoted to the development of techniques for the detection of gravitational radiation and to the application of these to the observation and study of signals from astronomical phenomena of various types.

Detection methods rely on sensing the change in the separation, $\Delta\ell$, of test masses separated by a distance, ℓ. As has been pointed out by A. Brillet in his review, likely sources are all characterised by the small value of the gravitational wave amplitude, $h = 2\frac{\Delta\ell}{\ell}$, at a detector on earth. Probably the most promising region of the frequency spectrum for initial ground based experiments is between a few hundred Hz and a few kHz. And recent estimates of source strength[1] suggest that while pulses of amplitude up to 10^{-16} are possible in such a frequency band, an amplitude sensitivity of approximately 10^{-21} or better should be aimed for. This could allow the detection of signals from stellar collapses and coalescing binary stars etc. at an estimated rate of one or two per month.

Perhaps the most promising method of eventually attaining the very high sensitivity required over a wideband is to use optical interferometry to sense changes in the spacing of nearly free test masses separated by relatively large distances (perhaps up to a few km)[2]. As already explained by A. Brillet, the test masses may be suspended to give two perpendicular baselines in which a gravitational wave signal will induce a differential length change. The relative length of the two arms can be monitored by a Michelson interferometer with its displacement sensitivity enhanced by the use of an optical delay line to make the light in each arm traverse the distance between the test masses many times[3]. (Experiments using such a scheme are discussed by A. Rüdiger in the preceding article). Alternatively, Fabry Perot cavities can be used in each arm of the interferometer and the relative phase of the light from the two cavities compared[2]. This cavity arrangement has similar sensitivity to the delay line scheme but has the advantage that smaller and potentially higher quality mirrors may be used.

A Prototype Optical Cavity Gravity Wave Detector

In order to develop the optical and electro mechanical techniques required for interferometric detectors with cavities in their arms, we have built and are currently developing a system at Glasgow University with an arm length of 10m.

The test masses for the detector, hung as pendulums for mechanical isolation, form 3 corners of a 10m square. They are placed in stainless steel vacuum vessels and the light paths between them are enclosed by vacuum pipes. High reflectivity dielectric mirrors are attached to the masses to form the two resonant optical cavities at right angles to each other, and the cavities are illuminated (via a polarising beam splitter) by an argon ion laser operating on a single longitudinal mode at a wavelength of 514.5 nm. After some initial experiments[4] in which 3 mirror ring cavities were used to avoid optical feedback effects we have now adopted two mirror cavities of higher finesse (presently ∿600). As is shown in fig. 1 an arrangement with a polarising beamsplitter and quarter wave plates is now used to separate input and output beams for the cavities, and to isolate the cavities from each other and the laser. Further optical isolation from the laser is provided by Faraday isolators placed in the input beam from the laser.

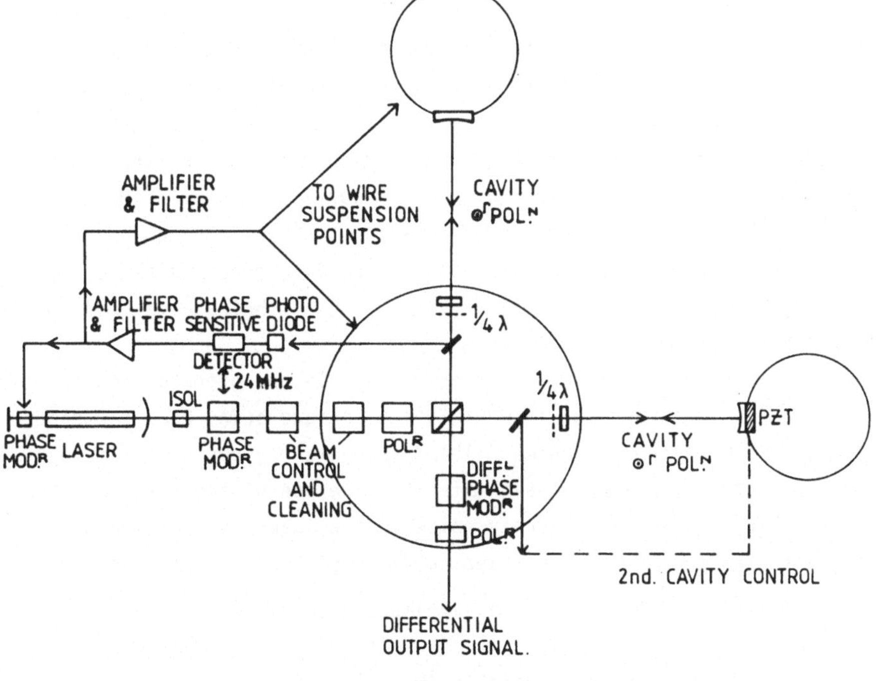

Fig. 1: Schematic diagram of prototype 10m gravitational radiation detector

10m long cavities of high finesse (600) have a bandwidth (25 kHz) considerably less than the fluctuations of the laser frequency and so a high degree of stabilisation of the laser frequency with respect to the cavity resonances is required. In most of our work to date we have stabilised the laser to one of the 10m cavities and then stabilised the second cavity to the laser light by means of a piezoelectrically driven mirror in the cavity. The feedback signal to this transducer gives a measure of the differential changes in length of the two baselines which could be induced by a gravitational wave.

The stabilisation schemes which are partly shown in schematic form in fig. 1 use our reflection r.f. sideband technique fully described in earlier publications [4,5]. Control of orientation of the cavity mirrors to an angular stability of better than 10^{-5} rad is achieved by applying signals from auxiliary optical systems using helium neon lasers to transducers which can move the supporting points of the wires suspending the test masses.

In our system provision has been made for reducing fluctuations in the geometry - position, direction, diameter and convergence - of the input laser beam. Beam direction and position relative to the optical system on the central mass are monitored by a pair of quadrant photodiodes mounted on the mass itself. The outputs from these position sensitive detectors control the orientation of four mirrors in the main laser beam path, by means of fast piezoelectric transducers and slower but wide-range moving coil elements, so that the final beam is kept stable in position and direction[6]. This system reduces fluctuations in the beam position and direction by a significant factor (Fig. 2). It has also proved very valuable in maintaining the long term alignment of the complete optical system and simplifying adjustments of optical components such as electro-optic modulators and mode matching lenses. Further reduction in beam direction and position fluctuations, and reduction in beam diameter and convergence fluctuations, may be achieved by passing the beam through an auxiliary optical cavity mounted on the central mass and arranged to suppress all but one resonant mode[7]. At present this auxiliary cavity is not being used while other noise sources are being investigated.

Provision is made for combining interferometrically the outputs from the two 10m cavities, a radio frequency phase modulation technique being used to measure the phase difference between the two beams. This arrangement gives better fringe visibility and potentially better sensitivity than the use of separate detectors for each cavity and allows residual frequency noise in the laser to be balanced out. Initial tests with the combined output have proved encouraging, some reduction in the effects of laser frequency noise artificially imposed on the system having been observed.

Considerable attention has been paid to improving some aspects of the mechanical design of the system. The end test masses in the system are made of aluminium of high quality factor; they are almost spherical in shape to keep their resonant frequencies high ($\sim$ 20kHz) and so thermal noise from their internal modes should not contribute significantly to the noise level of the system at present. However, the middle test mass is necessarily of more complicated structure as it has to carry some optical modulators and beam steering optics as well as the beam cleaning cavity; this will be discussed further in the next section.

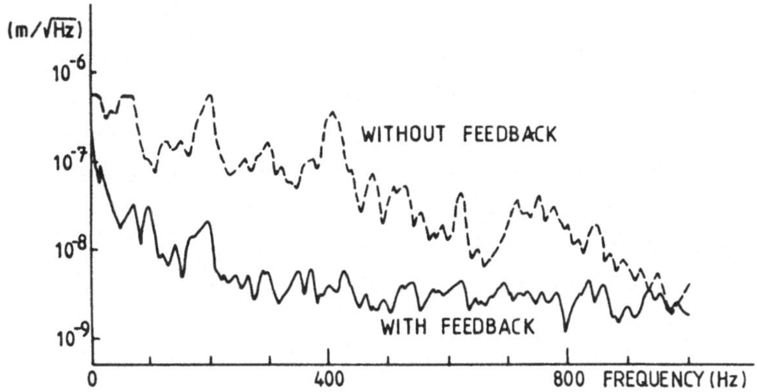

Fig. 2: Spectrum of positional fluctuations of the laser beam at a distance from the laser where the beam radius is 1 mm with and without the beam position stabilisation in operation.

Good mechanical isolation of the test masses from external vibration in the system has been achieved by the introduction of mechanical filters of lead and rubber between the ground and the suspension points of the masses, and by the removal of mechanically resonant components in the supporting structures.

Current Performance of our 10m Prototype System

The overall performance of our present system has been significantly improved over that obtained earlier[4)]. A spectrum of the feedback signal to the second cavity calibrated in terms of the spectral density of amplitude detectable is shown in Fig. 3.

At low frequency ($\sim$ 200Hz) the sensitivity of the system is 2 orders of magnitude better than we have observed before. In this region this has been brought about by better mechanical isolation, and reduction of electronic noise and mechanical resonances in the electro-mechanical systems controlling the suspension points of the test masses. Over a frequency band around 1800Hz the amplitude sensitivity approaches $4 \times 10^{-17}/\sqrt{Hz}$, (nearly an order of magnitude better than previously) and is within a factor of approximately 4 of the limit set by the photon noise in the relatively small amount of light used to control the laser and cavity stabilisation. (Note that most of the light in the detector is intended to be used in the beam recombination part of the system, and the photon noise in this output will be much less significant). However it does not quite reach photon noise in this frequency band. Evidence obtained by artificially exciting the mechanical resonances of the rather complicated central mass suggests that much of the noise in the spectrum may be due to excitation of these resonances. Fig. 4 shows a superposition of two system sensitivity spectra, with and without artificial excitation of the centre mass; we feel that the similar structure in both, especially between 1.5 kHz and 2 kHz is significant. When looked at over a wide frequency range the system does reach the photon noise limit at 17 kHz and above; again it appears that the excess noise below 17 kHz is mainly due to mechanical resonances.

As was mentioned in the last section, the central mass of the detector is of complicated construction, and it is not surprising that it has resonances within the

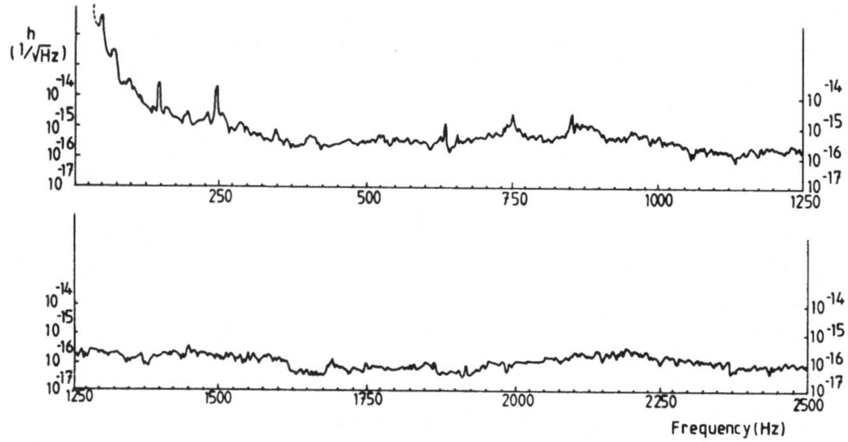

Fig. 3: Spectrum of noise in 10m interferometer, measured at piezoelectrically driven mirror in second cavity. Calibration is in terms of gravitational wave amplitude.

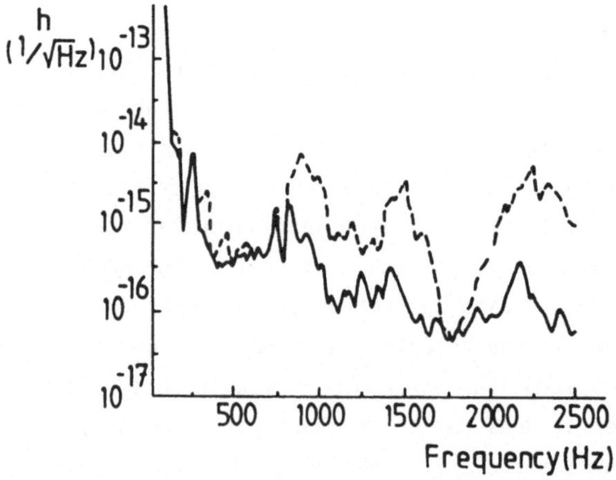

Fig. 4: Spectrum of noise in 10m interferometer. Solid curve, as Fig. 3, dotted curve with white noise excitation of central mass

frequency band of interest whose effects are important. While it is difficult to simplify this mass, a way round the problem is to move each of the two central cavity mirrors on to its own suspended test mass of good quality factor and high internal resonant frequency. Further servo systems are used to stabilise the relative positions of thse masses and the central mass and to control the orientations of the new masses. The possibility of using such an arrangement was envisaged from an early stage in the development of the detector, but it was felt to be more convenient to start operation of the system in the simplest way.

Currently, the new central test masses and servo systems are being added to our detector, and we expect to have further results in the near future†.

Further Related Work

In order to ease the dynamic range required of the laser stabilising system it was decided to rebuild one of our lasers (Coherent CR18 fitted with an Innova 2000 tube) with a separate silica rod resonator, isolated from vibrations of the laser cooling water. This resonator, similar in design to one built by A. Brillet, consists of three 2" diameter rods held in place by aluminium plates such that the ends of the rods form the tops and vertex of a V shape. These rods support end plates carrying the laser mirrors. Initial measurements of the frequency fluctuations of the laser operating single mode suggest that a reasonable improvement in the laser performance may be achieved in this way as is shown in Fig. 5.

Future Plans

Firstly we intend to continue development of our present prototype (including the installation of the new low loss laser gyro quality mirrors becoming available) to allow a search for pulsed gravitational waves to be carried out in conjunction with the low temperature bar detector at Stanford University and the laser interferometer detectors being developed at Caltech, Munich, and MIT. Amplitude sensitivities of around 10^{-18} are envisaged in such a search. Although the chances of finding a reasonable rate of signals at this level are not large, it is important that such new experiments (of higher sensitivity than any others previously done) are carried out.

Secondly, we are considering the possibility of building a larger detector, of arm length approximately 1 km, which could utilise the principles of light recycling suggested in Ref. 2 and discussed by A. Brillet in his review. Light recycling requires the use of very low loss laser gyro quality mirrors (losses less than 1 part in 10^4). Our ideas for this detector are currently evolving. We envisage the constru-

† Initial tests indicate that the new masses have made a considerable difference to the sensitivity spectrum which is now essentially flat at a level of better than $1.5 \times 10^{-17}/\sqrt{\text{Hz}}$ from approximately 500 Hz to 2.5 kHz. This is limited by the photon noise in the small amount of light used to control the laser and cavities.

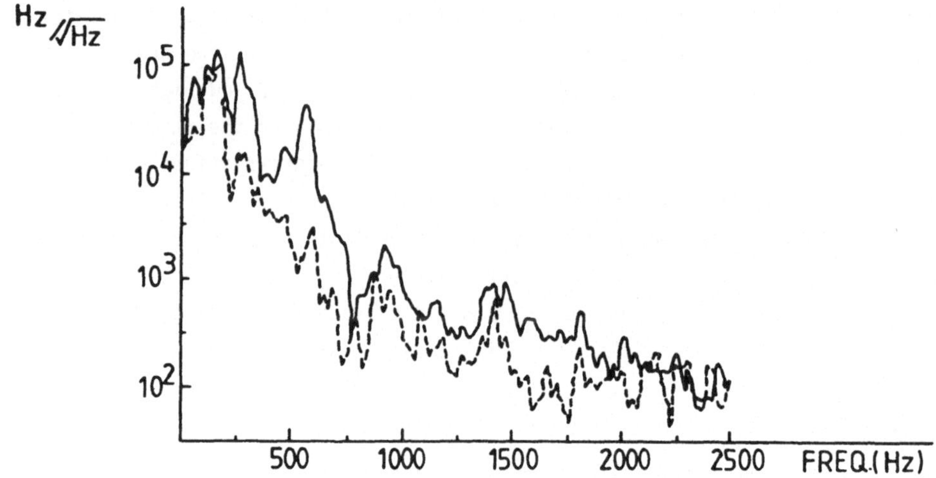

Fig. 5: Frequency noise of laser. Solid curve - standard laser. Dotted curve - mirrors mounted on separate resonator

ction of only one vacuum system. The Fabry Perot based optical systems, however, use near minimum optical beam diameters and thus a number of different interferometers may be operated within the same vacuum enclosure. This will allow some discrimination against local seismic disturbances and outbursts of gas from the vacuum pipes etc. With an arm length of 1 km and fairly simple suspension of the test masses we believe that the main limitation to sensitivity for short pulse detection should be photon noise in the detected light from the interferometer. Fig. 6 illustrates the limitation to sensitivity imposed by photon noise in such a system, illuminated by 20W of laser light, for the detection of pulses of length 10 ms to 1 ms. Two cases are shown - the limit for the situation where the storage time of the light in the cavities matches the timescale of the gravitational wave pulse and the limit when optical recycling is utilised with mirrors of a fractional loss of 10^{-4}. These sensitivity limits are superimposed on a series of curves which show the theoretically predicted range of signal strengths which might be expected for pulses arriving at the earth at a rate of 1/month. It can be seen that the possibilities for pulse detection are most encouraging.

As was outlined in ref. 2 there is a form of resonant recycling of light between the two arms of an interferometer which considerably reduces the limit set by photon noise to the detection of periodic sources of gravitational radiation. With an arm length of 1 km and low loss mirrors (10^{-4}) this could allow an amplitude sensitivity of 10^{-26} to be attained in a search with an integration time of 1 month. This is an interesting level of performance both for searches for the signals from known pulsars such as that in the Crab Nebula (estimated upper limit to signal $\sim 10^{-25}$)

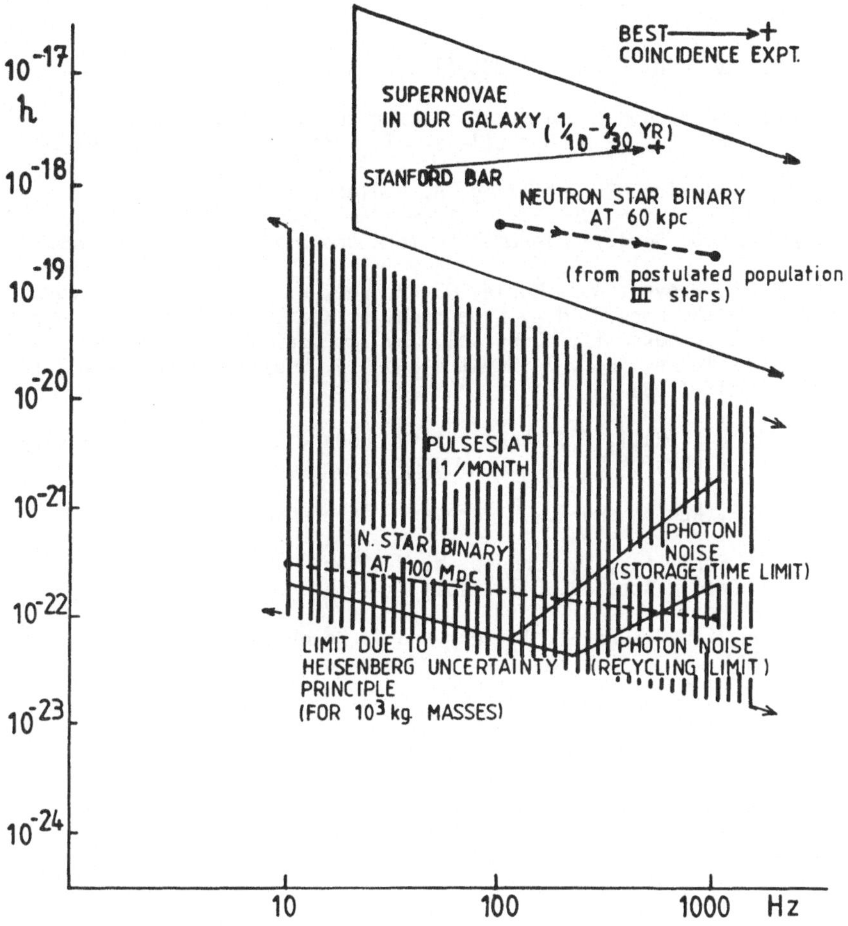

Fig. 6: Possible photon noise limited sensitivity (rms) of a 1 km interferometer illuminated by 20W of laser light. Signal strengths from some possible sources are also shown.

and for signals from unknown pulsars. This work, especially at frequencies below 100 Hz, might entail the use of active systems for isolating the suspensions of the test masses from ground noise [8].

Conclusion

The performance of the present interferometers using resonant cavities is continually improving and prospects for the detection of predicted sources of radiation with interferometers of longer baselines seem most encouraging.

References

1) K.S. Thorne, Rev. Mod. Phys. 52, 285 (1980)

2) R.W.P. Drever, *Gravitational Radiation,* Proc. of NATO Advanced Study Institute, Les Houches, 1982 (Ed. by N. Deruelle and T. Piran (North Holland))p.321

3) H. Billing, W. Winkler, R. Schilling, A. Rüdiger, K. Maischberger and L. Schnupp , *Quantum Optics, Experimental Gravity, and Measurement Theory* (Eds. P. Meystre and M.O. Scully) Plenum (New York and London) p.525, (1983)

4) J. Hough, R.W.P. Drever, A.J. Munley, S.-A. Lee, R. Spero, S.E. Whitcomb, H. Ward, G.M. Ford, M. Hereld, N.A. Robertson, I. Kerr, J.R. Pugh, G.P. Newton, B.J. Meers, E.D. Brooks III, Y. Gursel, *Quantum Optics, Experimental Gravity, and Measurement Theory* (Eds. P. Meystre and M.O. Scully) Plenum (New York and London) p.515,(1983)

5) R.W.P. Drever, J.L. Hall, F.V. Kowalski, J. Hough, G.M. Ford, A.J. Munley and H. Ward, Applied Physics B, 31, 97 (1983)

6) B.J. Meers, G.P. Newton and R.W.P. Drever - in preparation

7) A. Rüdiger, R. Schilling, L. Schnupp, W. Winkler, H. Billing and K. Maischberger, Optica Acta 28, 641 (1981)

8) N.A. Robertson, R.W.P. Drever, I. Kerr and J. Hough, J. Phys. E, Sci. Instrum., 15, 1101,(1982)

IMPROVED SENSITIVITIES IN LASER INTERFEROMETERS FOR THE DETECTION OF GRAVITATIONAL WAVES[1]

Roland Schilling, Lise Schnupp, David H. Shoemaker[2], Walter Winkler, Karl Maischberger, Albrecht Rüdiger

Max-Planck-Institut für Quantenoptik
D-8046 Garching bei München, Federal Republic of Germany

1. Introduction

The paper presented by Alain Brillet has given a thorough overview of gravitational wave detection using laser interferometry. He has discussed the principles of operation, some possible realizations, the major noise sources, and an estimate of possible future sensitivities.

What remains for me to do is to fill in some selected details of our experiments at Garching, and to show what best sensitivities have so far been reached. The emphasis will be mainly on the changes in the mechanical arrangement that led to encouraging reductions in the noise measured. It will become clear that despite the great efforts going into such experiments, the present sensitivities are still (orders of magnitude) short of the eventual goal.

This is not to be misunderstood as a pessimistic view. On the contrary, it has become clear from A. Brillet's talk that there is no reason to assume that the final goal of measuring Virgo cluster events cannot eventually be reached. But on the road to this goal a large number of obstacles have to be cleared away, and many problems, some of them not even known yet, have to be solved. This, I think, is a great challenge to inspired and devoted experimentalists.

2. The Garching Prototypes

The two prototypes currently in operation at Garching are Michelson interferometers (see Figure 1) in which the laser beam is first split into two perpendicular beams and then re-combined at the beam splitter. The interferometer is operated near the light minimum at one of the output ports, by keeping (via a Pockels cell) the light

---

[1] Presented by A. Rüdiger at Journées Relativistes, Aussois, May 1984

[2] Visiting Scientist from Massachusetts Institute of Technology

path difference constant to within much better than a wavelength. The voltage V_P required at the Pockels cell can be utilized as the output signal, which in an eventual antenna would be analyzed for possible gravitational waves.

3. The Delay Line

The interferometer is to be sensitive in a relatively wide frequency band, say from 0.5 to 2.5 kHz, which we will henceforth call our "frequency window". The sensitivity for gravity-wave-induced strains h increases with the total light path L, but only until L reaches half a wavelength of the gravitational wave. For a center frequency of 1.5 kHz, the optimum path length would thus be L = 100 km.

At Garching, we attempt to achieve the long optical path with the aid of so-called optical delay lines. Two spherical mirrors are placed at approximately the confocal spacing. A beam entering through an input hole in the near mirror will execute an even number of traversals (N "passes") between the two mirrors before it goes through the same hole again on its way back out (indicated for N = 4 passes in Figure 1). The total optical path L is simply N times the mirror distance.

Initial work at Munich, and later at Garching [1,2], had been done with a 3 m interferometer (arm length ℓ = 3.05 m), currently operated with N = 138 passes, for a total light path of L = 420 m. This is still more than two powers of ten below the desired length (of about 100 km). Increasing N would not provide better discrimination against seismic, acoustic, or thermal motions of the mirrors. On the other hand,

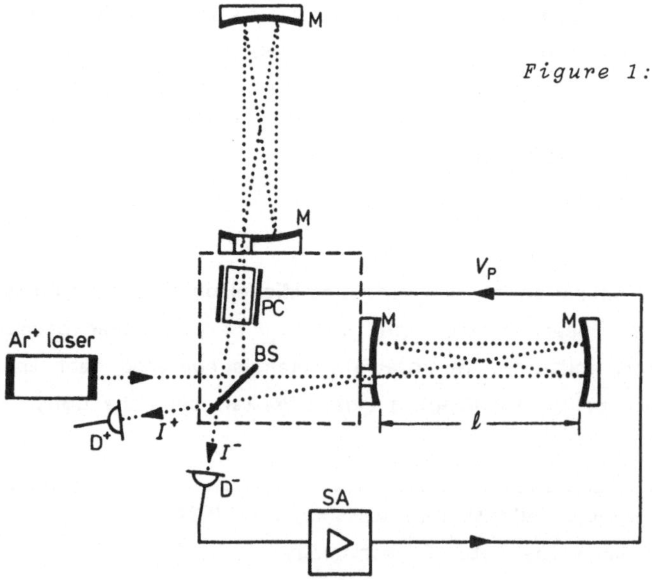

Figure 1: Michelson interferometer with optical delay lines of geometrical length ℓ, total light path $L = N \cdot \ell$

increasing the mirror separation ℓ, eventually to something of the order of several kilometers, linearly increases the sensitivity to strain without penalty.

As an intermediate step, an interferometer with an arm length of ℓ = 30 m has been put into operation. Only the central tank containing the beam splitter and the two near mirrors is situated inside the laboratory. The two 30 m vacuum pipes in earth-covered tunnels lead to the end houses containing the far mirrors.

4. Noise Spectral Densities

The gravitational-wave signals will have to compete with the noise prevailing in our interferometer set-ups. We can express this noise either as apparent fluctuations δL of the optical path length $L = N\ell$, or as an apparent strain $h = \delta L/L$. The noise (being mostly of a rather broad-band nature) is best described by its (linear) spectral density $\widetilde{\delta L}$ (or $\tilde{h}$), the square root of the noise power contributed per Hertz bandwidth. This is how we arrive at the (perhaps somewhat unfamiliar) units $m/\sqrt{Hz}$ for $\widetilde{\delta L}$, and $1/\sqrt{Hz}$ for $\tilde{h}$.

In this talk I will compare various such noise spectra, and this will give us a chance to discuss what sources seem to be responsible for this noise, and what can be done, or has already been done, to suppress it.

5. The 3 m Interferometer

Figure 2 shows noise spectra of our 3 m interferometer, at two stages of development (solid trace: September 1982; dotted: January 1981). The noise is expressed as fluctuation $\widetilde{\delta L}$ of the total path L. One can see that at high frequencies, above 10 kHz, the noise is almost down to the dashed line that indicates the fundamental limit due to shot noise (for 25 mW of available light power). The only feasible way to lower this level would be to increase the light power available in the interferometer. For want of more powerful lasers this can be done either by using a battery of several phase-locked argon lasers, or by using a scheme of feeding light from the unused interferometer port back in (so-called "recycling"). Both these approaches will require very delicate control schemes for the laser frequency, but they will have to be tackled.

Our current investigations are, however, concerned with understanding and eliminating the noise that is in excess of our present shot noise, particularly those contributions that are inside our frequency window.

We notice a considerable reduction, going from the dotted trace (1981) to the solid

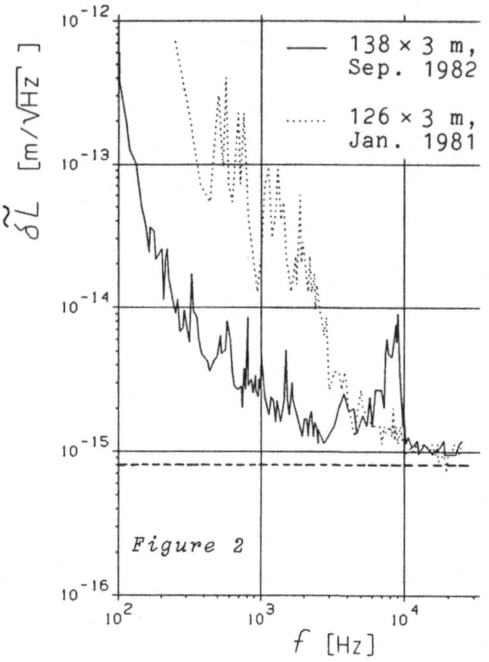

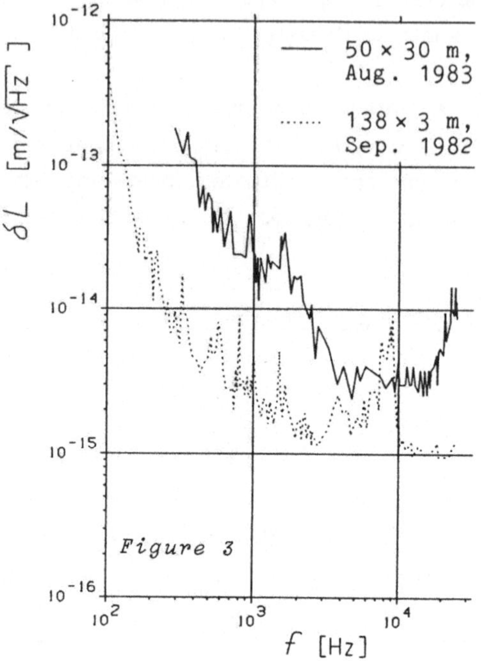

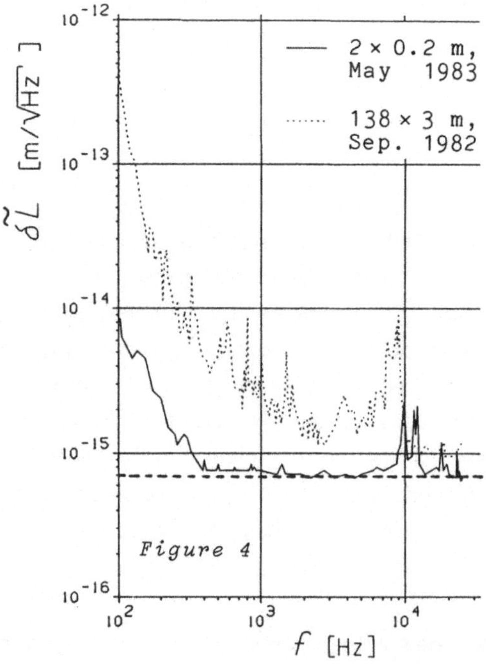

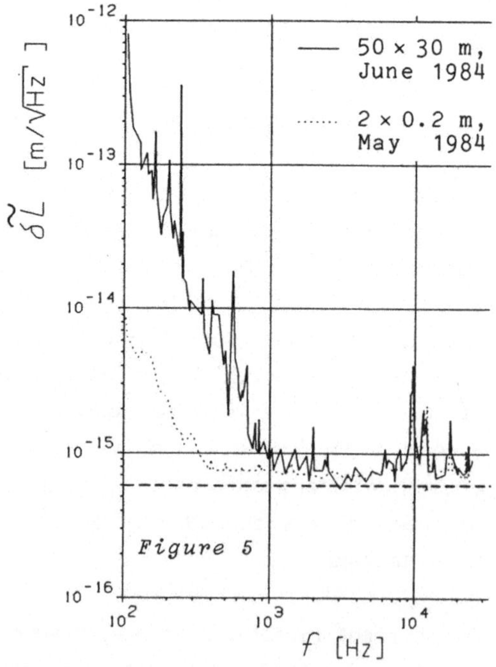

Figures 2, 3, 4, 5: Interferometer noise, expressed as fluctuation $\tilde{\delta L}$, for the different experimental set-ups at Garching. Dashed horizontal line indicates shot noise limit.

trace (1982). A large number of alterations had contributed to this improvement, e.g. more effective isolation against seismic noise and better stabilization of the absolute mirror distance. But the most decisive single improvement in the range from 1 to 5 kHz came from a re-design of the mirror mounts. Many of their mechanical resonances had been inside our frequency window, and even in the absence of seismic vibrations, they are always excited thermally. By mounting the mirrors more rigidly to the suspended metal blocks (several methods had been tried), and by simplifying some of the design features, most of these resonances had successfully been moved to above our frequency window.

In the solid trace we see these (thermally excited) resonances rising above a noise floor that down to a frequency of 2 kHz comes rather close to the shot noise limit. Some isolated resonances seem, however, to have remained, sticking out from a 1/f roll-off that extends from about 0.3 to 2 kHz.

It proved very difficult to establish the origin of the 1/f noise, and we hoped that switching to a longer mirror distance (30 m) would shed some light on the unidentified noise sources.

6. The 30 m Interferometer (Preliminary Run)

The construction of our 30 m interferometer was completed in mid-1983. For a crude operational test, it had been equipped with the old beam-splitter block (that had been put out of service in the 3 m experiment) and was illuminated with an old CR2 laser exhibiting high-frequency noise. The mirror distance (ℓ = 29.4 m) was chosen such that we had N = 50 passes in the delay line for a total path L of about 1.5 km.

The interferometer output, again expressed as $\widetilde{\delta L}$, is shown in Figure 3, in comparison with the 1982 trace from the 3 m set-up. It is evident that this preliminary 30 m test set-up is much noisier than the 3 m set-up of 1982. The flat noise floor from 3 kHz to 16 kHz, and the subsequent rise towards 25 kHz, were easily attributed to a combination of scattered light [1] and the very poor frequency stability of the CR2 laser. The low-frequency end, too, is significantly above the 3 m measurement of 1982. Actually, it shows some similarity with the old 3 m measurement of 1981 (Figure 2, dotted), not much to our surprise, because the central block, with its poor mechanical design, was common to these two experimental set-ups.

The positive result of this preliminary test was the lack of any new, unexpected noise sources in this larger interferometer. This allowed us to go ahead with an upgrade of this interferometer, utilizing the experience gathered in the 3 m set-up.

7. Suspension of Bare Mirrors

One design feature, which had already been tried out in the 3 m set-up, was a simplified suspension of the distant mirrors. Instead of suspending a metal block onto which the end mirror is mounted (with all the problems of mechanical resonances), now it is only the bare mirror (15 cm in diameter, 2.5 cm thick) that is suspended, by stringing a wire around the lower half of its perimeter. In this suspension scheme, the lowest resonances (bending modes of the mirrors) all occur well above 8 kHz, with a relatively high mechanical quality factor ($Q > 2000$).

The central block, however, was designed in such a way that the two near mirrors were fixed to a massive central block. As this block also carries the two Pockels cells and (what is particularly troublesome) the remote-control motors for the adjustment of the beam splitter, it had been the source of many undesirable resonances. The vibration of these resonant parts is, of course, also imparted to the near mirrors, where its influence on the light path is multiplied by N, the number of passes in the delay line.

To avoid these problems, it was proposed to suspend also the near mirrors separately (with the wire sling technique already used with the distant mirrors). Only the beam splitter and the Pockels cells would now have to be mounted on a central mass. Furthermore, the remote control (to adjust the beam splitter in the plane of symmetry between the two near mirrors) would no longer have to be done by motor-control on the central block, but rather the beam splitter block could be oriented as a whole, by proper adjustment in the suspension, or via coils and magnets, as is done to orient and damp the suspended bare mirrors.

It was hoped that with this new technique many of the bothersome resonances could either be completely avoided, or at least be moved to sufficiently high frequencies, well above our frequency window. It was, on the other hand, by no means clear how well the beam splitter could be kept in the plane of symmetry, and how high a price would have to be paid in the control of the mirrors and of the new beam-splitter block.

8. The 20 cm Test Interferometer

For a separate investigation of such questions, the 30 m set-up was radically simplified to a 20 cm interferometer: the two beams were immediately reflected by the near mirrors instead of letting them enter a delay line.

The orientation of the beam splitter was preset by a (motor-controlled) rotation of its suspension points and by (manual) adjustment of the points where the suspending wires leave the mirror rim. Fine adjustment of rotation and tilt was provided by offset currents in the coils, the fringe contrast being used as criterion, i.e. adjusting for the lowest value of the interference minimum. As it turned out, the adjustments remained at this optimum operating point for several days.

Some improvements in the laser illumination were made: installation of an Innova 90-5 laser and, in particular, better suppression of lateral beam jitter. The use of a single-mode fiber provided an excellent "mode cleaner" and also made the microphonic beam diverter obsolete. A series of noise measurements of this 2x20 cm interferometer were made, shown as a solid trace in Figure 4. For comparison, the 3 m result of September 1982 is shown as a dotted trace.

The range in which the measured noise is just the inevitable shot noise now extends as far down as 300 Hz. The steep rise, going roughly with $1/f^2$ at lower frequencies is not yet fully accounted for.

The encouraging result is the total lack of resonant peaks in a broad frequency window from 300 Hz to 8 kHz. It was hoped that this feature would be - at least partially - preserved if we went back to the 50x30 m interferometer.

Note: The results to be described in the next section did not become available until the end of June, 1984, but they are included here for completeness.

9. The New 30 m Interferometer

With the new beam-splitter block and the separate suspension of all four mirrors, the 30 m interferometer was put back into operation. We take as an indication of the high quality of our 30 m mirrors the fact that the interference minimum after N = 50 passes was nearly as good (1.0 % of the maximum) as in the 2x20 cm test (0.5 %).

A major improvement had to be made in the control of the absolute arm length. This problem had previously been tackled with an auxiliary (He-Ne) laser that measured the mirror spacing in one arm [1]. In the new scheme, the (average) mirror spacing in the two arms is measured (and subsequently controlled) by comparing a small portion of the ingoing main Ar^+ laser light with light returning from the interferometer. The arrangement, by the way, is very similar to what is needed to do "recycling" of the interferometer light.

The result of this new 50x30 m interferometer is shown as a solid trace in Figure 5, the 2x20 cm interferometer noise being shown dotted for comparison. Above 1 kHz, the 30 m spectrum is as close to the 2x20 cm one as could be hoped for. It is interesting to note that even the resonant peaks between 8 kHz and 15 kHz are almost exactly reproduced. This confirmed that all of these peaks are related to the beam splitter block, and not to the mirrors, which here are probed 25 times more often than in the 2x20 cm case. It appears necessary to pay even more attention to a sufficiently simple and rigid construction of this beam splitter block.

The rise towards low frequencies starts at about 1 kHz and exhibits a slope that is similar to those in the 138x3 m and 2x20 cm cases. The origin of this low-frequency noise is again not yet determined with certainty, although much of the evidence points to lateral beam jitter. We hope that the single-mode fiber will again help to reduce this noise contribution.

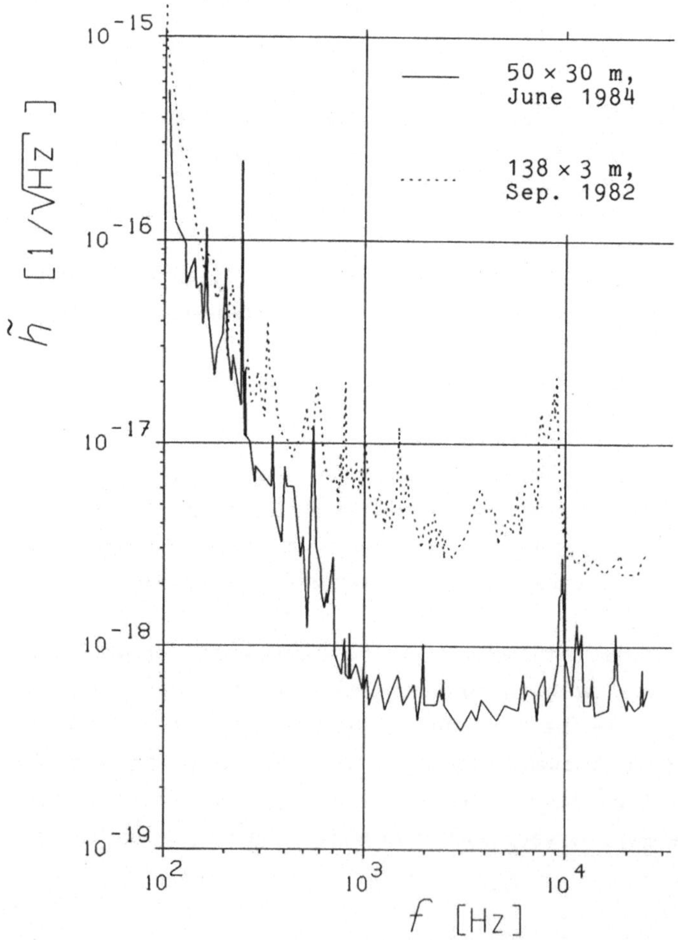

Figure 6: Interferometer noise, expressed as apparent strain h. Comparison of new 30 m results with best 3 m results.

10. Conclusion

The aim in our work is to provide high sensitivity with respect to the gravitational wave strain h. In a last plot (Figure 6), the spectral density $\tilde{h}$ of the new 30 m interferometer (solid trace) is compared with the previous optimum, the Garching 3 m interferometer (dotted). Close to a factor of ten has been gained (even though the light path L increased by only a factor of about 4), and even in a somewhat wider frequency window. With the noise spectral density $\tilde{h}$ of about $3 \cdot 10^{-19}/\sqrt{Hz}$, a sensitivity of 10^{-17} in a bandwidth of 1 kHz has been reached. Some improvements are yet intended to be made in the 30 m apparatus: a higher number of passes (perhaps about N = 100), an increase of the light power going into the interferometer, and possible attempts at recycling. But large increases in the sensitivity can only come from an increase in the geometric arm length ℓ. It is clear that plans for such a longer interferometer will have to be made in the very near future.

References

1. H. Billing, W. Winkler, R. Schilling, A. Rüdiger, K. Maischberger, L. Schnupp: "The Munich Gravitational Wave Detector Using Laser Interferometry", in Quantum Optics, Experimental Gravity, and Measurement Theory, Ed. P. Meystre, M.O. Scully, Plenum Publ. Corp. 1983, 525-566.

2. A. Rüdiger, R. Schilling, L. Schnupp, W. Winkler, H. Billing, K. Maischberger: "Gravitational Wave Detection by Laser Interferometry", in Lasers and Applications, Eds. I. Ursu, A.M. Prokhorov, CIPPress Bucharest 1983, 155-179.

INJECTION LOCKING AND COHERENT SUMMATION OF ARGON ION LASERS

C.N. Man and A. Brillet
Laboratoire de l'Horloge Atomique
C.N.R.S. E.R. 132
Bât. 221 - Université Paris XI
91405 Orsay Cedex, France

Introduction

The basis of the interferometric detection of gravitational wave radiation is the measurement of phase fluctuations as small as 10^{-10} fringe. So in order to lower the shot noise limit, most projects rely on the availability of very powerful ($\sim$100 W) and single frequency visible lasers [1]. In most lasers, single frequency operation is achieved by inserting many selective elements whose losses decrease the output power ; with argon ion lasers, the present commercial systems give less than 10 W on each line and this typical loss reaches 50%.

Injection locking [2] should be considered the solution to the problem of getting both a narrow-band and a high output power, because it allows one to deal separately with these two difficulties, and because the insertion of selective elements will eventually not be needed. This technique has been mainly used with pulsed lasers [3] or with low power lasers [4,5].

We show here that injection locking with argon ion lasers gives larger single frequency output power than with an intracavity etalon. Furthermore we verified that injection locking gives effectively a phase lock [6] which makes it very easy to coherently sum up the output beams from the two lasers.

Experimental configuration

The master oscillator is a single frequency laser which linewidth is reduced to 300 Hz when it is locked to a high finesse cavity. It delivers up to 0.7 Watt at 488.1 nm and 514.5 nm. The slave laser is a multimode laser (50 modes) with an output power of 1.3 Watt at 488.1 nm and 1.5 Watt at 514.5 nm. The injected beam is coupled to the slave laser through a directional coupler (acousto-optic crystal) and it is roughly mode-matched. (Figure 1).

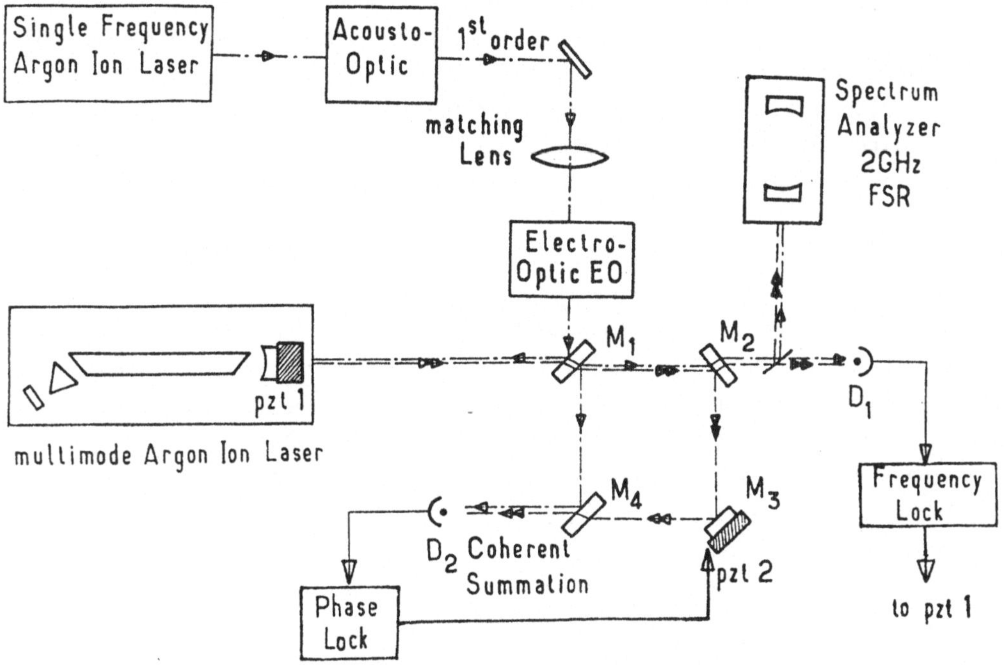

Figure 1 - Block diagram of the experiment.

Injection locking occurs when one of the resonances of the slave laser's cavity is close enough to the master laser's frequency. The locking range was typically of 1 MHz, i.e. much smaller than the frequency fluctuations of the slave cavity (which are induced by the water cooling). So in order to ensure a stable injection locked regime, we found it necessary to servo the length of the slave cavity by acting on its pzt driven mirror (pzt 1). The servo-loop also includes the electro-optic modulator for phase modulation and the detector D_1 monitoring the response of the slave cavity [7].

When the injection is realized and the servo-loop is closed, the whole multi-frequency power of the slave laser is converted into single freouency emission ; the typical injected power was 1/50th of the slave's laser power.

When the output power of the slave laser is increased, the behavior becomes different for the two lines. At 514.5 nm it was possible to keep it single frequency up to 0.4 Watt, but with further increasing power, we observed the appearance of a new regime, still locked and stable but with two or three modes, becoming gradually completely multimode and unstable. At 488.1 nm single frequency operation persisted up to the maximum current where it was possible to get 1.3 Watt with an injected power of 0.02 Watt. This difference can be explained by the fact that the gain at 488.1 nm is higher by a factor 2.5 and saturates more rapidly with increasing power than at 514.5 nm [8] ; so the 488.1 nm line is closer to be homogeneously broadened and more appropriate for single frequency operation.

In both cases, we must point out that the power available by injection locking is larger than the single frequency output power obtained from the same laser with an intracavity selective etalon, i.e. 0.35 Watt at 514.5 nm and 0.5 Watt at 488.1 nm.

A direct application of injection locking is the coherent summation of the two lasers which is shown in figure 1 by the interferometer M_1 M_2 M_3 M_4. Detector D_2 was used to monitor the relative phase or frequency fluctuations of the master and slave lasers. When the mirrors of the slave cavity were disconnected from the tube to get rid of the vibrations induced by the water cooling, the phase difference is a slow thermal drift which can be easily reduced to less than 10^{-2} rd (figure 2) with a servo-loop acting on pzt 2 (figure 1). This demonstrates that the two beams can be easily summed up in order to multiply the available power.

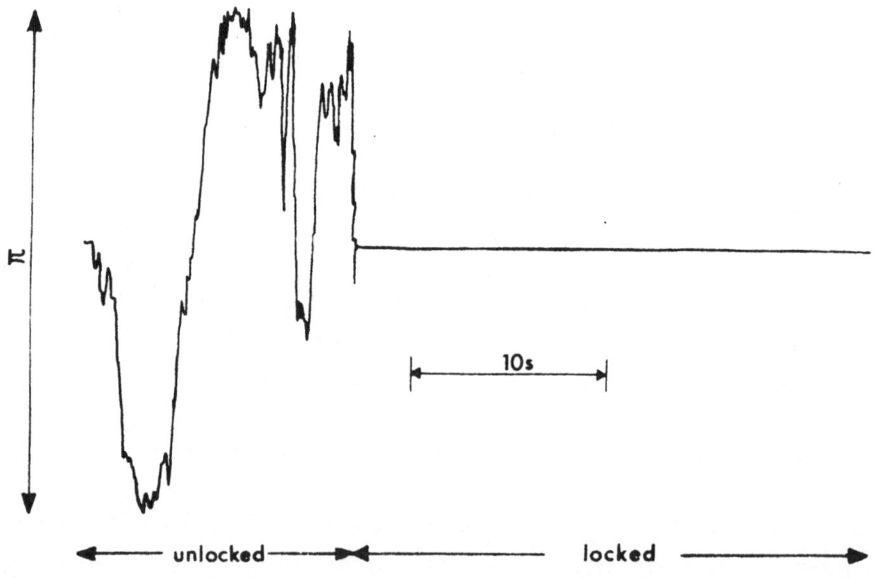

Figure 2 - Registration of the relative phase fluctuations of the master and slave lasers. Phase lock/loop is first off, then on (averaging time constant 40ms)

An obvious extension of this experiment is to phase lock a whole array of high power argon lasers. Rather than using a set of fixed reflectivity beamsplitters, one may choose a polarization technique, in which two orthogonally polarized beams are added on each polarizing beamsplitter. This would be a way of providing the tens of Watts which are needed for an interferometric gravity wave experiment.

REFERENCES

[1] A. Brillet, same issue.

[2] H.L. Stover and W.H. Steier, Appl. Phys. Lett., 8, 91 (1966).

[3] C.J. Buczek and R.J. Freiberg, IEEE J. Quantum Electron., QE-8, 641 (1972).

[4] B. Couillaud, A. Ducasse and E. Freysz, IEEE J. Quantum Electron., QE-20, 310, (1984).

[5] R.W. Dunn, S.T. Hendow, W.W. Chow and J.G. Small, Opt. Lett., 8, 319 (1983).

[6] C. Audoin in Metrology and Fundamental Constants, LXVIII Corso (Soc. Italiana di Fisica, Bologna, Italy, 1980).

[7] R.W. Drever, J.L. Hall and F.V. Kowalski, Appl. Phys. B, 31, 97 (1983).

[8] V.F. Kitaeva, A.N. Odintsov and N.N. Sobolev, Sov. Phys. Uspekhi, 99, 699 (1970).

CAN THE PHOTON NOISE BE REDUCED ?

A. Heidmann and S. Reynaud
Laboratoire de Spectroscopie Hertzienne
24, rue Lhomond - 75231 PARIS Cedex 05 - France

1. Introduction

The search for detection of gravitational waves has renewed the interest for the problem of quantum noise (which is of the same order than the weak expected signal [1,2]). We will here restrict the discussion to interferometric methods where the sensitivity is limited by the photon noise.

A Michelson interferometer (fig. 1) transforms a phase signal, produced for example by the passage of a gravitational wave, in an intensity signal recorded by photodetectors.

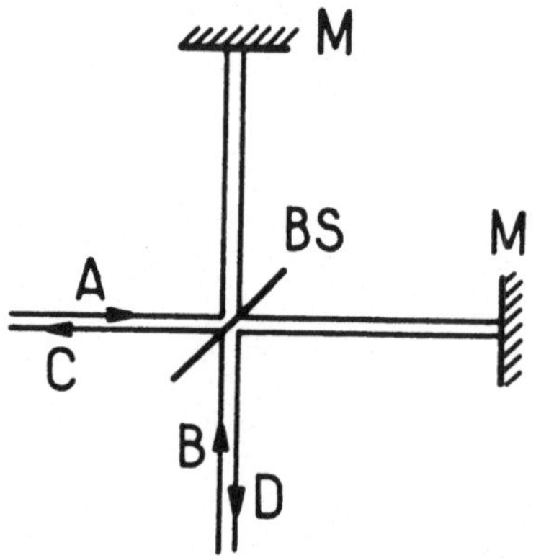

Fig. 1 : A Michelson interferometer : we have sketched the beam splitter BS, the two mirrors M , the two input ports A and B and the two ouput ones C and D . Usually, only one input port (A) is used.

In order to optimize this transformation, one usually measures the difference $\bar{N} = \bar{N}_C - \bar{N}_D$ between the mean number of photons in output channels C and D , with a fringe contrast equal to 1 , and a phase difference ϕ between the two arms such that $\bar{N} = 0$. A small variation $d\phi$ of ϕ thus produces an optimum response :

$$dN = \bar{N}_A \quad d\phi \tag{1.1}$$

($\bar{N}_A$: input intensity).

If all usual noise sources have been eliminated, the only remaining limitation is the fluctuation of N due to the corpuscular nature of light, i.e. photon or shot noise [3-5] :

$$\Delta N^2 = \Delta N_C^2 + \Delta N_D^2 = \bar{N}_C + \bar{N}_D = \bar{N}_A \qquad (1.2)$$

It has been recently realized [6] that the limit (1.2) can be reduced by entering "squeezed" states of the field [7] in the usually unused input port B of the interferometer (fig. 1). This property appears clearly in a wave interpretation of photon noise (see § 2). Some propositions have been done for generating such fields ([8] and references in [7]), leading to very interesting predictions concerning photon noise reduction (see § 3). But we will question these predictions, since the coherence properties of the squeezed field are not generally accounted for (§ 4). This will have important implications concerning the photon noise reduction (§ 5).

2. Photon noise in an interferometer

One usually consider only 4 modes, 2 input (A , B) and 2 output (C , D) modes, with annihilation operators a , b , c , d . It is easy to show that the output signal performs an heterodyne detection of the input mode B by the input mode A [3,5]. If the input field A is in a coherent state $|\alpha\rangle$ [9], the photon noise is thus given by :

$$\Delta N^2 = \bar{N}_A \, \Delta b_2^2 + \bar{N}_B \qquad (2.1)$$

where b_2 is the component of b in quadrature with α :

$$b_2 = i\,(b^+ - b) \qquad (2.2)$$

(we suppose α real).

The photon noise can thus be considered as resulting from the heterodyning of the fluctuations of b_2 with the classical field entering A (first term in equation 2.1 is dominant since $\bar{N}_B \ll \bar{N}_A$). If the field entering B is in the vacuum state, one has still to consider the vacuum fluctuations in mode b , with $\bar{N}_B = 0$ and $\Delta b_2^2 = 1$, so that one obtains the photon noise of equation (1.2). It appears clearly that the photon noise can be reduced if the field entering B is in a "squeezed" state such that :

$$\Delta b_2 < 1 \rightarrow \Delta N^2 < \bar{N}_A \qquad (2.3)$$

3. Generation of squeezed states

We will discuss the generation of squeezed light by four wave mixing [8] : an atomic medium is irradiated by two counterpropagating pump waves (fig. 2) and two weak counterpropagating probe waves. The two output waves (u and v on fig. 2) result from amplification and phase conjugation of the two probe waves. Interferences between these output waves then produces a field which can be described [8,11] by the annihilation operator :

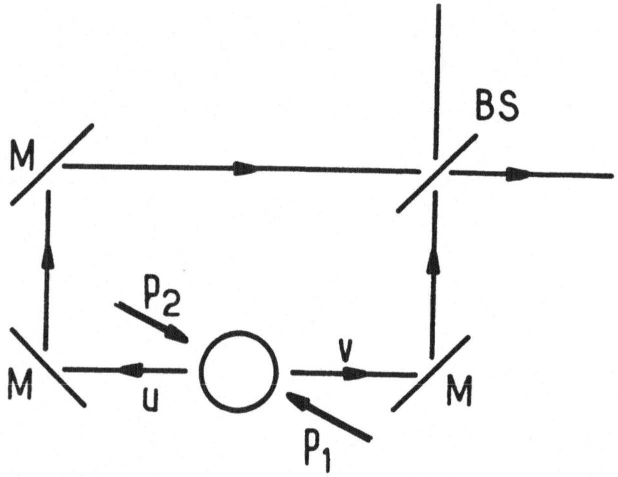

Fig. 2 : Generation of squeezed light by four wave mixing : P_1 and P_2 are the pump waves, u and v the probe ones. The two output probe waves are recombined using mirrors (M) and a beam splitter (BS).

$$b = \mathrm{ch}\,\xi\, b_0 + \mathrm{sh}\,\xi\, b_0^+ \qquad (3.1)$$

where b_0 is the sum of the two incident probe waves operators and ξ characterizes the phase conjugation efficiency (ξ is related to the interaction time, the pump amplitudes and the non-linear susceptibility of the atomic medium).

When the initial probe modes are in coherent states, the dispersions of the two field quadratures $b_1 = b + b^+$ and b_2 have the following values :

$$\Delta b_1 = e^{\xi} \quad , \quad \Delta b_2 = e^{-\xi} \qquad (3.2)$$

This device thus produces a squeezed field ($\Delta b_2 < 1$ for $\xi > 0$). It is important to note that the squeezing parameter obtained in equation

(3.2) does not depend on the amplitudes of the input probe waves. As a consequence, one gets squeezing even when there is no input probe waves (i.e. when the input modes are in the vacuum state).

4. Critical analysis of the few modes semiclassical treatment

We want now to question the validity of the model recalled in section 2 and 3, since it do not give a correct description of the coherence properties of the squeezed field.

First, the description of the atomic medium by a non-linear susceptibility consists in an identification between the dipole operators and their mean value. But statistical properties of the field (such as the photon noise) are related to field fluctuations which are generated by dipole fluctuations. A correct treatment of these quantum dipole fluctuations will give the temporal coherence properties of the field [10-12]. The coherence time thus obtained is of the order of the atomic lifetime :

$$\tau_c \simeq 1/\Gamma \tag{4.1}$$

(Γ spontaneous emission rate).

The treatment of section 3 can also be characterized as a few field modes analysis, involving only four field modes. But squeezing is independent of the presence of input probe waves which means that there is no privileged probe modes. In other terms, the radiated field has to be considered as a multimode field, by superposing all modes where phase conjugation process is efficient, i.e. all modes where phase matching condition is satisfied :

$$k_1 + k_2 = k_u + k_v \tag{4.2}$$

(k_1 and k_2 are the pump wave vectors, k_u and k_v the probe ones). The vector ($k_1 + k_2$) can be slightly different from zero because of the divergence of the pump waves, allowing four wave mixing to be efficient for directions k_u and k_v not exactly opposite. One obtains in this manner [11,12] a coherence solid angle Ω_c equal to the divergence solid angle of the pump laser beams :

$$\Omega_C \simeq \lambda^2/w^2 \tag{4.3}$$

(λ wavelength, w beam waist).

5. Effect of coherence on photon noise

This discussion has important implication concerning photon noise. As a matter of fact, the photon noise ΔN^2 is a photodetection signal, and it is well known from the photodetection theory [9] that such a signal is very sensitive to the coherence properties of the field. Calculating ΔN^2 in the photodetection theory, one finds [5,12] :

$$\Delta N^2 = \bar{N}_A \ (1 + Q) + \bar{N}_B \tag{5.1}$$

This expression is similar to (2.1) but the factor Q is now a double integral over the detection volume (detection area S and time T) of two points correlation functions of the squeezed field. Since these functions are naught for two points not in the same coherence volume, Q is in fact proportional to the coherence volume of the squeezed field [5]. We have performed a calculation of factor Q with the squeezed field source of section 3 [12]. In the best conditions, Q reaches the optimum value :

$$Q \simeq - .02 \tag{5.2}$$

One can thus expect that the photon noise is reduced in the best conditions to :

$$\Delta N = .99 \sqrt{\bar{N}_A} \tag{5.3}$$

6. Conclusion

We have recalled how photon noise can be reduced by using squeezed states, and how such states can be theoretically generated. We have then shown that it is necessary to take into account the coherence properties of the squeezed field. By doing so, we have calculated that the reduction factor was 1% in the best conditions.

The point is that the source of squeezed field considered here is an incoherent source. The problem is now to understand how to design a coherent source of squeezed field [11,13].

Acknowledgements

We are grateful to C. Cohen-Tannoudji and J. Dalibard for fruitful discussions.

References

1. "Quantum Optics, Experimental Gravitation and Measurement Theory", ed. P. Meystre, M.O. Scully (Plenum 1983).
2. This volume.
3. C.M. CAVES, Phys. Rev. Lett. 45, 75 (1980).
4. R. LOUDON, Phys. Rev. Lett. 47, 815 (1981).
5. S. REYNAUD and A. HEIDMANN, to be published in Ann. Phys. Fr.
6. C.M. CAVES, Phys. Rev. D-23, 1693 (1981).
7. D.F. WALLS, Nature 306, 141 (1983).
8. H.P. YUEN and J.H. SHAPIRO, Optics Letters 4, 334 (1979).
9. R.J. GLAUBER in "Quantum Optics and Electronics", ed. C. de Witt, A. Blandin, C. Cohen-Tannoudji (Gordon and Breach, 1965).
10. R. LOUDON, Optics Comm. 49, 24 (1984).
11. S. REYNAUD and A. HEIDMANN, Optics Comm. 50, 271 (1984).
12. A. HEIDMANN, Thèse de 3ème Cycle (Paris VI, 1984, unpublished).
13. M.D. LEVENSON ; R.E. SLUSHER, B. YURKE and J.F. VALLEY ; M.D. REID and D.F. WALLS in "Thirteenth International Quantum Electronics Conference", JOSA B1, 525 (1984).

The problem of the optical stability of a pendular Fabry-Perot

Nathalie Deruelle and Philippe Tourrenc

Laboratoire de Physique Théorique
Institut Henri Poincaré
11 rue P.et M. Curie, 75005, Paris.

In order to obtain a sufficient signal-to-noise ratio in the interferometers designed for detecting gravitational radiation, very large systems ($\sim$1 km long) and very large effective laser powers ($\sim$ 1 kwatt) must be considered [1]. The problem of the optical stability of such systems must then be addressed. The dynamics of the mobile mirrors may indeed become complex, firstly because of the importance of the radiation pressure which implies that the force acting on the mirrors is highly non-linear and secondly because of the large size of the apparatus which implies that the round trip travel time of light in the cavity may not be negligible.

In a very different context, the optical multistability and the possibility of bifurcation to chaos in a Fabry-Perot of fixed geometry but filled with a non-linear medium have been extensively studied theoretically as well as experimentally [2]. The equations describing an empty Fabry-Perot with a mobile mirror being similar, one can expect similar behaviours. Indeed the optical bistability and the associated hysteresis induced by the radiation pressure on the mirror of a small pendular Fabry-Perot has already been observed experimentally by Dorsel et al.[3].

In the prospect of pursuing the comparison between such systems, we shall state here the general problem of the optical stability of a pendular Fabry-Perot. Some conclusions concerning the gravitational wave detectors will be drawn. A detailed analysis of the motion of the mirror in the different regimes will be given elsewhere.

I-The conditions of multistability.

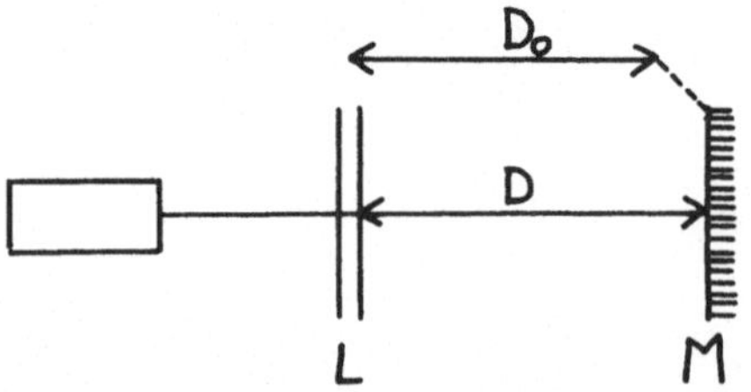

Fig.1 : The parameters of the Fabry-Perot cavity. P is the laser power, λ its wavelength. $R=e^{i\varphi}\cos\theta$ is the reflectivity of the fixed mirror (L). The mobile mirror (M) of mass M and angular pulsation Ω is suspended to a wire anchored at D_0.

In the static case the force $\mathcal{F}$ acting on the mobile mirror of a Fabry-Perot cavity (see fig.1) is the sum of the mechanical restoring force:

$$(1) \qquad \mathcal{F}_{mec} = -M\Omega^2(D-D_0)$$

and the radiation pressure force which, when the length of the Fabry-Perot is constant, is the Airy function:

(2) $$\mathcal{F}_{rad} = \frac{2P}{c} \frac{\sin^2\theta}{1+\cos^2\theta + 2\cos\left(\frac{4\pi}{\lambda}D+\theta_\ell\right)}$$

where c is the speed of light. Introducing:

(3) $$x = \frac{4\pi}{\lambda}(D-D_m) \qquad ; \qquad x_o = \frac{4\pi}{\lambda}(D-D_o)$$

where D_m is a value of D such that $\mathcal{F}_{rad}$ is maximum, $\mathcal{F}$ reads:

(4) $$\frac{4\pi}{M\lambda\Omega^2}\mathcal{F} = -x - x_o + \frac{A\sin^2\theta}{1+\cos^2\theta - 2\cos\theta\cos x} \equiv F_{exa}$$

where $A=8\pi P/(c\lambda M\Omega^2)$. For small x (and actually for all $|x|<\pi$ when θ is small), $\mathcal{F}_{rad}$ is well approximated by a Lorentzian and $\mathcal{F}$ reduces to:

(5) $$\frac{4\pi}{M\lambda\Omega^2}\frac{\sqrt{\cos\theta}}{(1-\cos\theta)}\mathcal{F} \simeq -z - \frac{1}{\alpha}\left(\beta - \frac{1}{1+z^2}\right) \equiv F_{app}$$

with

(6) $$\alpha = \frac{(1-\cos\theta)^2}{(1+\cos\theta)}\frac{1}{\sqrt{\cos\theta}}\frac{1}{A} \; ; \quad \beta = \frac{(1-\cos\theta)}{(1+\cos\theta)}\frac{x_o}{A} \; ; \quad z = \frac{\sqrt{\cos\theta}}{(1-\cos\theta)}x$$

When θ is small:

(6a) $$\alpha \sim \frac{\theta^4}{8A} \qquad ; \quad \beta \sim \frac{\theta^2 x_o}{4A} \qquad ; \quad z \sim \frac{2x}{\theta^2} .$$

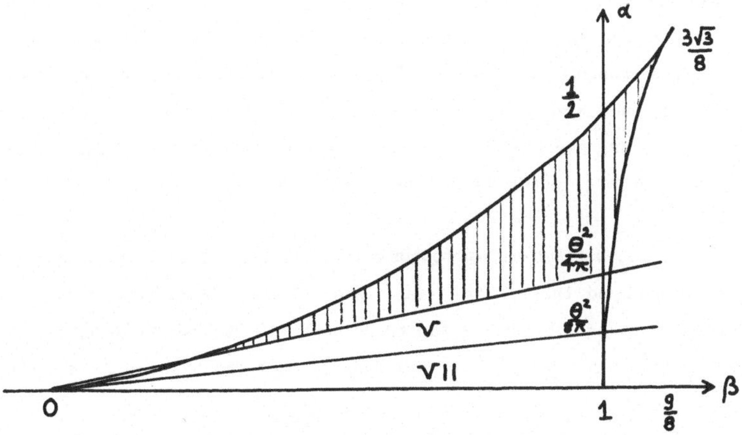

Fig.2 : The conditions of multistability. When α and β range in the hatched area the system is bistable. In region V the mirror has 5 positions of equilibrium, etc. The axis β = 1 corresponds to a system such that D_m , where the radiation pressure is maximum, is a position of equilibrium. For given θ and x_o , $\alpha/\beta \simeq \theta^2/2x_o$ is constant for all P.

The system is multistable if there is more than one position of equilibrium for the mirror that is if $\mathcal{F}=0$ has more than one solution. Fig.2 gives the values of the parameters α and β in which range eq.(5) has three real roots. When θ is small, eq.(4) has then at least three real roots and the system is multistable. When $\beta=1$, bistability occurs for $\alpha < 1/2$. The positions of equilibrium then are:

$$(7) \qquad z_m = 0 \ ; \quad z_i = \frac{1}{2\alpha}\left(-1+\sqrt{1-4\alpha^2}\right) \overset{\alpha\to 0}{\sim} -\alpha \ ; \quad z_s = \frac{1}{2\alpha}\left(-1-\sqrt{1-4\alpha^2}\right) \overset{\alpha\to 0}{\sim} -\frac{1}{\alpha} \ ;$$

D_m, where the radiation pressure is maximum, is then a metastable position of equilibrium.

The force acting on the mirror derives from a potential $\mathcal{V}(D)=-\int dD\, \mathcal{F}(D)$. Its explicit expression is:

$$(8) \qquad \left(\frac{4\pi}{\lambda}\right)^2 \frac{1}{M\Omega^2}\mathcal{V} = \frac{x^2}{2} + x_0 x - 2A\left[\arctan\left(\frac{1+\cos\theta}{1-\cos\theta}\, tg\,\frac{x}{2}\right) + \pi E\left(\frac{x+\pi}{2\pi}\right)\right]$$

which, when the Airy function is approximated by a Lorentzian, reduces to:

$$(8a) \qquad \left(\frac{4\pi}{\lambda}\right)^2 \frac{1}{M\Omega^2}\frac{\cos\theta}{(1-\cos\theta)^2}\mathcal{V} \simeq \frac{z^2}{2} + \frac{1}{\alpha}\left(\beta z - \arctan z\right).$$

Let us consider the case when the laser power P slowly increases (α then decreases), θ and x_o remaining constant (α/β is then a constant chosen to be less than $1/\sqrt{3}$ -see fig.2). For small P the potential $\mathcal{V}$ exhibits only one well (A) whose minimum corresponds to the position of equilibrium D_e of the mirror. When P increases a second well (B) appears in the potential but the mirror remains in well (A) by continuity. Eventually the well (A) disappears and the mirror falls to the minimum of well (B). When P decreases back to small values the mirror will remain in well (B) and therefore the function $D_e(P)$ will exhibit an hysteresis cycle. Such hysteresis cycles giving the experimental evidence for the existence of optically bistable regimes in a pendular Fabry-Perot were observed by Dorsel et al. [3].

*

Values of the parameters appropriate to gravitational wave detectors are: $\lambda=0.5\ 10^{-6}$m, M=100 kg, $\Omega=2\pi$ rd/sec (so that $A/P \sim 4\ 10^{-5}$ watt^{-1}) and $\theta=0.1$. In these detectors D_m is chosen to be a position of equilibrium for the mirror ($\beta=1$). Multistability then occurs for $\alpha < 1/2$ that is for $P > 0.6$ watt. For $P > (400\ n + 0.6)$ watts, the mirror has at least (2n+3) positions of equilibrium.

When the system is multistable, the equilibrium position of the mirror at D_m is metastable. The height of the potential barrier preventing the mirror from falling to the stable position D_s is, when θ is small:

$$(9) \qquad \mathcal{V}(z_i) \simeq M\Omega^2\left(\frac{\lambda}{4\pi}\right)^2 \frac{\theta^4}{4}\frac{\alpha^2}{6} = \frac{(M\Omega^2)^3}{P^2}\left(\frac{\lambda}{4\pi}\right)^4 \frac{c^2\theta^{12}}{3\times 2^{11}} \ ;$$

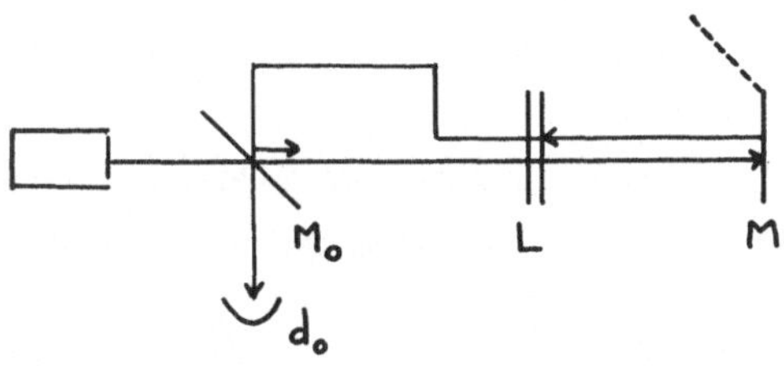

Fig. 3 : The recycling of the light.

Large effective laser powers may be reached using Drever's idea of recycling the light [4]: a suitable servo continuously adjusts the position of the mirror M_0 (see fig.3) so that the power on the detector d_0 is zero. When the adjustment is perfect the system is equivalent to an ordinary Fabry-Perot (without M_0) except that P is now an effective laser power $P=P_{laser}/T_0^2$ where T_0 is the transmittivity of M_0. For P=1000 watts: $V(z_i)/kT^\circ \sim 0.5\ 10^{-3}$ at room temperature. The stabilization of the mirror will therefore require a good control of its position.

In practice the adjustment will not be perfect and in the extreme case when M_0 is held fixed at a position such that the power on d_0 is zero only when the mirror is at D_m, the system is equivalent to a cavity inside a cavity: the effective laser power is P/2 and θ must be replaced by θT_0 so that the height of the potential barrier decreases considerably. The control of the position of M_0 will therefore be crucial.

II-The equations of motion of the mirror.

When the response time of the cavity to a variation of its length is ignored, the equations of motion of the mirror are:

(10) $$M\frac{d^2D}{dt^2} + \frac{M\Omega}{Q}\frac{dD}{dt} = \mathcal{F}_{mec} + \mathcal{F}_{rad} + \mathcal{F}_{ext}(t)$$

where Q is the quality factor of the mirror and where $\mathcal{F}_{ext}(t)$ is any external force. They also read:

(11) $$\frac{d^2x}{d\tau^2} + \frac{1}{Q}\frac{dx}{d\tau} = F_{exa} + \frac{4\pi}{M\lambda\Omega^2}\mathcal{F}_{ext}(\tau)$$

with $\tau=\Omega t$ and where F_{exa} is given by eq(4). When the Airy function is approximated by a Lorentzian they reduce to:

(12) $$\frac{d^2z}{d\tau^2} + \frac{1}{Q}\frac{dz}{d\tau} = -z - \frac{1}{\alpha}\left(\beta - \frac{1}{1+z^2}\right) + F_{ext}(\tau)$$

with $F_{ext}(\tau)=(8\pi/M\lambda\Omega^2\theta^2)\,\mathcal{F}_{ext}(\tau)$ and where α, β and z are defined by eq(6).

For $\beta=1$ (D_m, where the radiation pressure is maximum, is then a position of equilibrium) and for $|z| \ll \alpha$, a case in which the gravitational wave detectors fall if the thermal noise of the mirror is sufficiently reduced, the radiation force [the second term in the rhs of eq(12)] can be ignored. The equations of motion then reduce to those of a forced harmonic oscillator.

A numerical analysis of (12) when F_{ext} is sinusoidal is underway [5]. For given α and β the solutions, as the amplitude of F_{ext} increases, are expected to follow a cascade of subharmonic bifurcations similar to the route towards chaos observed in the case of the Duffing oscillator.

*

When the response time of the cavity is no longer neglected, the radiation force on the mirror is no longer the Airy function. Let then $\mathcal{E}_i(t)$ be the field entering the cavity and $\mathcal{E}_m(t)$ the field on the mirror at time t (see fig.4). We have:

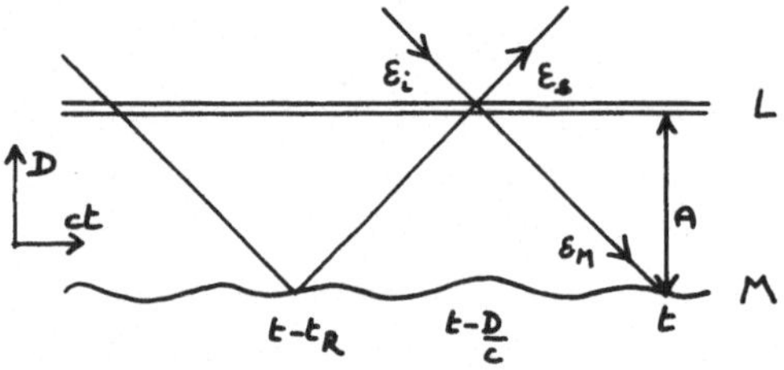

Fig.4 : Space-time diagram of the system. The mirror (M) is supposed to be perfectly reflective. The reflectivity of (L) is $R=e^{i\theta_L}\cos\theta$ and its transmittivity is $T=ie^{i\theta}\sin\theta$

$$\mathcal{E}_m(t) = T\mathcal{E}_i\left[t-\frac{D(t)}{c}\right] - R\,\mathcal{E}_m(t-t_r) \tag{13}$$

where $t_r=[D(t)+D(t-t_r)]/c$ is the round trip travel time of light in the cavity. Intoducing a new function $\hat{D}(t)$ we write $\mathcal{E}_m(t)$ as:

$$\mathcal{E}_m(t) = \frac{T e^{\frac{2i\pi}{\lambda}D(t)}}{1+R e^{\frac{4i\pi}{\lambda}\hat{D}(t)}}\,\mathcal{E}_i(t) \tag{14}$$

so that the total force on the mirror (the mechanical force plus the radiation pressure force) reads:

$$\frac{4\pi}{M\lambda\Omega^2}\hat{\mathcal{F}}(t) = -x - x_0 + \frac{A\sin^2\theta}{|1-\cos\theta\, e^{i\hat{x}(t)}|^2} \equiv \hat{F}_{exa} \tag{15}$$

where we have set $\hat{x}(t)=(4\pi/\lambda)[\hat{D}(t)-D_m]$. The function $\hat{x}(t)$ is deduced from (13-14); $\mathcal{E}_i(t)$ being a plane wave, $\hat{x}(t)$ is a solution of the time difference equation:

$$e^{i\hat{x}(t)} = \frac{e^{ix(t-t_r)}}{1+\cos\theta\left[e^{ix(t-t_r)} - e^{i\hat{x}(t-t_r)}\right]} \tag{16}$$

with $t_r=2D_m/c+[x(t)+x(t-t_r)]\lambda/4\pi c$.

For small x the equations of motion reduce to:

$$\frac{d^2z}{d\tau^2} + \frac{1}{Q}\frac{dz}{d\tau} = -z - \frac{1}{\alpha}\left(\beta - \frac{1}{1+\hat{\mathcal{Z}}^2(\tau)}\right) + F_{ext}(\tau) \tag{17}$$

with $\tau=\Omega t$, α, β and z defined by eq(6), and where $\hat{\mathcal{Z}}^2(\tau)=\hat{z}_r^2(\tau)+2(1-\cos\theta)\hat{z}_i(\tau)$, $\hat{z}_r$ and $\hat{z}_i$ being the real and the imaginary parts of $\hat{z}(\tau)=\sqrt{\cos\theta}\,\hat{x}(\tau)/(1-\cos\theta)$. The expansion of (16) then leads to:

$$\begin{cases} \hat{\mathcal{Z}}^2(\tau) = \cos\theta\,\hat{\mathcal{Z}}^2(\tau-\tau_r) + (1-\cos\theta)\,\hat{z}_r^2(\tau-\tau_r) \\ \hat{z}_r(\tau) = \cos\theta\,\hat{z}_r(\tau-\tau_r) + (1-\cos\theta)\,z(\tau-\tau_r) \end{cases} \tag{18}$$

where $\tau_r \simeq 2D_m\Omega/c$, D_m being the length at rest of the cavity.

When τ_r can be neglected, $\hat{\mathcal{Z}}^2=\hat{z}_r^2=z^2$: the equations of motion (17-18) reduce to eq (12). When τ_r is small compared to the evolution time of the solution z(t), the time difference equations (18) that we shall write generically as:

(18a) $$\phi(\tau+\tau_r) = \cos\theta\,\phi(\tau) + (1-\cos\theta)f(\tau)$$

can be expanded in Taylor series in τ_r. At first order they read:

(19) $$\phi(\tau) + \tau_r \frac{d\phi}{d\tau}(\tau) \simeq \cos\theta\,\phi(\tau) + (1-\cos\theta)f(\tau).$$

In order to eliminate the runaway solutions, eq(19) is solved iteratively:

(20) $$\phi(\tau) \simeq f(\tau) - \frac{\tau_r}{1-\cos\theta}\frac{df}{d\tau}(\tau).$$

That is the so-called first order predictive solution of (18a) [6]. At first order in τ_r the equations of motion therefore reduce to the ordinary differential equation (17) where:

(21) $$\hat{\mathcal{Z}}^2(\tau) \simeq z^2(\tau) - \frac{2\tau_r}{1-\cos\theta}\frac{d}{d\tau}\left[z^2(\tau)\right]$$

which, when z $\ll$1 further reduce to:

(22) $$\frac{d^2z}{d\tau^2} + \frac{dz}{d\tau}\left[\frac{1}{Q} - \frac{4\tau_r}{1-\cos\theta}z\right] = -z - \frac{1}{\alpha}\left(\beta-1+z^2\right) + F_{ext}(\tau).$$

Taking into account the response time of the cavity therefore introduces a "damping" force in the equations of motion.

For larger τ_r it may occur that the solution of (18a) can no longer be obtained as in (19-20) via a Taylor series expansion associated with an iterative reduction of order [7]; the existence of retarded effects may then lead to chaotic solutions for the equations of motion [5].

*

We aknowledge fruitful discussions with L.Bel, Ch.J. Bordé and B.Mosca.

References

[1] see e.g. A.Brillet in these proceedings
[2] A.Szöke et al., Appl. Phys. Lett. 15, 376 (1969)
H.M.Gibbs et al., Phys. Rev. Lett. 36, 1135 (1976)
K.Ikeda et al., Phys. Rev. Lett. 45, 709 (1980)
H.M.Gibbs et al., Phys. Rev. Lett. 46, 474 (1981)
[3] A.Dorsel et al., Phys. Rev. Lett. 51, 1550 (1983)
[4] R.Drever in "Gravitational Radiation", N.Deruelle, T.Piran Edts, North Holland (1983)
[5] In preparation
[6] L.Bel, X.Fustero, Ann. IHP A25, 411 (1976) and references therein
[7] J.M.Aguirregabiria, L.Bel, in "Frontiers of non-equilibrium statistical Physics", G.T.Moore Edt, Plenum, to be published

Much ado about Geminga

Nathalie Deruelle
Laboratoire de Physique Théorique, Institut H.Poincaré
11 rue P.M.Curie, 75231 Paris

Resume: On October 11th 1983, the Centre National de la Recherche Scientifique (CNRS) announced that gravitational radiation had been observed. The origin of this claim goes back almost ten years. In 1976 an oscillation of the Sun at 160 minutes was observed and has resisted since then all theoretical explanations. At about the same time a powerful, nearby but mysterious gamma- ray source, "Geminga", was discovered. Could these facts be two pieces of the same puzzle? Could the solar oscillation be excited by gravitational waves emitted by Geminga? That was the idea of P.Delache (Observatoire de Nice), J.Paul (CEA, Saclay), G.Bignami (Milano) and G.Isaak (Birmingham). They then searched for a 160 min periodicity in the Geminga gamma data and claimed they found one. "Thus", as the CNRS communique puts it, "for the first time a causal relationship between an emitter and a receptor comes to confirm the reality of gravitational waves predicted by Einstein's theory of General Relativity".

As one might suspect this announcement started a lively controversy, a first aspect of which concerned the possibility of a coupling between the Sun and Geminga via gravitational radiation. An answer was quickly given. When considering standard models of the Sun such a coupling is impossible. A second problem concerned the significance of the result, that is of the 160 min periodicity in the Geminga data. This issue in May 1984 seemed still unclear.

In the following we review the main aspects of this controversy.

Un mode du soleil inexpliqué

Des oscillations du soleil de l'ordre de 5 minutes avaient été observées il y a une vingtaine d'années mais ce n'est qu'en 1975 qu'on put les interpréter comme des oscillations globales de toute la sphère solaire. On comprend l'importance de telles observations, qui permettent d'obtenir des renseignements sur les couches profondes du soleil puisque les prédictions des modèles sur les modes propres dépendent des hypothèses faites sur l'intérieur du soleil [1-2].

Ces modes propres se scindent en deux catégories. Les "modes-p", pour lesquels la force de rappel est la pression, sont les mieux connus. Leur identification aux prédictions des modèles est à peu près satisfaisante et fixe l'abondance initiale d'hélium à environ 23% en masse. Ce résultat est un des succès de la sismologie solaire. Néanmoins l'accord avec les prédictions des modèles standards n'est qu'approximatif.

Quant aux "modes-g", pour lesquels la force de rappel est la gravité, ils ont dans leur ensemble à peu près répondu aux prédictions des modèles [3] à condition d'y inclure une hypothèse de mélange entre les régions radiative et convective. Le spectre de fréquence de ces modes est alors dans ses grandes lignes explicable, mais leur amplitude, de l'ordre de 30cm/sec, est deux ordres de grandeur plus grande que prévu. D'après P.Delache, cela indique qu'il faut probablement tenir compte d'effets non linéaires, à déterminer.

Le modèle du soleil est donc loin d'être définitif et la sismologie solaire, bien que très prometteuse, laisse des problèmes en suspens. De plus cette discipline possède ses "monstres", qui, eux, ne rentrent dans aucun cadre.

*

Celui qui nous intéresse ici est la présence dans le spectre d'une oscillation de 160 minutes découverte en 1976 dès les débuts de la sismologie solaire, observée depuis dans les domaines optique, infra-rouge et radio en Crimée, à Stanford et au pôle sud, et qui a résisté à toute explication [4-5].

Sa période tout d'abord a posé problème. En effet elle est proche du 1/9ème de jour et on a pensé pendant un temps qu'elle était un artefact lié à la rotation de la terre sur elle-même. La polémique semble maintenant terminée avec la mesure plus précise de cette période: 160.0095$\pm$ 0.001 minutes, qui ne peut être confondue avec le 1/9ème de jour. Il semble donc assuré qu'il s'agit d'un mode de gravité, probablement quadrupolaire ($l=2$) [6-8].

Son amplitude aussi a posé problème. En 1976 les mesures donnaient une vitesse de 2m/sec soit une amplitude de l'ordre de 10 km. On donne actuellement plutôt la valeur 0.5m/sec ce qui peut signifier ou que l'amplitude a décrû en 10 ans ou que les premières mesures étaient imprécises. En tous cas cette amplitude est beaucoup plus grande que ce que les modèles standards prévoyaient. Mais depuis que l'on a découvert d'autres modes de gravité dont l'amplitude est tout aussi extravagante, le mode de 160 min n'est plus exceptionnel en cela.

Ce qui rend ce mode de 160 min unique c'est en fin de compte sa cohérence de phase. Cette oscillation est observée depuis près de 10 ans et elle n'a bougé ni en fréquence, ni (probablement) en amplitude et n'a subi aucun sursaut ("glitch"). En cela elle diffère considérablement des autres modes de gravité dont la durée de vie est de l'ordre de quelques mois (c'est ce qui les rend d'ailleurs beaucoup plus difficiles à observer).

Les tentatives d'explication de ce mode furent passées en revue lors d'une conférence qui eut lieu à Catania en Juin 1983 [2]. Celle de G.Isaak est que le système solaire serait plongé dans des ondes gravitationnelles venant d'un système binaire proche qui exciteraient un mode de pression (et non de gravité) au centre du soleil. Ce mode, en se propageant vers l'extérieur du soleil, s'amplifierait jusqu'à 100 000 fois peut-être, comme le "claquement d'un fouet" (à cause de la diminution de densité).

P.Delache, présent à cette réunion, fut séduit par cette idée d'Isaak. Sans apparemment questionner la possibilité théorique du phénomène, et nonobstant le fait que ce mode semblait de période trop longue pour être un mode de pression, il se mit alors en quête d'un excitateur possible en dehors du système solaire. C'est ainsi qu'il se trouva sur la piste de Geminga.

Une mystérieuse source gamma

L'astronomie gamma débute avec l'ère spatiale. Deux satellites lui furent exclusivement consacrés: SAS-2, lancé par la NASA en 1972, qui expira au bout de huit mois et COS-B mis sur orbite par l'Agence Spatiale Européenne; géré par un cartel de 6 laboratoires baptisé "Collaboration Caravane", il fut lancé en août 1975 et fonctionna pendant 7 ans. En plus d'un fond diffus, 25 sources ponctuelles furent répertoriées. Sur ces 25 sources 4 seulement sont actuellement identifiées: le quasar 3C273 et le nuage Rho Ophiuchi ainsi que les pulsars du Crabe et de Véla [9-12].

*

Pour identifier une source gamma à un pulsar on ne procède pas uniquement par coïncidence positionnelle vue la taille considérable de la boite d'erreur gamma (qui a au mieux 1/2° de côté). On cherche plutôt s'il existe dans les données gamma la même périodicité que dans le domaine radio. Or les données gamma consistent en quelques centaines de photons recueillis dans un petit nombre de fenêtres d'observation d'un mois chacune environ, réparties sur 7 ans. Pour trouver une périodicité dans ces données on divise donc les fenêtres d'observations en intervalles égaux à la période cherchée (33 msec pour le Crabe et 89 msec pour Véla); comme un mois compte environ 10^7 telles périodes on doit aussi tenir compte de leur variation avec le temps, $\dot{P}$, qu'il faut par conséquent également connaître par ailleurs. On répartit ensuite dans ces intervalles les photons détectés (leurs temps d'arrivée sont connus à mieux de 0.25 msec près). Puis on effectue un "repliage" des périodes les unes sur les autres en aditionnant, modulo la période, les photons reçus. On voit bien que s'il n'y a pas de périodicité dans les données gamma ou si les valeurs de P ou de $\dot{P}$ sont inexactes on obtiendra après repliage une distribution uniforme de photons. En revanche si une structure de pulse apparaît c'est signe de l'existence d'une période. Quelques essais autour des valeurs de P et de $\dot{P}$ de départ permettent alors de faire apparaître cette structure plus franchement, une opération que l'on rend quantitative par un test χ^2 qui mesure l'écart entre la structure considérée et une distribution uniforme.

Il est capital de noter que cette méthode de repliage n'a de sens que si l'on connaît a priori la période que l'on cherche. En effet chercher une période au hasard entre une msec et une min par exemple implique un grand nombre d'essais (d'autant plus qu'il faut déterminer $\dot{P}$ également); la probabilité de trouver, due au hasard, une structure de pulse devient alors grande et trouver une structure n'est plus significatif. A cause de ces délicats problèmes de signification statistique des résultats il n'y a que deux cas seulement où l'identification source gamma-pulsar se soit faite sans problème. Il s'agit des pulsars du Crabe et de Véla [13-17].

*

Geminga, primitivement appelé Gamma 195+05 ou 2CG 195+04 se trouve dans l'anticentre de la galaxie, à 14° du Crabe. Après Véla c'est la source gamma la plus brillante du ciel et elle fut donc découverte très tôt, dès 1975 par l'équipe de SAS-2. COS-B la détectait en 1976 et en recueillit en tout 913 photons. Mais alors que l'identification du pulsar du Crabe était quasi immédiate, ce que l'on connut de Geminga se limita pendant 10 ans à son spectre gamma [9,12].

En comparant ce spectre à ceux des pulsars du Crabe et de Véla on en déduisit que cet objet était probablement aussi un pulsar ou du moins une étoile à neutrons. Par ailleurs parce que les photons semblaient s'espacer de nombres entiers de minutes, on chercha une périodicité de l'ordre de la minute dans les temps d'arrivée. L'équipe de SAS-2 prétendit en 1977 avoir effectivement trouvé un résultat positif, marginalement significatif; ils obtinrent également une valeur pour la dérivée, $\dot{P}$, très supérieure aux P habituels des pulsars ($\sim 2\ 10^{-9}$ au lieu de 10^{-13}-10^{-14}). Ces résultats, confirmés dans un premier temps, furent ensuite rejetés par la Collaboration Caravane; mais comme nous le verrons, ce débat a été relancé récemment.

Pendant ces 10 années on chercha aussi une contrepartie à cette source gamma. Sans succès. D'où le nom que l'équipe milanaise lui a donné: "Geminga", une contraction de gamma et de Gémeaux (la constellation où se trouve cette source) qui signifie "il n'y a rien" en milanais. Mais en Septembre 1979 et Mars 1981 le satellite Einstein était pointé vers Geminga et détectait une source de rayons X: 1E 0630+18, que l'on identifia à Geminga. G.Bignami et ses collaborateurs pouvaient enfin écrire "the quest for elusive Geminga (...) comes here to an end". En Septembre 1983 le satellite européen Exosat confirmait l'existence et les propriétés de cette source X [18-19, 37].

Cette identification X qui réduisait la boite d'erreur où se trouvait Geminga à 3 sec d'arc rendit possible la recherche d'une contrepartie optique. H.Sol et M.Tarenghi avaient déjà, sur des plaques prises à la Silla (Chili), des indications de l'existence d'une étoile à cet endroit mais des résultats plus concluants furent obtenus par Caraveo et al. à partir de photos prises à Hawaï en Avril 1983. Dans la boite d'erreur X se trouve un objet de magnitude m_b=21.7±0.3 et m_r=20.5±0.2 (trop faible donc pour avoir été répertorié dans le catalogue du mont Palomar). L'argument le plus convaincant pour identifier cette source optique à Geminga n'est pas tant sa coïncidence positionnelle que le fait qu'elle est très différente des 70 autres sources du champ: si l'on trace le diagramme m_r en fonction de m_r-m_b la source optique considérée se démarque nettement des autres. L'existence de cette source optique fut depuis confirmée par Halpern et Grindlay (observations de Septembre 1983 à l'aide du télescope de Hale) et par H.Sol et al. après des observations en fin 1983 et Janvier 1984 au Chili et à Hawaï [20-22].

Ces identifications apportèrent la preuve que Geminga était un objet proche, situé sûrement à une distance inférieure à 100 psec car les X mous du spectre ne sont pas absorbés. Tant que seule l'identification X fut sûre (car l'identification optique et l'absence de parallaxe élimina cette possibilité) on pensa même que Geminga pourrait former avec le soleil un système double et ainsi rendre compte des perturbations inexpliquées dans les trajectoires des planètes extérieures. Cette hypothèse hardie et frappante pour l'imagination- on parla du "compagnon noir" du soleil- courait au moment (été 1983) où P.Delache cherchait un objet exotique et proche qui ferait vibrer le soleil à 160 min. On comprend que Geminga l'ait intéressé.

L'annonce d'une coïncidence de périodes entre le soleil et Geminga

Songeant donc à Geminga pour expliquer l'oscillation à 160 min du soleil, P.Delache demanda à J.Paul et G.Bignami de chercher dans les données gamma de Geminga une périodicité aux environs de 160 min.

Pour que le résultat soit statistiquement significatif, il fallait que l'éventuelle périodicité dans les données gamma de Geminga soit trouvée indépendamment de l'existence de l'oscillation solaire. Or les astronomes gamma, pour vérifier l'existence d'une période, avaient besoin, comme on l'a vu, qu'on la leur donne a priori. Leur idée fut donc de chercher la valeur de période à trouver dans les données gamma elles-mêmes puisqu'à l'époque, l'été dernier, Geminga n'était connu de façon précise que dans le domaine gamma. Il calculèrent donc la transformée de Fourier de la distribution temporelle des temps d'arrivée des photons. Comme les 913 photons émis par Geminga en 7 ans étaient répartis sur 5 fenêtres d'observation d'un mois chacune environ il fallait tenir compte des effets liés à la présence de ces fenêtres. Ils divisèrent donc les 7 ans en environ 10^5 intervalles de 45 min chacun et définirent une fonction, nulle en dehors des fenêtres et égale, dans chaque intervalle, au nombre de photons reçus dans cet intervalle diminué du nombre moyen de photons reçus dans la fenêtre d'observation.

La transformée de Fourier de cette fonction présente aux alentours de 160 min un pic au milieu d'une structure complexe, structure qui est due à l'existence des fenêtres. En effet si on calcule la transformée de Fourier d'un signal sinusoïdal modulé par ces fenêtres on obtient dans ce cas également un pic central encadré de fantômes. Afin de tester la réalité de ce pic les séries temporelles dans chaque fenêtre d'observation furent inversées sans changer l'ordre de ces fenêtres; si le pic était dû à la présence des fenêtres uniquement, cette modification ne changerait pas le résultat; or le pic disparaît. Enfin pour être sûr que ce pic n'était pas dû à des effets inhérents au satellite, les photons issus d'une autre région du ciel furent étudiés de la même façon. Aucun pic ne fut observé près de 160 min. J.Paul et G.Bignami en conclure que ce pic à 159.96 min était réel, sa hauteur indiquant que la probabilité que sa présence soit due au hasard est de 5.10^{-3}.

L'étape suivante dans l'identification d'une période dans les données gamma de Geminga fut, elle, standard. Connaissant dorénavant la période à identifier, 159.96 min, J.Paul et G.Bignami utilisèrent la méthode des repliages pour affiner le résultat, un test χ^2 mesurant à quel point l'histogramme obtenu diffèrait de l'hypothèse de distribution uniforme. Le maximum est obtenu pour T_G=159.9588 min; la probabilité pour que cette structure soit due au hasard est de 5.10^{-4} [23].

La période de l'oscillation solaire est T_S=160.0095min. La différence entre les deux fréquences, $1/T_G-1/T_S$, est à 10^{-5} près, de 1/1an, différence dont on conçoit qu'elle puisse s'expliquer par un artefact lié à la rotation de la terre autour du soleil. La coïncidence entre ces deux périodes devient alors troublante.

En Septembre 1983, Delache, Paul, Bignami et Isaak s'étaient donc convaincus que ces deux périodes, celle du soleil et celle de Geminga, devaient être reliées causalement, et que ce lien devait être les ondes gravitationnelles émises par Geminga. Une découverte qui fut annoncée avec un certain bruit [24-25].

Geminga peut-il faire vibrer le soleil?

La réaction des astrophysiciens relativistes à cette nouvelle ne se fit pas attendre. En une quinzaine de jours la réponse tombait sous forme d'une série de préprints: impossible; le mode de 160 min du soleil ne peut, en aucune manière, être excité par une onde gravitationnelle [26-33].

Une première question concerna le mécanisme d'émission par Geminga des ondes gravitationnelles. Est-ce un système double qui perd de l'énergie sous forme de rayonnement gravitationnel? L'hypothèse semblait gratuite: rien n'indiquait que Geminga fût un système double. En admettant néanmoins que ce soit le cas, une périodicité de 160 min pour les ondes émises implique que la période orbitale du système est le double, soit 320 min. Comment alors expliquer une période de 160 min dans l'émission gamma? En supposant, comme le suggère P.Delache, que les deux membres du système émettent dans le domaine gamma (on verrait alors passer deux phares gamma toutes les 320 min, phénomène équivalent à un phare toutes les 160 min)? C'est là un modèle peu plausible, tout comme celui d'un seul objet émetteur dont l'axe des zones émissives s'alignerait deux fois par période avec l'axe de visée.

Comment par ailleurs expliquer la différence de 1/1an entre les deux fréquences? Par un artefact lié à la rotation de la terre autour du soleil? Dans l'article du "Monde" du 11 Octobre 1983, M.Arvonny (et P.Delache) suggèrent que le mode solaire est propagatif: c'est une "vague" qui tourne autour du soleil; pour un observateur sationnaire cette vague a une fréquence $1/T_G$; pour un observateur terrestre qui tourne autour du soleil dans le même sens que la vague en 1 an, la fréquence est de $1/T_G-1/1an=1/T_S$; c'est l'effet "Phileas Fogg" comme l'a baptisé M.Arvonny dans son article, en hommage au héros de J.Verne qui fit le tour du monde en 80 jours pour ses amis londoniens et en 81 à sa montre. A cette explication, les astrophysiciens relativistes objectèrent qu'une onde gravitationnelle (à l'approximation linéaire du moins) interagissant avec une sphère de fluide parfait excite le mode quadrupolaire ($l=2$); tenir compte de la rotation de la terre autour du soleil entraîne une levée de la dégénérescence des modes azimuthaux ($m=\pm2,\pm1,0$); on devrait donc observer non pas un mode solaire mais 5; pourquoi n'en observerait-on qu'un seul? En réponse P.Delache argue que traiter le soleil comme un fluide parfait est une grossière approximation; à cause d'effets non-linéaires et de la rotation différentielle du soleil, on peut concevoir que des modes autres que quadrupolaires soient excités. Soit, mais quel pourrait être le modèle?

L'objection la plus décisive cependant provint d'arguments énergétiques (présentés par S.Bonazzola, B.Carter, J.Heyvaerts, J.P.Lasota de l'Observatoire de Meudon, par P.Tourrenc de l'Institut H.Poincaré et par J.Fabian et D Gough de l'Université de Cambridge). Considérons le soleil comme un oscillateur soumis à une force excitatrice résonnante, en l'occurrence une onde de gravitation périodique provenant d'un système double, Geminga. Supposons que cette onde, dont on connaît l'amplitude maximale puisque les mesures de détection ont jusqu'à présent été négatives [27], rende compte de l'amplitude observée du mode solaire; ceci implique que le facteur de qualité du soleil est très élevé, mais admettons que le soleil soit un résonateur de premier ordre. On en déduit alors que la bande passante du soleil, c'est-à-dire l'intervalle de fréquences autour du mode pour lequel le soleil entre en résonance, est très étroite. Or la période des ondes gravitationnelles émises par Géminga ne peut rester accordée au mode solaire très longtemps; en effet le système supposé double de Géminga perd de l'énergie, ne serait-ce que parce qu'il émet des ondes gravitationnelles; les deux objets spiralent l'un vers l'autre de plus en plus vite, et la période des ondes émises décroît en proportion. En combinant ces deux contraintes, on s'aperçoit que pour que le soleil et Geminga restent en phase, il faudrait que Geminga se trouve quelque part entre Neptune et Pluton...Une conclusion qui semble sans appel.

P.Delache récuse cet argument qui repose sur un modèle extrêmement simplifié du soleil, répondant linéairement à une excitation extérieure. L'onde gravitationnelle n'a pas en effet à rendre compte de l'amplitude du mode solaire comme l'implique le modèle ci-dessus mais seulement de sa cohérence de phase- seul point qui le distingue des autres modes de gravité. P.Delache suggère alors de traiter le soleil comme un oscillateur répondant à une force générée en son centre, qui rende compte de l'amplitude et de la durée de vie des modes de gravité en général (notons que l'on ignore encore l'origine et la forme de cette force puisque l'amplitude au moins des modes de gravité est inexpliquée); supposons maintenant que l'effet de l'onde gravitationnelle en provenance de Geminga soit de moduler la fréquence de l'oscillateur solaire de sorte que l'équation du mode de 160 min devienne une équation de Matthieu (on peut imaginer que l'onde gravitationnelle modifie la pression entraînant, via des phénomènes non linéaires, une variation de la vitesse du son et donc une modulation de la fréquence du mode); on sait que l'on peut alors voir apparaître des phénomènes d'amplification paramétrique bien plus efficaces que dans l'approximation linéaire.

Les objections de P.Delache au modèle simplifié proposé par ses détracteurs sont pertinentes, mais le modèle qu'il suggère, à supposer qu'il soit mis en forme et explique comment une onde gravitationnelle modulerait la fréquence d'une oscillation solaire, dépendra a priori d'un grand nombre de paramètres que l'on peut fixer à son gré et sera donc nécessairement ad hoc.

La conclusion de cette polémique semble donc claire. Si l'impossibilité absolue d'un couplage par ondes gravitationnelles entre le soleil et Geminga ne peut, bien sûr, être démontrée, il n'en reste pas moins que tant qu'un modèle du soleil n'aura pas été fermement établi, cette coïncidence entre une période solaire et une période de Geminga, aussi frappante soit-elle, doit rester au rayon des curiosités.

Y a-t-il une période dans les données gamma de Geminga?

Si le débat sur la possibilité d'exciter un mode solaire par des ondes gravitationnelles semble clos, la polémique sur la réalité même des faits présentés reste vive. Il est en effet bien évident que si le soleil ne vibre pas à 160 min, ou si les données gamma de Geminga n'exhibent pas, en fin de compte, de périodicité à 160 min, alors l'édifice s'effondre.

Or la Collaboration Caravane dans son ensemble, mis à part J.Paul et G.Bignami, a nié l'existence d'une période de 160 min dans les données gamma de Geminga ou du moins ne l'a pas trouvée significative, ainsi que l'a fait savoir sans détours et sans ménagements le télex de Scarsi du 16 Février 1984 à la revue "Nature" :"...the Caravane Collaboration wishes it to be known that, using their well tried analysis procedure they do not find the result to be statistically significant".

Le désaccord porte sur la façon dont G.Bignami et J.Paul ont analysé les données. Pour trouver une période dans les données de Geminga J.Paul et G.Bignami ont procédé, comme on l'a vu, en deux étapes distinctes. Dans un premier temps, en analysant la transformée de Fourier des données, ils ont obtenu une période. Cette période étant alors donnée, il ont effectué un repliage et ont trouvé un résultat qu'ils estiment significatif. Mais, et c'est là l'objection qui leur est faite, les deux étapes de l'analyse ne sont pas indépendantes puisque le stock de données est le même pour chaque étape. La période que Paul et Bignami recherchent par repliage n'est pas donnée "par ailleurs": elle est tirée du même lot de données. Et prendre, ainsi qu'ils l'ont fait, une transformée de Fourier suivie d'un repliage comportant peu d'essais est équivalent à un repliage, non précédé d'une transformation de Fourier, comportant beaucoup d'essais et le résultat, est beaucoup moins significatif qu'ils ne l'affirment. Il aurait fallu scinder les données en deux groupes distincts et statistiquement indépendants.

Quelle que doive être l'issue de ce débat, il fut rapidement clair à tous les protagonistes que le plus sûr moyen de ne pas s'y enliser était de chercher une périodicité de 160 min dans Geminga "par ailleurs" au sens strict du terme, c'est-à-dire dans les contreparties X et optique nouvellement découvertes de cet objet.

Résultats récents sur la contrepartie X de Geminga

Très récemment (Avril 1984), Bignami, Caraveo et Paul ont étudié les données de la contrepartie X de Geminga pour y chercher une période de l'ordre de la minute. Le résultat fut positif [34].

La première conclusion des auteurs est double. Tout d'abord l'existence d'une périodicité aux alentours de 1 min dans les données X prouve que sa présence dans les données gamma, suspectée par l'équipe de SAS-2 en 1977, était significative contrairement aux conclusions finales de la Collaboration Caravane. Par ailleurs si cette périodicité existe bien à la fois en X et en gamma, alors on tient là le meilleur argument pour identifier ces deux sources.

Mais les auteurs vont plus loin. En effet la période X, en 2 ans, semble avoir varié. Collectant toutes les indications existant sur cette périodicité ils obtiennent $\dot{P}$ en fonction du temps. Un meilleur ajustage leur permet d'en déduire que jusqu'en 1979 environ $\dot{P}$ valait $\sim 2.4\ 10^{-9}$

mais a ensuite quasiment doublé: $P \sim 4.68\ 10^{-9}$. Les auteurs considèrent comme une confirmation de leurs conclusions le fait que d'après les observations du satellite HEAO 3 analysées par Durouchoux et al., Geminga soit apparu comme un puissant émetteur de gamma mous (photons d'énergie 1Mev) en 1980 mais pas en 1979.

Par ailleurs une variation du flux X sur une échelle de temps de quelques milliers de secondes a été notée mais les résultats, bien que suggestifs, ne semblent pas encore suffisants pour que leur signification statistique soit claire [37-38].

Résultats récents sur la contrepartie optique de Geminga

Le premier résultat nouveau concernant la contrepartie optique de Geminga, découverte rappelons-le en 1983, fut obtenu par H.Sol et al. et concerne son mouvement propre et sa parallaxe. Aucune parallaxe n'ayant été observée, la précision des mesures implique que Geminga est au moins à 10 psec du système solaire. Cette mesure est la première qui donne une limite inférieure à la distance de Geminga. Quant au mouvement propre il est inférieur à 0.2 sec d'arc par an. On en déduit la distance de Geminga, en faisant des hypothèses sur sa vitesse tangentielle: une vitesse de 10 km/sec mettrait l'objet à une distance de 10 psec, une vitesse de 200 km/sec à 200 psec; les pulsars ayant une vitesse tangentielle moyenne de 130 km/sec, on voit qu'un modèle qui fait de Geminga une étoile à neutrons à une centaine de psec du système est cohérent avec ces observations [35].

Mais un résultat annoncé par Vigroux et al. peut, s'il est confirmé, donner une nouvelle impulsion au débat: l'intensité du rayonnement optique émis pourrait être modulée...à 160 min [36]. Cette indication fut obtenue par repliage d'une série d'une trentaine de clichés pris en 2 nuits en Janvier 1984 à Hawaï, couvrant ainsi une dizaine de périodes de 160 min. Les auteurs considèrent que la modulation obtenue est compatible avec une périodicité de 160 min, à condition que l'existence d'un telle périodicité soit démontrée par ailleurs.

Un modèle pour Geminga?

En admettant les conclusions de Bignami et al. sur la contrepartie X (existence d'une période de 59 sec de dérivée $\dot{P} \sim 10^{-9}$) et en supposant que les indications de Vigroux et al. sur la contrepartie optique (existence d'une période de 160 min) soient confirmées, un modèle pour Geminga émerge [34,39]: ce serait une étoile à neutrons freinée par la présence d'un compagnon, ce qui expliquerait la valeur de $\dot{P}$, beaucoup trop importante pour être celle d'un pulsar isolé (pour lesquels $\dot{P} \sim 10^{-13}$). Mais le freinage ne s'effectuerait pas par transfert de matière, car alors le disque d'accrétion serait visible, mais par un phénomène de marée ou de frottement électromagnétique à élucider. La possibilité qui semblerait la plus plausible est que Geminga soit une "variable cataclysmique", c'est-à-dire un système formé d'une étoile à neutrons et d'une naine blanche qui en l'occurrence serait assez peu lumineuse. Pour corroborer ces dires et leur donner un certain piment, une étude des données astronomiques chinoises montre qu'en 435 après J.C. une nouvelle étoile apparut dans le ciel, visible en plein jour, au point de coordonnées 195+04 (à 1°

près), c'est-à-dire là où se trouve Geminga. On ne pense pas que cet événement ait été une supernova; il aurait pu en revanche marquer la naissance d'une variable cataclysmique...

Les objections que l'on peut opposer au modèle de Bignami et al., mis à part le fait qu'il repose sur des résultats pour le moins statistiquement peu concluants, sont nombreuses. L'une d'elles consiste à remarquer que les rapports de luminosité L_γ/L_X et L_X/L_{opt} ne sont pas ceux habituellement observés pour les étoiles à neutrons. En réponse J.Paul remarque que la source observée en optique n'est peut-être pas l'étoile observée en gamma et en X mais son compagnon. Mais H.Sol et al. en tirent, eux, une conclusion beaucoup plus drastique: la source optique actuellement identifiée à Geminga (ou à son compagnon) ne serait ni Geminga ni son compagnon. La contrepartie optique de Geminga serait en fait un autre objet G', de magnitude 24, présent à la fois sur les plaques d'H.Sol et al. et sur celles de L.Vigroux et al., objet dont l'intensité dans le domaine optique serait mieux adaptée à un modèle de pulsar pour Geminga. Si cette conclusion est correcte, la modulation à 160 min de l'intensité de l'objet G suggérée par Vigroux et al., dans la mesure où elle est significative, n'est plus qu'une coïncidence de plus...Mais l'existence même de G' n'est pas encore certaine: il semblerait en effet qu'il ne soit pas à la même place sur les plaques d'H.Sol et al. et sur celles de Vigroux et al.; est-ce parce qu'il aurait bougé, auquel cas G' serait très proche de nous??

*

Cette série d'interrogations montre bien que si l'"affaire" Geminga concernant l'excitation d'un mode solaire par des ondes gravitationnelles est un dossier clos pour l'instant, le problème Geminga, lui, reste entier. En deux ans une mine de renseignements a été obtenue sur cet objet, mais on ignore toujours ce qu'il est.

Remerciements

Cette revue n'aurait pas été possible sans de nombreuses discussions et sans avoir la possibilité d'avoir accès à de nombreux documents non publiés. Je remercie tout spécialement S.Bonazzola, B.Carter, P.Delache, J.P.Lasota, J.Paul, H.Sol, P.Tourrenc et C.Vanderriest.

Références

[1] J.Christensen-Dalsgaard, D.O.Gough, Nature 259,87 (1976)

[2] D.Gough, Nature 304,689 (1983)

[3] P.Delache, P.H.Scherrer, Nature 303,651 (1983)

[4] A.B.Severny, V.A.Kotov, T.T.Tsap, Nature 259,87 (1976)

[5] J.R.Brookes, G.R.Isaak, H.B. van der Raay, Nature 259,92 (1976)

[6] A.B.Severny, V.A.Kotov, T.T.Tsap, A&A 88,317 (1980)

[7] S.Koutchmy, O.Koutchmy, V.A.Kotov, A&A 90,372 (1980)

[8] F.Cavallini, G.Ceppatelli, A.Righini, Mem. Soc. Astron. Ital. 51,611 (1980)

[9] D.J.Thompson et al., Ap.J. 213,252 (1977)

[10] H.A.Mayer-Hasselwander et al., Proc. 9th Texas Meeting, Annals New York Ac. Sc. (1980)

[11] K.Pinkau, Proc. 9th Texas Meeting, Annals New York Ac. Sc. (1980)

[12] B.N.Swanenburg et al., Ap.J. 243,269 (1981)

[13] J.P.Leray et al., A&A 16,443 (1972)

[14] H.Ogelman et al., Ap.J. 109,584 (1976)

[15] K.Bennett et al., A&A 61,279 (1977)

[16] G.Kanbach et al., A&A 90,163 (1980)

[17] R.D.Wills et al., Nature 296,723 (1982)

[18] G.F.Bignami, P.A.Caraveo, R.C.Lamb, Ap.J. 272,L9 (1983)

[19] P.A.Caraveo, G.F.Bignami, P.Giommi, Bull. Am. Astron. Soc. 15,908 (1983)

[20] G.F.Bignami et al., Bull. Am. Astron. Soc., 15,908 (1983)

[21] J.P.Halpern, J.E.Grindlay, Bull. Am. Astron. Soc., 15,908 (1983)

[22] P.Caraveo et al., Ap.J., 276,L45 (1984)

[23] J.Paul et al., preprint

[24] M.Arvonny, Le Monde, 12039,1,14 (1983)

[25] R.Walgate, Nature, 305,665 (1983)

[26] A.J.Dean, Nature, 308,113 (1984)

[27] J.D.Anderson et al., Nature, 308,158 (1984)

[28] A.C.Fabian, D.O.Gough, Nature, 308,160 (1984)

[29] S.Bonazzola et al., Nature, 308,163 (1984)

[30] J.R.Kuhn, S.P.Boughn, Nature, 308,164 (1984)

[31] B.W.Carroll et al., Nature, 308,165 (1984)

[32] J.M.Moffat, preprint

[33] P.Tourrenc, preprint

[34] G.F.Bignami, P.A.Caraveo, J.A.Paul, à paraître dans Nature

[35] H.Sol, M.Tarenghi, C.Vanderriest, L.Vigroux, G.Lelièvre, preprint

[36] L.Vigroux, J.A.Paul, P.Delache, G.F.Bignami, P.A.Caraveo, C.Salotti, preprint

[37] P.A.Caraveo, G.F.Bignami, P.Giommi, S.Mereghetti, J.A.Paul, Nature, in press

[38] P.L.Hertz, J.E.Grindlay, Ap.J., in press

[39] J.A.Paul, J.Astr.Franç., à paraître

THE 3K BACKGROUND RADIATION: OBSERVATIONAL AND THEORETICAL STATUS

R. Fabbri
Istituto di Fisica Superiore, Università di Firenze,
Via S. Marta 3, I-50139 Firenze, Italy

Introduction

About twenty years have elapsed since the discovery of the cosmic background radiation (CBR) by Penzias and Wilson (1965). Whereas early experimental efforts were intended to verify the agreement of its gross features with the idealized (homogeneous and isotropic) version of the general-relativistic hot-big-bang picture, at present most cosmologists believe that the standard model is sufficiently well established, at least as a zero-order approximation to the real universe (see however the next section). As a result, most experimental and theoretical efforts are directed to the details of the CBR properties, i.e. spectral distortions, angular anisotropies and polarization, as sensitive probes of the large scale structure of the universe.

To appreciate this basic point, let us begin recalling that the idealized version of the standard model is the homogeneous and isotropic Friedmann cosmology, whose metric can be written as

$$ds^2 = g_{ik}^{iso}\, dx^i dx^k = -c^2 dt^2 + a^2(t)\left[d\chi^2 + \Sigma^2(\chi)\,(d\theta^2 + \sin^2\theta\, d\varphi^2)\right]; \qquad (1)$$

we use here the notation of Misner et al. (1973). The evolution of the cosmic geometry is simply described by the scale factor a(t), in terms of which the cosmological redshift is given by $1+z = a(t_o)/a(t)$ with t_o the present epoch. The model gives also a quite simple description of the cosmic medium, which in the redshift interval $10^8 \gtrsim z \gtrsim 10^3$ is schematized as a uniform expanding plasma (mainly ionized hydrogen and helium) in thermal equilibrium with a radiation field at the thermodynamic temperature

$$T = T_o(1+z). \qquad (2)$$

While the temperature decreases with time according to Eq. (2), the average photon occupation number $N = \left[\exp(h\nu/kT) - 1\right]^{-1}$ is conserved, with the ratio ν/T constant along each photon geodesic path between any two scatterings. Later, when T drops to about 4000 K, the cosmic plasma becomes a neutral gas and the radiation begins thereby to propagate freely; thus the notion of a last-scattering (hyper)surface placed at z $\simeq 10^3$ is introduced. The photon occupation number being since conserved for each mode, a redshifted blackbody is left over for today's observers.

Applying ordinary physics to the cosmic medium of the Friedmann cosmology gives the prediction of a structureless CBR as a relic of the primeval fireball. Any variation of this simple picture should be ascribed to one of the following complications.

(a) Non-equilibrium processes. The spectrum may become non-Planckian when matter and radiation are not in strict thermal equilibrium but some interaction is still allowed. This situation occurs in the idealized model, too, during the decoupling around $z \simeq 10^3$. However the resulting spectral distortion is placed at observation wavelengths $\lesssim$ 0.2mm, where the CBR is masked by dust emission from our Galaxy; see the calculation of Peebles (1968) and the experiment of DeBernardis et al. (1984). More relevant distortions arise in the "generalized" standard model which allows the dissipation of inhomogeneities; also, apparent distortions may be generated by extragalactic emissions superposed to the CBR. For spectral distortions of any type, we can define a brightness temperature $T(\nu)$, in terms of which $N = \left[\exp(h\nu/kT(\nu)) - 1\right]^{-1}$. Note that non-equilibrium processes or spurious emissions are usually calculated in the metric of Eq. (1), but in fact they imply inhomogeneities at some length scale. As an example, Zeldovich and Sunyaev (1969) found that a distortion is produced by Compton scattering at $z \sim 10$ if the cosmic medium experienced a secondary ionization; such a reheating, although it is schematized as homogeneous, could be powered only by strongly inhomogeneous processes.

(b)Perturbations of the geometry of spacetime. In this case the metric coefficients of Eq. (1) are perturbed, so that $g_{ik} = g_{ik}^{iso} + h_{ik}$. The perturbations are in general spatially inhomogeneous, although homogeneous anisotropic models are also considered in the literature (MacCallum 1979). When the inhomogeneities are associated with matter density fluctuations (which is not the case for pure gravitational waves), it is often attempted to connect them to the observed structure, namely to the origin and properties of galaxies, clusters and superclusters; see e.g. Silk (1981). In models of this type, the thermodynamic equilibrium of matter and radiation at early epochs is usually maintained locally, the temperature being a function of all spacetime coordinates. As a result, a cosmological observer sees an angle-dependent thermodynamic temperature $T(\theta,\varphi)$ However, in models where the anisotropic distribution is intrinsically related to spectral distortions, it clearly becomes frequency-dependent.When the CBR is made anisotropic by large-scale perturbations, it is also expected to be partially polarized.

(c) Peculiar motion of the observer. If we are not cosmological observers, namely, our frame of reference does not coincide locally with the cosmological frame, a Doppler shift affects the angular distribution (but not the polarization state) of the CBR. Obviously local motions are related to the matter distribution which is strongly inhomogeneous on scales $\lesssim$ 50 Mpc; in practice we shall ascribe to point (c) any motion within the Local Supercluster, and consider length scales $\gtrsim 10^2$ Mpc as cosmological, see point (b).

(d) Finally, we mention the possibility of a breakdown of some piece of local physics, or of general relativity, or of the hot-big-bang scenario. Solutions based on one or more of these assumptions have been proposed for the spectral-distortion problem which will be considered in the next section.

In conclusion we can state that most theoretical models try to link the CBR properties to the cosmic structure within the framework of a refined standard model or to the details of the thermal history of the universe. We should also mention that, owing to the present collaboration of cosmology and high energy physics, even the properties of the CBR are often linked to the properties of particles. For instance, the hypothesis of galactic halos made of massive neutrinos or more exotic particles affects the predictions about the CBR anisotropies (e.g. Silk 1981), while a non-vanishing photon mass may explain spectral data (Georgi et al. 1983).

In the following sections we give a short summary of the experimental knowledge of the CBR and of the current theoretical work. For lack of space, we shall skip technical problems of experimental interest (as discussed by Weiss 1980 or Fabbri et al. 1981), and only give some hint of the physical input of models.

The spectrum

A large number of measurements of the brightness temperature have been performed over the years. While complete lists of earlier results are given by many review papers (Weiss 1980, Bussoletti et al. 1980), here we select a limited number of experiments which suffice to cover the wavelength region 50 cm $\gtrsim \lambda \gtrsim$ 1 mm, mostly with claimed accuracies of about 0.2 K. Table I gives the results of some radio-antenna measurements down to $\lambda \simeq$ 3 mm. (For indirect data based on interstellar molecular clouds, see the above reviews.) Figure 1 summarizes the data of the Berkeley infrared experiment (Woody and Richards 1981) which seems to imply a peculiar spectral distortion in the milli-

Table I. Selected radio measurements of the brightness temperature.

λ (cm)	T (K)	Reference
49.2	3.7 ± 1.2	Howell and Shakeshaft (1967)
20.7	2.8 ± 0.6	Howell and Shakeshaft (1966)
12.0	2.62 ± 0.25	Smoot et al. (1983)
6.3	2.71 ± 0.20	Smoot et al. (1983)
3.2	$2.69^{+0.16}_{-0.21}$	Stokes et al. (1967)
3.0	2.91 ± 0.19	Smoot et al. (1983)
1.6	$2.78^{+0.12}_{-0.17}$	Stokes et al. (1967)
0.9	3.16 ± 0.26	Ewing et al. (1967)
0.9	$2.56^{+0.17}_{-0.22}$	Wilkinson (1967)
0.9	2.87 ± 0.21	Smoot et al. (1983)
0.33	2.61 ± 0.25	Millea et al. (1971)

metric region. While the average of many radio results gives a brightness temperature of 2.7 K, there seems to be a radiation excess ($T = 3.1$ K) at $\lambda = 3$ mm and a decline to about 2.8 K at $\lambda = 1$ mm. The reality of this distortion has been a topic of debate for several years. In fact, it is difficult to reconcile such a spectral shape with the standard hot-big-bang ideas and standard physics.

Distortions can arise in the conventional scenario from an injection of energy in the radiation field. It turns out that the distorted spectrum does not depend on the source (which may be density waves or turbulent motions, an electron temperature higher than the radiation temperature, or exotic processes such as matter-antimatter annihilation), but rather on the interaction transferring the energy excess to the radiation. The relevant interactions at redshifts $z \lesssim 10^7$ (earlier distortions could not survive down to $z = 0$) are well-known processes, namely, Compton scattering and emission-absorption processes (double Compton, bremsstrahlung, radiative recombination, line emission). The resulting spectrum is easily calculated by solving the transport equation for unpolarized radiation in the Friedmann metric.

If the energy injection occurred at $z_i > z_a \sim 10^4$ multiple Compton scatterings were dominant, except at very long wavelengths where bremsstrahlung was more important, and

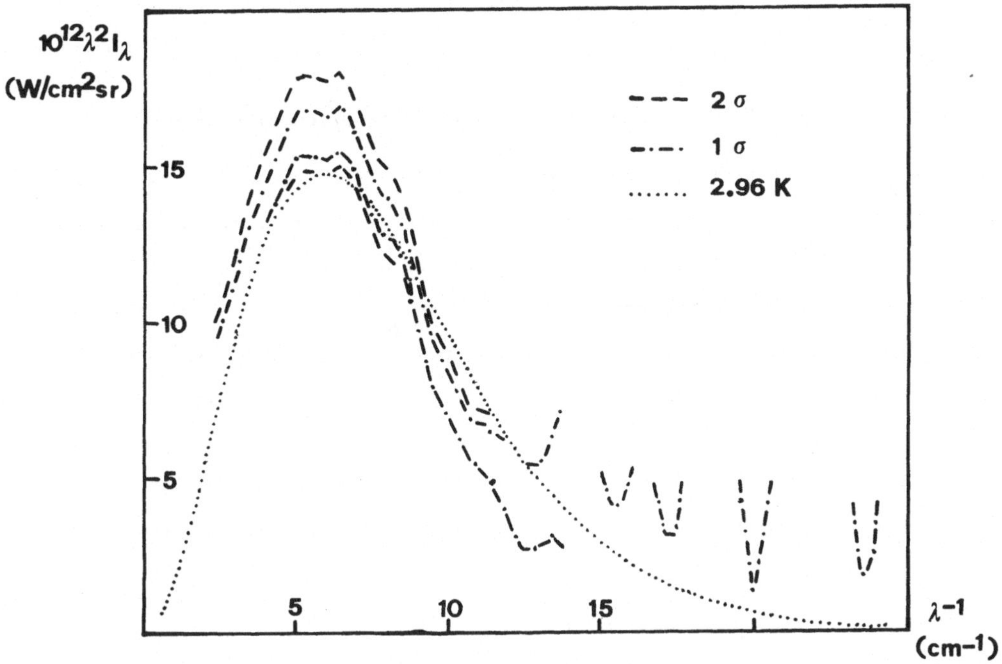

Fig. 1. The millimetric spectrum from the Berkeley experiment (Woody and Richards 1981).

produced a Bose distribution (Sunyaev and Zeldovich 1970, Illarionov and Sunyaev 1975)

$$N = \left[\exp\,(h\nu/kT + \mu) - 1\right]^{-1}. \tag{3}$$

Because of bremsstrahlung, μ in fact depends on the frequency and generates a radiation defect at some point of the Raileigh-Jeans region. At large times, the maximum value of μ is

$$\mu_{max} = (\Delta\rho/\rho_{rad})\,F(z_i) \tag{4}$$

with $\Delta\rho/\rho_{rad}$ the percent energy excess transmitted to the radiation field and $F(z_i)$ a dimensionless function which is 1.4 at $z_i = z_a$ and decreases at larger z_i. When the distortion is produced at $z_i < z_a$, the ordinary Compton scattering dominates and gives (Zeldovich and Sunyaev 1969, Chan and Jones 1975)

$$\frac{T(\nu) - T}{T} = u\left(\frac{h\nu}{kT}\coth\frac{h\nu}{2kT} - 4\right); \tag{5}$$

the parameter u is $(1/4)\,\Delta\rho/\rho_{rad}$, or, when heat transfer is due to an electron temperature $T_e > T$,

$$u = \int k(T_e - T)\,(m_e c^2)^{-1}\, d\tau \,, \qquad (6)$$

with τ the plasma optical depth. Now $T(\nu)$ is an increasing function of ν.

Both (3) and (5) give millimetric excesses of radiation with respect to the Raileigh-Jeans region, but fail to predict the subsequent decline around 1 mm. Although they have been used to fit radio and infrared data, the resulting fits were poor, and we can only set upper limits on the parameters: thus $\mu_{max}(z_a) < 2 \cdot 10^{-2}$ (with less stringent limits for more primordial z_i) and $u < 3 \cdot 10^{-2}$ (cfr. Danese and DeZotti 1977, Bussoletti et al. 1981). The limits on μ_{max} give constraints on the spectra of primordial perturbations (density waves and vortex motions, see Sunyaev and Zeldovich 1970), on the amount of cosmic antimatter (Stecker and Puget 1972) and exotic particles (Dolgov and Zeldovich 1980). The limits on u apply to the Compton distortion from a reheated plasma at $z \sim 10$, constraining thereby the strength of the secondary ionization.

Although the above limits are of clear interest for the conventional scenario, the Woody-Richards distortion remains unexplained. Thus new ideas have been set out. A few proposals are based on theories inconsistent with general relativity, such as the chronometric cosmology (Jakobsen et al. 1979). Another approach considers the distortion as not intrinsic to the CBR, but originated by the superposition of radiation from rather primordial ($z \simeq 200$) dust grains. This scenario, investigated by Aiello et al. (1980) and Rowan-Robinson et al. (1979), can be stretched out to the unconventional view that the entire CBR is non-primordial (Rees 1978, Carr 1981). In any case, it requires an early generation of pregalactic stars, usually referred to as Population III, as the basic powerhouse, and optically-thick interstellar dust as a thermalizing medium. We finally mention the proposal of Georgi et al. (1983), who maintain the standard cosmological scenario but modify the foundations of electrodynamics and cast doubts on Grand Unified Theories. The authors assume the existence of two photon states, say a_1 and a_2, the latter possessing a non-zero rest mass m_γ. The ordinary photon, which is coupled to matter, is described by the state $\cos\Phi\, a_1 + \sin\Phi\, a_2$, and the state orthogonal to it is uncoupled and unobserved. Photon oscillations then arise, quite analogous to $K\bar{K}$ or neutrino oscillations, and can reproduce the Woody-Richards distortion for $m_\gamma \sim 10^{-18}$ eV.

More certain data are required before one can assess the question of the spectral distortions. The Berkeley Group will probably supply new millimetric information at CIRP 3 in Zurich, July 1984; part of the discussion in this section might become obsolete in the near future.

Anisotropies

The distribution of the CBR in the sky has been measured so far at angular scales ranging from about 20" to the largest (dipole) scale. Small-scale surveys usually have a statistical character and provide rms angular fluctuations; if the CBR exhibits an angle-dependent temperature $T(\theta,\varphi)$ a typical double-beam experiment is able to measure $\Delta^2(\alpha,\beta) = \langle(T_1 - T_2)^2\rangle$, where the subscripts 1 and 2 denote two sky positions separated by an angular distance α, and both T_1 and T_2 are temperatures averaged over the beamwidth β. $\Delta(\alpha,\beta)$ is connected to the smeared angular correlation function $\Gamma(\alpha,\beta)$

$$\Delta^2(\alpha,\beta) = 2\left[\Gamma(0,\beta) - \Gamma(\alpha,\beta)\right]. \tag{7}$$

Table II summarizes a number of upper limits available on $\Delta(\alpha,\beta)$. For a better interpretation of the Table, we specify that in each experiment the beamwidth β is somewhat smaller than the quoted value of α, or coincides with its minimum when a range of values is reported. The quoted results of the Soviet Group (Parijskij et al. 1977, Berlin et al. 1982) have been criticized sometimes and appear to conflict with the confusion limit of discrete sources calculated by Danese et al. (1982). However, more recent data at 1.5 cm, where the source confusion limit is a less serious problem, vindicate earlier claims that the CBR is isotropic up to the level of a few parts in 10^5 in the arcminute range. Analogous limits are now available at scales $> 1^\circ$. The fluctuations detected at 6° and 2° (Melchiorri et al. 1981, Fabbri et al. 1984) are at present identified as millimetric emission ($\lambda \simeq 0.5$ mm) from Galactic dust. The large-scale distribution of the Galactic dust emission has been recently measured by DeBernardis et al. (1984).

For the result given by Fixsen et al. (1983) at scales $\geq 10^\circ$ (last entry in Table II) the large-scale (dipole) contribution has been subtracted. As a matter of facts, on large angular scales it is more convenient to use a deterministic approach: the aim of experiments is then to measure the first few coefficients $a_{\ell m}$ of the well-known harmonic expansion

$$\frac{\Delta T}{T} = \frac{T(\theta,\varphi) - \langle T\rangle}{\langle T\rangle} = \sum_{\ell,m} a_{\ell m} Y_{\ell m}(\theta,\varphi). \tag{8}$$

Table II. Selected upper limits on the rms angular fluctuations.

α	λ (cm)	$10^5\Delta(\alpha,\beta)$	Reference
18"-1'	6	100-50	Fomalont et al. (1982)
4!5	1.5	2	Uson and Wilkinson (1984)
5'-1°5	4	8-2	Parijskij et al. (1977)
4!5-1°	7.6	1-3	Berlin et al. (1982)
2°	0.05-0.3	3	Fabbri et al. (1984)
6°	0.05-0.3	3	Melchiorri et al. (1981)
≥ 10°	1.2	3	Fixsen et al. (1983)

Table III. Recent data on the dipole anisotropy.

λ (cm)	T_x	T_y	T_z	Reference
1.2	-3.07±0.17	0.67±0.09	-0.45±0.09	Fixsen et al. (1983)
0.3	-2.94±0.15	0.55±0.10	-0.28±0.09	Lubin et al. (1983)+
0.05-0.3	-3.5 ±0.7	0.0 ±0.2	-0.2 ±0.2	Ceccarelli et al. (1982)
0.05-0.3	-2.9 ±0.7	0.5 ±0.2	-0.3 ±0.2	Fabbri et al. (1984)

+Lubin et al. (1983) gives this result in terms of antenna temperature.

In practice experimentalists use real quantities instead of the complex $a_{\ell m}$; thus they describe the dipole anisotropy (ℓ = 1) by means of the vector T_n defined by

$$- T_x + i\,T_y = (3/2\pi)^{\frac{1}{2}} \langle T \rangle\, a_{11}, \qquad T_z = (3/4\pi)^{\frac{1}{2}} \langle T \rangle\, a_{10}. \tag{9}$$

Table III lists some recent results concerning the dipole anisotropy. The reported T_n are in mK, and refer to the thermodynamic temperature, except the data of Lubin et al. (1983) which refer to the antenna temperature (cfr. Fabbri et al. 1981). The same experiments have provided upper limits on the quadrupole anisotropy (ℓ = 2).

The radio data (Fixsen et al. 1983, Lubin et al. 1983) imply

$$\langle T\rangle \left(|a_{2m}|^2/4\pi\right)^{\frac{1}{2}} < 0.2 \text{ mK}, \tag{10}$$

while the infrared data at present give a corresponding upper limit of 0.5 mK (Ceccarelli et al. 1982, Fabbri et al. 1984). Also, A.A. Starobinsky informed us that a Soviet group (Strukov and Skulachev 1984) has performed the first satellite measurement of the large scale anisotropy. Their data imply a dipole amplitude about 20% less than the corresponding American radio data, while similar upper limits are given on the quadrupole terms.

We can conclude that at present only the detection of the dipole anisotropy is undisputed, while earlier results (Fabbri et al. 1980, Boughn et al. 1981) which were tentatively ascribed to a CBR quadrupole, are now recognized to be affected by Galactic emission. This is a pity, because current cosmological models often predict a richer structure in the CBR anisotropy.

As remarked in the Introduction, the angular distribution is a powerful means to probe the large-scale structure of the universe. The connection between such a cosmic structure and the anisotropy is founded on simple physical mechanisms. One mechanism consists of the spatial temperature fluctuations at the last scattering epoch; these spatial fluctuations (which arise from inhomogeneities in $\rho_{rad} \propto T^4$) are reflected on angular variations at observation. A further mechanism is the combined Doppler-gravitational redshift along the photon geodesics; when the metric (1) is perturbed, we have a redshift modulation depending on the line of sight (Sachs and Wolfe 1967). Assuming a definite last-scattering epoch, this effect can be written simply as (Fabbri 1984)

$$\frac{\Delta T}{T} = \frac{1}{2}\int_{t_o}^{t_1} \frac{\partial}{\partial t} h_\chi^\chi + v^\chi(t_o) - v^\chi(t_1). \tag{11}$$

Here v^χ is the radial component of the velocity perturbation ($v^\chi = 0$ in the Friedmann cosmology), h_χ^χ the radial component of the metric perturbation, and t_1 is the time of the last scattering of photons. In general, since the last scatterings occupy a finite time interval, the fluctuation of ρ_{rad} and Eq. (11) are not sufficient to calculate the observed $\Delta T/T$; one should incorporate such effects in a relativistic transfer equation for the radiation, which in fact was done by Peebles and Yu (1970), Wilson and Silk (1981) and Wilson (1983). However, for perturbations whose length scale was larger than the thickness of the last-scattering interval at $t \sim t_1$, Eq. (11) is sufficient (Fabbri et al. 1983).

Most models appearing in the literature are fully linearized, namely, not only h^i_k but also the density fluctuation $\Delta\rho/\rho$ is small. (Thus they cannot apply to galactic scales for $t \simeq t_o$.) In the linear regime, perturbations are classified as density, velocity and gravitational waves (Lifshitz and Khalatnikov 1963). Refined treatments of density perturbations distinguish adiabatic and isothermal waves, which have different temporal evolutions at redshifts $z > 10^3$. A limiting case is provided by homogeneous anisotropic models, where $\Delta\rho/\rho = 0$ but the Hubble expansion is anisotropic and a homogeneous velocity field may be allowed.

Even in this limiting case, in spite of the fairly small number of degrees of freedom, we encounter many complicated anisotropy patterns. When the harmonic expansion (8) is considered, however, the situation becomes somewhat simpler for homogeneous models, because we find that the dominant contribution comes from either the quadrupole or high-order harmonics. In the latter case, which applies to models where the cosmic density parameter is $\Omega_o < 1$, there appears a peculiar hot spot or ring of fire whose angular size α_o is of the order of Ω_o. This feature, studied by Novikov (1968) and Collins and Hawking (1973), is due to the dominance of harmonics with $\ell \sim \pi/\alpha_o$. For a more complete analysis of the harmonic expansion, see Fabbri et al. (1984).

Homogeneous models never give anisotropies with characteristic scales less than 1°, since $\Omega_o \gtrsim 0.1$. On the other hand, inhomogeneous models were in the past connected mainly to small-scale anisotropies (e.g., Peebles and Yu 1970, Doroshkevich et al. 1978). We shall not discuss here the constraints set on the vortex theory of galaxy formation (Kurskov and Ozernoi 1978), since the present interest is focused on adiabatic density waves. It turned out that models of galaxy formation based on such waves typically predicted anisotropies of order 10^{-4} at some preferred scale of order 10'. The limits on $\Delta(\alpha,\beta)$ becoming more and more stringent over the years, theoreticians worked hard to lower the predicted anisotropies; a possible solution was to introduce massive neutrinos in the theory of galaxy formation. (Neutrinos had been invoked for the problem of invisible galactic halos.) To make more precise statements, we recall that the adiabatic theory considers primordial density fluctuations whose Fourier spectrum $\delta(\vec{k})$ satisfies

$$|\delta(\vec{k})|^2 \propto k^n \tag{12}$$

with $4 \geq n \geq -3$. The experimental limits on the rms fluctuations in the arcminute range can be satisfied taking $n \simeq 4$ in the "standard" adiabatic theory or $n \gtrsim 1$ in in the extended theory with massive neutrinos.

In the Eighties the interest for the large-scale anisotropy has increased, and the same models have been employed to calculate $\Delta(\alpha,\beta)$ for large values of α, and the expectation values of $a_{\ell m}$ for low ℓ. Silk and Wilson (1981) found a dipole anisotropy of order 10^{-3} for adiabatic waves with $n \gtrsim 2$ for a cosmic density parameter (of the unperturbed Friedmann cosmology) $\Omega_o = 1$, so that the observed dipole might have a substantial cosmological component. Attempts were made to explain the quadrupole and the 6° data interpreted as actual detections; see for instance Silk (1982), and also the model of Peebles (1981) which assumes a spectrum like (12) at the present epoch (not primordially). However, interpreting the above results as upper limits, the adiabatic theory does not meet problems for the large-scale anisotropy (predicting a quadrupole of order 10^{-5} for several values of n and Ω_o), but it is only marginally consistent with small-scale data and is in serious difficulty also at intermediate scales for $\Omega_o = 0.1$ (Wilson 1983). The situation would be more favourable for isothermal waves (Davis and Boynton 1980, Silk and Wilson 1981).

No small-scale problem arises if instead of a power spectrum one considers a large-scale matter bump. Models of this kind include a spherical overdensity (Raine and Thomas 1981, Goicoechea and Sanz 1984, Fabbri 1984) and a long plane wave or its curved-space analog (Fabbri et al. 1983). The calculations show that for length scales much less than the Hubble radius, $L \ll c/H_o$, the dipole and harmonics with $\ell \sim c/H_oL$ dominate.For $L \gg c/H_o$ either the dipole or the quadrupole prevails, depending on the type of perturbation and the curvature of 3-dimensional space. In the intermediate case, $L \sim c/H_o$, a rich structure appears, and a carefull choice of the length scale and the bump geometry could reproduce any large-scale pattern. Models of this kind, however, are not founded on a theory of galaxy formation (this is the case, of course, also for homogeneous models and gravitational wave models).

As far as only a dipole anisotropy is detected, one can adopt the view that this is entirely due to the local dynamics (motion of the observer inside the Local Supercluster). However, in such a case we should find an agreement between the dipole parameters and the velocity field of galaxies on scales $\lesssim$ a few tens of Megaparsec. At present people tend to believe that such an agreement exists, owing to the recent astronomical data of Hart and Davies (1982). However, previous results by deVaucouleurs et al. (1981) and Tammann et al. (1979) do not agree with Hart and Davies' (and with each other), and limiting ourselves to the most recent results may be too optimistic. Further, the claimed agreement of the dipole and Hart and Davies' results holds for the direction of motion but not for the amplitude. The velocity of our Local Group inside the Local Supercluster is claimed to be (436 ± 55) km/sec, while its velocity against the CBR as calculated by Fixsen et al. (1983) is (600±50) km/sec. The result of Lubin et al. (1983a), when expressed in terms of thermodynamic temperature, seems to imply a still higher velocity; only the result of the Soviet Group (Strukov and Skulachev 1984) may weaken the discrepancy. If alternatively we adopt the view that the dipole anisotropy is not completely explained by the local dynamics, we may ask which length scale is involved. We still may invoke motions of the cosmological matter on scales of $\simeq 10^2$ Mpc or slightly more, with corresponding quadrupole amplitudes less than 10^{-5} (cfr. Silk and Wilson 1981); but if we make a further step and consider scales of some thousand Megaparsec, a quadrupole amplitude of 10^{-5}-10^{-4} becomes unavoidable (Fabbri et al. 1983). Therefore, we cannot exclude that cosmological effects are hidden in the non-dipolar 0.1 mK residuals of the current experiments.

Before concluding this section, we briefly discuss the possibility of having frequency dependent anisotropies. In fact, one example is known and fairly popular among the experimentalists, namely, the Sunyaev-Zeldovich (1972) effect which arises from the Compton spectral distortion in the hot gas clouds of clusters of galaxies. At present microwave dips (as predicted by Eq. (5) adapted to the present context) appear to have been detected in three clusters (Birkinshaw et al. 1984). Also, strongly frequency-dependent anisotropies would naturally arise all over the sky in unorthodox models Hogan(1982). But an important, although often overlooked, fact is that even in the standard scenario the anisotropies may strongly vary with frequency if the spectrum is not Planckian (Danese and DeZotti 1981). In particular, DeBernardis et al. (1984) have noticed that the spectral modulation predicted by Georgi et al. (1983) would imply spectacular oscillations of the dipole amplitude in the radio region, so that measurements of the low-frequency anisotropy can accurately determine the photon mass in the a_2 state.

Other properties of the CBR

The polarization properties of the CBR, too, have been investigated. Partially polarized radiation is fully described by the Stokes parameters (I, Q, U, V), the first of which denotes the total intensity; the linear polarization state is described by $S_1 = Q/I$ and $S_2 = U/I$, and the circular polarization by $S_3 = V/I$ (Chandrasekhar 1950). Selected upper limits on S_1 and S_2 (on the order of 10^{-4} or 10^{-5}) are listed in Table IV. A much weaker upper limit on S_3 is given by Lubin et al. (1983b).

Table IV. Selected upper limits on the linear polarization.

λ(cm)	Angular Scale	S_1, S_2	Reference
3.2	Large	$5\cdot10^{-4}$	Nanos (1979)
0.9	$\ell = 2$	$3\cdot10^{-5}$	Lubin et al. (1983b)
	$\ell \leq 3$	$7\cdot10^{-5}$	
	7°	$7\cdot10^{-5}$	
0.05-0.3	$0^{\circ}\!.5$-40°	$(50-6)\cdot10^{-4}$	Caderni et al. (1978)

Theoretical models, too, have been mainly concerned with linear polarization. As shown by Rees (1968), this is coupled to the CBR anisotropy by Compton scattering. The study of the CBR polarization requires the use of a transport equation for the Stokes parameters and is fairly complex. Fully general-relativistic calculations in the cosmological context have been performed so far only for homogeneous anisotropic models (Anile 1974, Basko and Polnarev 1980, Negroponte and Silk 1980, Matzner and Tolman 1982). The predicted linear polarization is $\sim 10^{-2}$ times the large-scale anisotropy $\Delta T/T$ for a last scattering around $z \simeq 10^3$, and of the order of $\Delta T/T$ if the CBR was scattered by the reheated plasma at $z \simeq 10$. The polarization pattern is quadrupolar for such models; since however I+iV and Q-iU have spin weights 0 and 2 respectively, the linear polarization is expanded in spin-weighted harmonics. Contrary to the case of the CBR anisotropy, the polarization is very sensitive to magnetic fields. The question was raised by Ceccarelli et al. (1982b), who concluded that an intergalactic magnetic field of realistic strength may affect the polarization in the radio region but not around 1 mm. The transport equation for homogeneous models with a cosmic magnetic field has been written by Milaneschi and Fabbri (1984), who analyse also some models and find that, when the magnetic field effects are important, the polarization pattern shifts from quadrupolar to octupolar.

The CBR polarization in inhomogeneous models has been only studied, by means of a simplified approach, by Kaiser (1982), who finds a linear polarization of about 20% the small-scale anisotropy. Only one small-scale experimental result is available (Caderni et al. 1978): this is an upper limit of $5\cdot10^{-3}$ on a scale of 30'.

Finally, a CBR property which has only begun to be investigated is the temporal fluctuations. The pioneering experiment of Dall'Oglio et al. (1982) has provided an upper limit on the CBR noise, which is equivalent to setting $T(\nu) \leq 3.1$ K in the millimetric region. To the best of our knowledge, no theoretical work has considered the CBR noise in the proper cosmological context, and only the standard equations for the blackbody or greybody quantum noise can be compared to the experiment.

We thank A.A. Starobinsky for communicating to us the results of the Soviet experiment on the large-scale anisotropy, and F. Melchiorri for providing work from his group before publication.

References

Aiello, S., Cecchini, S., Mandolesi, N. and Melchiorri, F. (1980). Lett. Nuovo Cimento 27, 472.
Anile, A.M. (1974). Astrophys. Space Sci. 29, 415.
Basko, M. and Polnarev, A.G. (1980). MNRAS 191, 207.
Birkinshaw, M., Gull, S.F. and Hardebeck, H. (1984). Nature 309, 34.
Berlin, A.B., Bulaenko, E.V., Vitkovsky, V.V., Kononov, V.K., Parijskij, Yu.N. and Petrov, Z.E. (1982). In Early evolution of the Universe and its present structure, IAU Symposium.
Boughn, S.P., Cheng, E.S. and Wilkinson, D.T. (1981). Astrophys. J. (Lett.) 243, L113.
Bussoletti, E., Fabbri, R. and Melchiorri, F. (1980). In Infrared Astronomy. Eds. P. Bernacca and R. Ruffini. Dordrecht: Reidel.
Caderni, N., Fabbri, R., Melchiorri, B., Melchiorri, F. and Natale, V. (1978). Phys. Rev. D17, 1908.
Carr, B.J. (1981). MNRAS 195, 669.
Ceccarelli, C., Dall'Oglio, G., DeBernardis, P., Masi, S., Melchiorri, B., Melchiorri, F., Moreno, G. and Pietranera, L. (1982a). XVII Rencontre de Moriond, p. 175.
Ceccarelli, C., Dall'Oglio, G., DeBernardis, P., Masi, S., Melchiorri, B., Melchiorri, F., Moreno, G., Pietranera, L. and Pucacco, G. (1982b). XVII Rencontre de Moriond, p. 191.
Chan, K.L. and Jones, B.J.T. (1975). Astrophys. J. 200, 461.
Chandrasekhar, S. (1950). Radiative Transfer. Oxford: Clarendon Press.
Collins, C.B. and Hawking, S.W. (1973). MNRAS 162, 307.
Dall'Oglio, G., DeBernardis, P., Masi, S. and Melchiorri, F. (1982). In Early evolution of the Universe and its present structure, IAU Symposium.
Danese, L. and DeZotti, G. (1977). Riv. Nuovo Cimento 7, 277.
Danese, L., DeZotti, G. and Mandolesi, N. (1982). XVII Rencontre de Moriond, p. 205.
Danese, L. and DeZotti, G. (1981). Astron. Astrophys. 94, L33.
Davis, M. and Boynton, P. (1981). Astrophys. J. 237, 365.
DeBernardis, P., Masi, S., Melchiorri, B., Melchiorri, F. and Moreno, G. (1984a). Astrophys. J. 278, 150.
DeBernardis, P., Masi, S., Melchiorri, F. and Moleti, A. (1984b). Astrophys. J. Lett. (in press).
De Vaucouleurs, G., Peters, W., Bottinelli, L., Gauguenheim, L. and Paturel, G. (1981). Astrophys. J. 248, 408.

Dolgov, A.D. and Zeldovich, Ya.B. (1980). Rev. Mod. Phys. 53,1.
Doroshkevich, A.G., Zeldovich, Ya.B. and Sunyaev, R.A. (1978). Sov. Astron. 22, 523.
Ewing, M.S., Burke, B.F. and Staelin, D.H. (1967). Phys. Rev. Lett. 19, 1251.
Fabbri, R. (1984). Astron. Astrophys. (in press).
Fabbri, R., Guidi, I., Melchiorri, F. and Natale, V. (1980). Phys. Rev. Lett. 44, 1563.
Fabbri, R., Guidi, I. and Natale, V. (1983). Astron. Astrophys. 122, 151.
Fabbri, R., Guidi, I., Natale, V. and Ventura, G. (1984). Convegno GIFCO, L'Aquila.
Fabbri, R., Melchiorri, B. and Melchiorri, F. (1981). Adv. Space Res. 1, 19.
Fabbri, R., Pucacco, G. and Ruffini, R. (1984). Astron. Astrophys.(in press).
Fixsen, D.J., Cheng, E.S. and Wilkinson, D.T. (1983). Phys. Rev. Lett. 50, 620.
Fomalont, E.B., Kellerman, K.I. and Wall, J.V. (1984). Astrophys. J. (Lett.) 277, L23.
Georgi, H., Ginsparg, P. and Glashow, S.L. (1983). Nature 306, 765.
Goicoechea, L.J. and Sanz, J.L. (1984). Phys. Rev. D29, 607.
Hart, L. and Davis, R.D. (1982). Nature 297, 191.
Hogan, C. (1982). Astrophys. J. 256.
Howell, T.F. and Shakeshaft, J.R. (1966). Nature 210, 1318.
Howell, T.F. and Shakeshaft, J.R. (1967). Nature 216, 753.
Illarionov, A.F. and Sunyaev, R.A. (1975). Sov. Astron. 18, 691.
Jakobsen, H.P., Kon, M. and Segal, I.E. (1979). Phys. Rev. Lett. 42,1788.
Kaiser, N. (1982). MNRAS 198, 1033.
Kurskov, A.A. and Ozernoi, L.M. (1978). Astrophys. Space Sci. 56, 67.
Lifshitz, E.M. and Khalatnikov, I.M. (1963). Adv. Phys. 12, 185.
Lubin, P.M., Epstein, G.L. and Smoot, G.F. (1983a). Phys. Rev. Lett. 50, 616.
Lubin, P.M., Melese, P. and Smoot, G.F. (1983b). Astrophys. J.(Lett.) 273, L51.
MacCallum, M. (1979). Lecture Notes in Physics 109, 1.
Matzner, R.A. and Tolman, B.W. (1982). Phys. Rev. D26, 2951.
Melchiorri, F., Melchiorri Olivo, B., Ceccarelli, C. and Pietranera, L. (1981). Astrophys. J. (Lett.) 250, L1.
Millea, M.F., McColl, M., Pederson, R.J. and Vernon, F.L. (1971). Phys. Rev. Lett. 26, 919.
Milaneschi, E. and Fabbri, R. (1984). Preprint.
Misner, C.W., Thorne, K.S. and Wheeler, J.A. (1973). Gravitation. San Francisco: Freeman.
Nanos, G.P. (1979). Astrophys. J. 232, 341.
Negroponte, J. and Silk, J. (1980). Phys. Rev. Lett. 44, 1433.
Novikov, I.D. (1968). Sov. Astron. 12, 427.
Parijskij, Y.N., Petrov, Z.E. and Cherkov, L.N. (1977). Pisma Astron. Zh. 50, 453.
Peebles, P.J.E. (1968). Astrophys. J. 153, 1.
Peebles, P.J.E. (1981). Astrophys. J. (Lett.) 243, L119.
Peebles, P.J.E. and Yu, J.T. (1970). Astrophys. J. 162, 815.
Penzias, A.A. and Wilson, R.W. (1965). Astrophys. J. 142, 419.
Raine, D.J. and Thomas, E.G. (1981). MNRAS 195, 649.

Rees, M.J. (1968). Astrophys. J. (Lett.) 153, L1.
Rees, M.J. (1978). Nature 275, 35.
Rowan-Robinson, M., Negroponte, J. and Silk, J. (1979). Nature 281, 635.
Sachs, R.K. and Wolfe, A.M. (1967). Astrophys. J. 147, 73.
Silk, J. (1981). Ann. New York Acad. Sci. 375, 188.
Silk, J. (1982). Scr. Varia Pont. Acad. Sci. 48.
Silk, J. and Wilson, M.L. (1981). Astrophys. J. (Lett.) 244, L37.
Smoot, G.F., DeAmici, G., Friedmann, S.D., Witebsky, C., Mandolesi, N., Sironi, G., Partridge, R.B., Danese, L. and DeZotti, G. (1983). Phys. Rev. Lett. 51, 1099.
Stecker, F.W. and Puget, J.L. (1972). Astrophys. J. 178, 57.
Stokes, R.A., Partridge, R.B. and Wilkinson, D.T. (1967). Phys. Rev. Lett. 19, 1199.
Strukov, I.A. and Skulachev, D.A. (1984). Pisma Astron. Zh. 10, 1.
Sunyaev, R.A. and Zeldovich, Ya.B. (1970). Astrophys. Space Sci. 7, 20.
Sunyaev, R.A. and Zeldovich, Ya.B. (1972). Comments Astrophys. Space Phys. 4, 173.
Tammann, G.A., Yahil, A. and Sandage, A. (1979). Astrophys. J. 234, 775.
Uson, J.M. and Wilkinson, D.T. (1984). Astrophys. J. (Lett.) 277, L1.
Weiss, R. (1980). Annu. Rev. Astron. Astrophys. 18, 489.
Wilkinson, D.T. (1967). Phys. Rev. Lett. 19, 1195.
Wilson, M.L. (1983). Astrophys. J. 273, 2.
Wilson, M.L. and Silk, J. (1981). Astrophys. J. 243, 14.
Woody, D.P. and Richards, P.L. (1981). Astrophys. J. 248, 18.
Zeldovich, Ya.B. and Sunyaev, R.A. (1969). Astrophys. Space Sci. 4, 129.

CLOSE-UP ON GRAVITATIONAL LENSING : THE GRAVITATIONAL MIRAGES.

C. VANDERRIEST
Groupe d'Astrophysique Relativiste
Observatoire de Meudon
92195 Meudon Principal Cedex (FRANCE)

Summary :

Among the various manifestations of gravitational *lensing*, we emphasize the interest of gravitational *mirages* (defined as situations where several images of a single source are produced).

We first recall the principle of this phenomenon and the conditions for its occurence. Then, we discuss a few possible uses of gravitational mirages, particularly in cosmology.

Finally, we review the present results on the 5 known cases and speculate on the observational developments expected in the near future.

A few words about terminology :

In many papers, the expression "gravitational lens" is improperly or ambiguously used.

For example, Weymann et al. (1980) announced the discovery of the so-called "triple quasar" as follows : "The triple Q.S.O. P.G. 1115 + 080 : Another probable gravitational lens". Such a title can be misleading, because the quasar is not a gravitational lens. In fact, its peculiar aspect results from the interposition of a gravitational lens ... which remains to be found.

On the other hand, gravitational lenses always alter the appearence of a distant source, but very rarely to the point of producing several images.

For the sake of clarity and precision of language, we will thus distinguish between :

1 - Gravitational lenses :

In a broad sense, a gravitational lens can be identified with any local departure from a model Universe (for example, the homogeneous Friedmann-Robertson-

Walker Universe). Excess of matter with respect to this reference Universe can be called a *positive* lens, depletion a *negative* one.

Local curvatures :

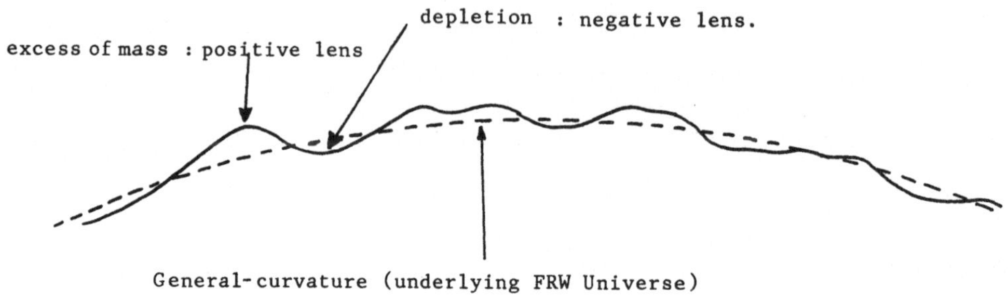

2 - Gravitational lensing effects :

They consist in displacement and distortion (and thus magnification or demagnification) of the observed images with regard to what they would be in the reference Universe. In this sense, they exist everywhere.

3 - Gravitational mirages :

We will reserve this phrase to situations where several images of a single source are produced (which is an extreme case of distortion). Strictly speaking, it encompasses two distinct concepts :

- It designates the phenomenon of multiple imaging by a gravitational lens.
- It designates also the observable images resulting from this phenomenon.

The same is true for atmospheric mirages which present many analogies with gravitational mirages. Both phenomena are deceiving, but they are only the rare and spectacular consequence of an omnipresent mechanism : atmospheric refraction and gravitational deflection, respectively.

Two more remarks :

A - The mere concept of gravitational lens is not necessary if we can deal with the propagation of light in the real (highly inhomogeneous) Universe. Attempts in this way, based on the Optical Scalar Equation (Sachs, 1961), have been made for example by Zeldovitch (1964) and Nottale (1983).

B - We will adopt, however, the more widespread method of using a FRW underlying metrics (or even a simple euclidian space for qualitative demonstrations).

- The following is entirely devoted to gravitational mirages defined in this sense.-

I) The mechanism of gravitational mirages :

A - A long story :

The emergence of gravitational mirages, first as a theoretical concept, and more recently as an observational reality, is marked out by a few significant dates and by long periods of fallow.

- 1704 : Newton suggests the possibility of the deflection of light by a gravitational field.
- 1916 : The General Relativity predicts a deflection of light by 1.75" at the solar limb.
- 1919 : Eddington observes such a deviation. He suggests at once the possibility of gravitational mirages : if the Sun were a point-mass, it would be possible to observe 2 images of a background star.
- 1936 : Einstein is amused by some aspects of gravitational mirages, but he is very skeptical about the possibility of observing the phenomenon among stars.
- 1937 : Zwicky proposes to search for gravitational mirages among galaxies. The source would be, for example, the nucleus of a distant galaxy, and the lens a nearer intervening galaxy.

Link describes the main observable features of gravitational mirages. The subject is then considered as a mere theoretical curiosity for almost 25 years.

- 1963 : Discovery of the first quasars. Owing to their distance, brightness and point-like aspect, they can be ideal sources for a gravitational mirage phenomenon.
- 1964 : Refsdal exposes the possible use of gravitational mirages for cosmological tests.
- 1965 : Barnothy try to explain <u>all</u> quasars as Sy I nuclei amplified by gravitational lensing effect.
- 1979 : Discovery of the "double quasar" 0957 + 561.

Since then, an impressive amount of works on gravitational mirages, both observational and theoretical, has flourished.

B - Conditions for occurence of a gravitational mirage :

Let us consider a distant source S, an observer O and a spherically symmetric deflecting body near the line of sight (for example, galaxy G). The following figure shows the situation in euclidian space.

The light rays emitted by S and deflected by G can reach by several optical paths (1, 2, 3,...) the observer O, which will see several images (A, B, C,...) of the source. For a light ray reaching the observer, we have the condition :

$$\alpha = (\beta - \beta_o) \frac{D_{SO}}{D_{SG}}$$

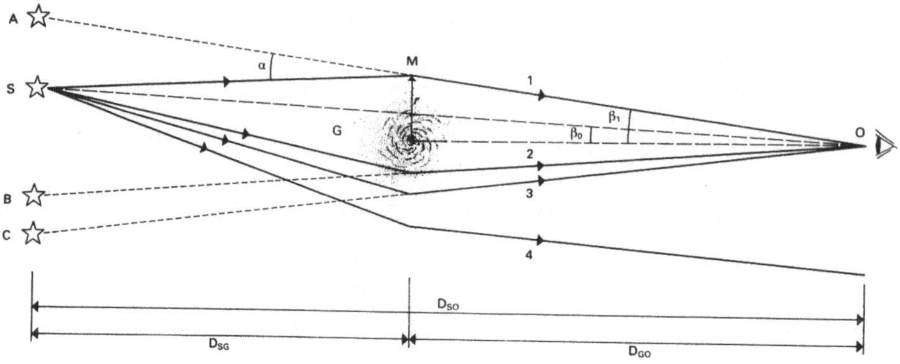

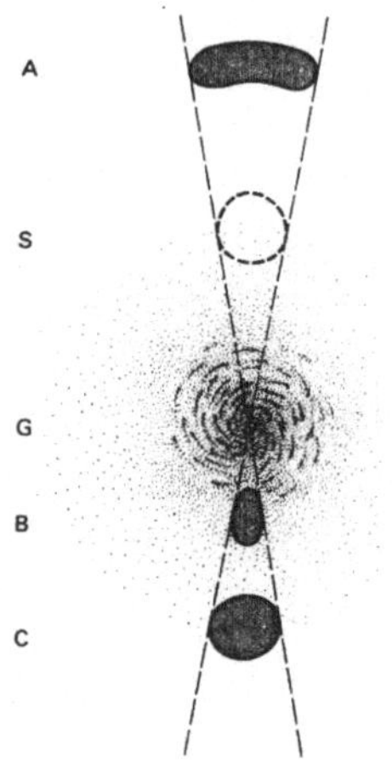

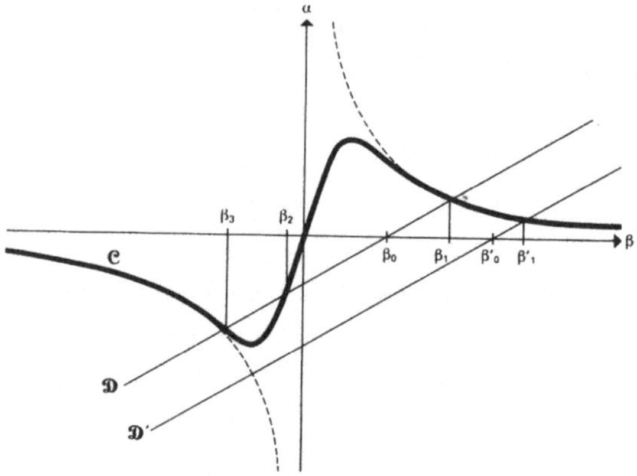

which, in the plane (α, β), describes a straight line D. The deflection angle is also given by Einstein's formula :

$$\alpha = \frac{4G}{c^2} \frac{M(r)}{r} = \frac{4G}{c^2} \frac{1}{D_{GO}} \frac{M(\beta)}{\beta}$$

which is the equation of a certain curve C.

[$M(\beta)$ is the mass inside a cylinder of angular radius β] For a point-mass, $M(\beta) = M$ (a constant) and C is an hyperbola. The solutions given by the intersections of D and C are always to the number of 2.

In the case of a mass distribution representative of a galaxy, $M(0) = 0$ and $\frac{M(\beta)}{\beta} \to 0$ when $\beta \to 0$.

Thus, $\alpha(0) = 0$ and the curve C behaves as shown in the figure. Provided that the slope of D is less than in the inner part of C (which means, for a given galaxy and a given source, a minimum value of D_{GO}), we see that there can be 3 solutions, and thus 3 images, when β_o is less than a critical value. When β_o is too large, there is only 1 solution (position D'). Note also that, when 3 images are produced, at least one is seen through rather dense parts of the deflecting galaxy.

This simple discussion in euclidian space can be easily transposed in FRW metrics. We just have to use the angular diameter distances, i.e. :

$$D_{SO} = \frac{c}{H_o} \cdot \frac{Q_oZ_S + (Q_o-1)\ [(2Q_oZ_S + 1)^{\frac{1}{2}} - 1]}{Q_o^2\ (1+Z_S)^2}$$

$$D_{GO} = \frac{c}{H_o} \cdot \frac{Q_oZ_G + (Q_o-1)\ [(2Q_oZ_G + 1)^{\frac{1}{2}} - 1]}{Q_o^2\ (1+Z_G)^2}$$

$$D_{SG} = \frac{c}{H_o} \cdot \frac{Q_o(Z_S-Z_G) + (Q_o-1)\ [(2Q_oZ_S+1)^{\frac{1}{2}}(1+Z_G) - (2Q_oZ_G+1)^{\frac{1}{2}}(1+Z_S) + (Z_S-Z_G)]}{Q_o^2\ (1+Z_S)^2\ (1+Z_G)^2}$$

Here, the direct observables Z_S and Z_G become apparent, as well as the parameters of the FRW model H_o and Q_o.

General Case :

For a non-spherical galaxy (but still in the weak field and thin deflector approximation), we have to deal with 2 dimensions. The displacement vector field $\vec{\alpha}$ in the plane of the sky is given by :

$$\vec{\alpha} = -\frac{2}{c} \int_S^0 \vec{\nabla}\, \Phi \ . \ dt$$

where Φ is the (classical) gravitational potential of the deflector and $\vec{\alpha}$ links the "image (s)" positions to the "source" positions.

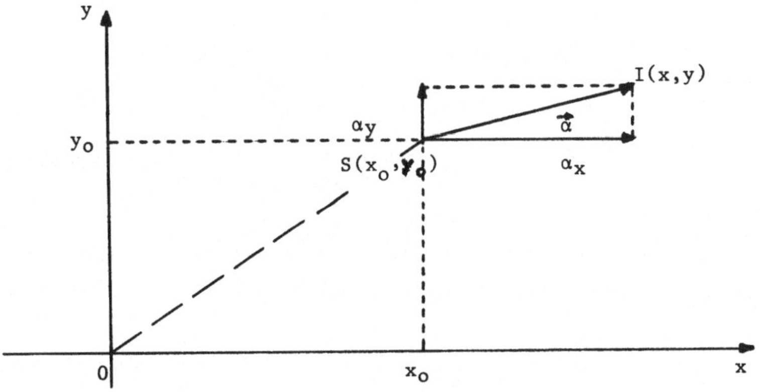

This vector field is evaluated in the image plane.

An elegant way to resolve the problem uses the complex notation introduced by Bourassa and Kantowski (1975).

To $\vec{\alpha}\,(x,y)$ corresponds the complex number $\alpha = \alpha_x + i\ \alpha_y$, and to $\Phi\ (x,y)$, the "scattering function"

$$I = \frac{c}{2G} \int_{-\infty}^{+\infty} \frac{\partial \Phi}{\partial x} \cdot dt - i \int_{-\infty}^{+\infty} \frac{\partial \Phi}{\partial y} \cdot dt$$

So, we have only to resolve :

$$\alpha = -\frac{4G}{c^2} \; I *$$

an equation very similar, formally, to that of the spherical case.

It can be easily shown that, for a transparent lens, there is always an odd number of images (with eventual merging in degenerate cases). Of the $(2n + 1)$ images, $(n + 1)$ have the parity of the source (positive amplification), while n are mirror-inverted images (negative amplification).

Etherington (1933) demonstrated that surface brightness is conserved during the propagation of light in a gravitational field. Thus, image magnification depends only on the ratio of area of the image to that of the source (area of the source in absence of the deflector). For each image, the magnification is thus the value of the Jacobian of the transformation $(x,y) \rightarrow (x_o, y_o)$.

C - Optical simulation of a gravitational lens :

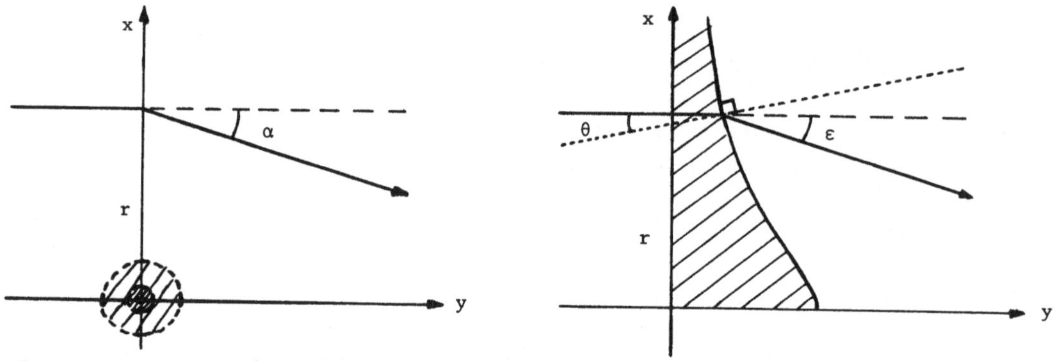

Gravitational lens (Mass M(r)) with spherical symmetry.

equivalent optical lens (profile y (x); glass of index n).

It is interesting to simulate a gravitational lens by a classical glass lens (see figure above). We have the correspondance :

$$\alpha(r) = \frac{4G}{c^2} \; \frac{M(r)}{r} = K. \; \frac{M(r)}{r}$$

$$\varepsilon(r) = (n-1) \; \theta(r) = (n-1) \; \left.\frac{dy}{dx}\right]_r$$

Thus : $\frac{dy}{dx} = K' \; \frac{M(x)}{x}$

1 - Point mass $M(x) = M$:

The profile of the optical lens should be :

$$y(x) = \ln x + cst$$

2 - Simulation of a Spiral galaxy :

$$M(x) = 2\pi \int_0^x \sigma(x) \; x \; dx$$

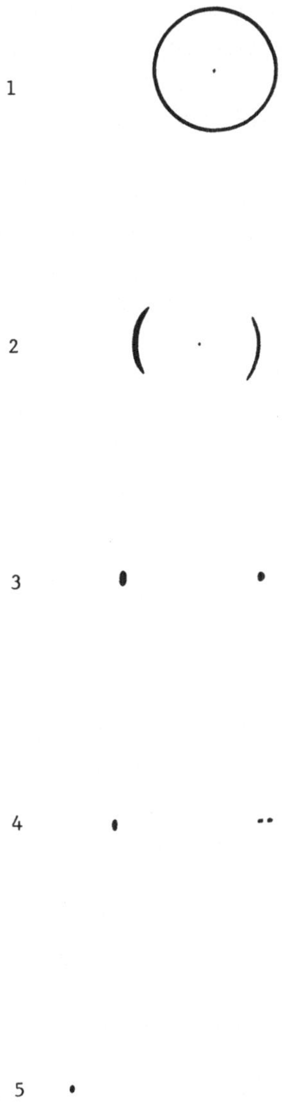

with the surface mass distribution of a Spiral:

$$\sigma(x)=e^{-\frac{x}{x_o}} \Rightarrow M(x) = M_t \left[1 - e^{-\frac{x}{x_o}} \left(1 - \frac{x}{x_o} \right)\right]$$

with $z = \frac{x}{x_o}$, the profile should thus be :

$$y(z) = \int_0^z \frac{dz}{z} - \int_0^z \frac{e^{-z}}{z} dz - \int_0^z e^{-z} \, dz$$

i.e. : $y(z) = \ln z - EI(z) + e^{-z} + cst$

where EI (z) designates the Exponential Integral function.

Such a lens has been manufactured and gives a good idea of the real aspect of gravitational images produced by a galaxy. In the joined figure, picture 1 shows a small circular source seen through the lens when the alignement is perfect ($\beta_o = 0$).

Numbers 2 to 5 correspond to increasing values of β_o. We can see that the "third" image is always much fainter than the 2 "principal" ones, except near merging (picture 4). The 2 principal images are of comparable intensities, and their separation is almost constant, nearly equal to the diameter of the degenerate circle of picture 1. In picture 5, there is no more gravitational mirage, but the remaining image is still amplified (elliptical shape).

All these features show the drastic differences with the point-mass case.

D - Effects of clustering :

A model involving only a source, a deflecting galaxy and an observer is not fully realistic ; it is well known that galaxies belong generally to more or less important groups or clusters. This can affect seriously the appearance of a gravitational mirage.

Let us consider the lensing of a quasar (Q) by a main galaxy (G) belonging to a cluster (C). For the sake of simplicity, we can assume that the mass distributions in both C and G are spherically symmetric and that the true position of the quasar lies on the axis joining their centers. The problem is therefore reduced to one dimension and can be again discussed graphically. The bending angles due to G and C being additive, the presence of the cluster can change noticeably the curve α (β).

Even near the core of a rich cluster, the gradient $(d\alpha/d\beta)_c$ (which can be considered as constant) is generally not sufficient to produce, by itself, multiple imaging ; but it can help (or inhibit) a gravitational mirage induced mainly by the galaxy G. The allowed interval in β for multiple imaging is shifted and increased. A full 2-dimensional treatment would have shown that the multiple imaging region (the interior of the critical circle in the isolated spherical galaxy case) is badly distorted and its area noticeably increased.

It can be shown that the main effect of clustering is to shift the *separation histogram* of the images towards high values.

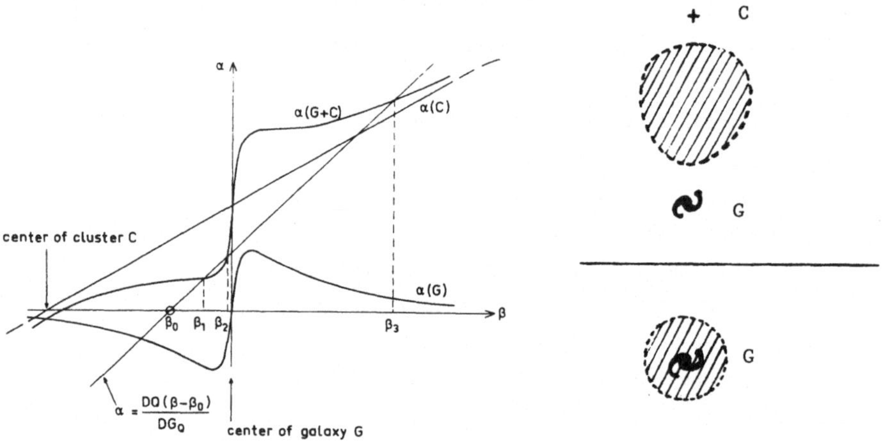

Action of a cluster of galaxies on the image separation and on the allowed region for multiple imaging.

II) Astrophysical uses of gravitational mirages :

A - Miscellaneous possible uses :

A gravitational mirage involves 4 elements :

- the source
- the deflecting gravitational field
- the propagation space of the light
- the observer.

It is remarkable that a careful analysis of the phenomenon can yield illuminating data on these 4 elements.

- concerning the observer :

A gravitational mirage is certainly a challenge for observing skill and theoretical imagination. The problem is to make the best of a very unusual situation : *the possibility to observe an object by several different light paths in the Uni-*

verse.

- concerning the source :

Each image shows the source from a slightly different angle of view and at a slightly different time. This opens the opportunity to analyse the fine structure of the regions emitting the continuum or the broad emission lines.

The flux ratio of these images can be measured with high accuracy (it is relative photometry and the images have perfectly identical spectra). For example, a variation of 0.03 magnitude has been detected between 0957 + 561 A and B in 10 hours (proper time) ; it would have been undetectable on any other object of 17.5 magnitude in absence of multiple imaging (Vanderriest et al., 1982).

- concerning the deflector :

If we find a gravitational mirage produced by an isolated galaxy, or if the influence of other cluster members can be properly disentangled, the total mass of this galaxy (luminous and not luminous) could be precisely measured.

The mass distribution inside the galaxy can be reached via the observed amplification factors A (β).

At much smaller scale, a gravitational mirage can present the startling opportunity of detecting solar or sub-solar mass bodies in the deflecting galaxy, whatever its redshift may be (see, e.g., Gott, 1981 ; Young, 1981).

- concerning the space between source and observer :

A gravitational mirage makes it possible to set at least upper limits on the sizes of intergalactic absorbing clouds or even to evaluate directly their sizes and masses. Such clouds have been detected on high resolution spectra of the first gravitational mirages (cf. infra).

But the most important application, perhaps, is the possible direct measurement of the distance to the source, and thus of the Hubble constant Ho.

B - Cosmological tests :

The values of the classical parameters Ho, Qo and Λ which give the best fit to the real Universe can be found by comparison of apparent quantities (m,θ,...) with the corresponding absolutes ones (M, D, ...).

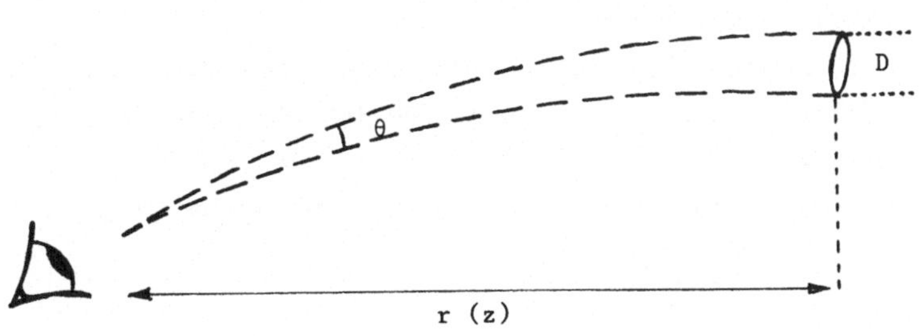

The relations have the form :

$\theta = D.\ f\ (r) \quad [\ f\ (r) \equiv \frac{1}{r}$ in euclidian space]

$f_\nu = \Phi_{\nu_0}\ .\ f'\ (r) \quad [f'\ (r) \equiv \frac{1}{r^2}$ in euclidian space]

All the cosmologically relevant informations are included in the functions $f_{Ho,Qo,\Lambda}(r)$ or $f'_{Ho,Qo,\Lambda}(r)$.

There are, thus, 2 basically different strategies :

a) Statistical methods :

In these approaches, we have to select populations of objects having absolutes quantities (D, Φ_{ν_0},...) as constant as possible. Then, the set of data points at different redshifts is fitted by the theoretical curve. A well-known example is the Hubble diagram.

Such methods are difficult to use because of the difficulties to find good "standard candles" or "standard yardsticks" and to take into account evolution effects.

b) Non-statistical methods :

If, for a given object, we can determine exactly (by mean of some model) the absolute quantity involved, we do not have to make statistics.

3 kinds of "objects" seem attractive :

1 - Supernovae.

2 - "Superluminal" radio-sources.

3 - Gravitational mirages.

For discussion and references, see for example Shaver (1983).

A common characteristic of these 3 methods is that we obtain a *length* from a certain *velocity* :

1 - $V = V_{photosphere}$

2 - $V \geqq 4.45\ C$ in the dipole model (Bahcall, 1980)

3 - $V = C$

In the last case, there is no uncertainty on the velocity. The method is based on the fact that, in a gravitational mirage, the different light paths joining the source to the observer do not have, generally, the same length. To a difference of length Dl_{ij} between the paths for images i and j, corresponds a *time delay* ΔT_{ij}. This term is purely geometrical and proportional to $\frac{1}{Ho}$.

In fact, the observed time delay would be the sum of this geometrical term and of a relativistic time delay Δt_p due to the difference of the gravitational potential along the two paths.

The expressions of the geometrical and potential terms are :

$$C\,\Delta T_g = (1 + Z_G)\ \frac{D_{GO}.\ D_{SG}}{D_{SO}}\ \frac{\alpha^2}{2}$$

$$C\,\Delta\,T_p = (1 + Z_G)\int_{-\infty}^{+\infty} \frac{2\Phi}{c^2}\,dl$$

[the factor $(1+Z_G)$ corresponds to the transformation from the rest frame of the deflecting galaxy to the frame of the observer].

If we can find, observationally, the global value $\Delta t_{i,j}$ of the time delay between 2 images i and j, we thus have to subtract the potential term to have any hope of finding Ho. It can be calculated from the mass distribution of the deflector.

Without detailing the calculations, let us look at the feasability of the method .

- Observable quantities :

  the position angles of the images : $\beta_i = \beta_i(\beta_o, M(\beta), Z_S, Z_G, H_o, Q_o, \Lambda)$

  the time-delays : $\Delta t_{ij} = \Delta t_{ij}(\beta_o, M(\beta), Z_S, Z_G, H_o, Q_o, \Lambda)$

  the amplification ratios : $\frac{A_i}{A_j} = \frac{A_i}{A_j}(\beta_o, M(\beta), Z_S, Z_G, H_o, Q_o, \Lambda)$

and , of course, the redshifts Z_S and Z_G.

- Unknowns :

  β_o, $M(\beta)$, H_o, Q_o, Λ

If $M(\beta)$ is sufficiently simple (for example if the deflector is an isolated galaxy with spherical symmetry, described by 2 parameters : ρ_o, r_e), the problem seems over-determined for a gravitational mirage with 3 images.

We have, effectively, 7 observables (β_1, β_2, β_3, $\Delta T_{1,2}$, $\Delta T_{1,3}$, $\frac{A1}{A_2}$, $\frac{A1}{A_3}$) and 6 unknowns (β_o, ρ_o, r_e, H_o, Q_o, Λ).

In fact, even in this quasi-ideal case, the equations are inequally sensitive to the various parameters and observational errors can be unacceptable.

For example, we need *very* high accuracy on the angle measurements to see the effect of Q_o (Lacroix and Schneider, 1982). It is believed that, with a single gravitational mirage with 3 images, only H_o can be precised (Young et al., 1981).

C - A crucial point : the measurement of $\Delta T_{i,j}$:

The knowledge of $\Delta T_{i,j}$ is necessary (but certainly not sufficient) to find H_o.

For this task, time variations of any observable parameter of the source can be used. Practically, the observable can be morphology, flux or polarisation (variations of the spectrum are probably too slow).

1 - Morphology :

The most favourable case occurs when the source is a radio-lond quasar showing superluminal motions. It is then theoretically possible to find $\Delta T_{i,j}$ from only 2 observations if the motion is linear (Vanderriest, 1982). V.L.B.I. observations of the first gravitational mirage 0957 + 561 are currently undertaken in hope to detect this effect (Gorenstein et al. 1984).

2 - Flux (or polarisation) :

The variations being random, many observations are needed. If τ is a characteristic variation time, the sampling frequency f_S should verify : $f_S^{-1} \ll \tau \ll \Delta T_{i,j}$.

ΔT would then be found from cross-correlation of the data over a time interval $> \Delta T$.

III) The presently available gravitational mirages :

During the last 5 years, 5 cases of possible or certain gravitational mirages have been discovered.

A - The "double" quasar 0957 + 561 :

The story of the discovery of the first gravitational mirage (29/03/1979)and of its successful interpretation within a few months will probably become archetypal in the near future. It is unnecessary to dwell on it.

Let us just remind that it was found during a routine programme of optical identification of radio-sources. Two objects, separated by 6.17" showed identical spectra of quasars at Z_{em} = 1.405 (Walsh, Carswell, Weymann, 1979). The interpretation in term of gravitational mirage was quickly conforted : identity of the spectra at U.V., visible, I.R. and radio wevelengths (Gondhalekar and Wilson, 1980 ; Lebofsky et al., 1980 ; Pooley et al., 1979), presence in both images of similar absorption lines at Zabs = 1.391 and Zabs = 1.125 (Young et al., 1981), detection of the main deflecting galaxy at Z_G = 0.36 (Young et al., 1980 ; Stockton, 1980 ; Young et al., 1981), compatibility of the VLA radio map and of the VLBI data with the gravitational mirage model (Walsh, 1983 and references therein ; Gorenstein et al. 1984 and references therein) etc...

Unfortunately, the main deflecting galaxy belongs to a rather dense group and the modelling will be complex. Optical, radio (VLA) and VLBI data already put severe constraints on the usable model. For example, the VLA map shows that the true position of the quasar is very near the critical curve : the radio core is imaged twice (and thus probably 3 times), but the extensions are detected only once, near image A (see figure).
Despite the complex imaging situation, 0957+561 is still the best (or the less unfavourable) candidate for measurement of H_o. $\Delta T_{A,B}$ is of the order of 1-5 years and several teams are actively trying to determinate its exact value by photometric monitoring. Data on the third image (not yet detected with certainty) are necessary for a full modeling. It is expected that the Space Telescope will yield them. But the measurement of $\Delta T_{A,B}$ is a long term affaire which should be pursued for a few years more.

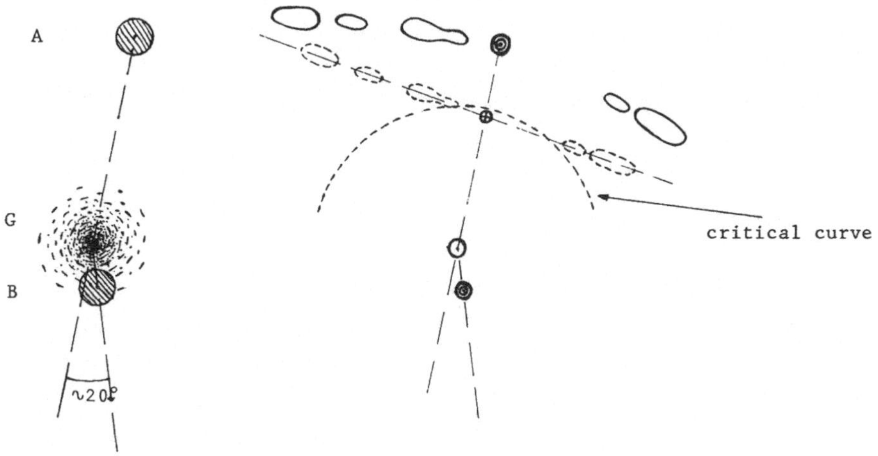

The region of 0957 + 561 seen at optical and radio wavelengths.

B - The "triple" quasar P.G. 1115 + 080 :

The second gravitational mirage was also found by chance (Weyman et al., 1980).
Even if, because of the compactness of the configuration, it is difficult to check rigorously the identity of spectra of the different images, it was quickly accepted as a *bona fide* gravitational mirage. In fact, it is not a "triple" quasar, but more probably a "quintuple" one, the brightest image being a close double. This duplicity is suspected on pictures obtained by good seeing conditions (joined figure), and was confirmed by speckle interferometry (Hege et al., 1981). The separation A_1-A_2 is about 0.54". As for 0957+561, a common absorption system as been found on the spectra of the different images

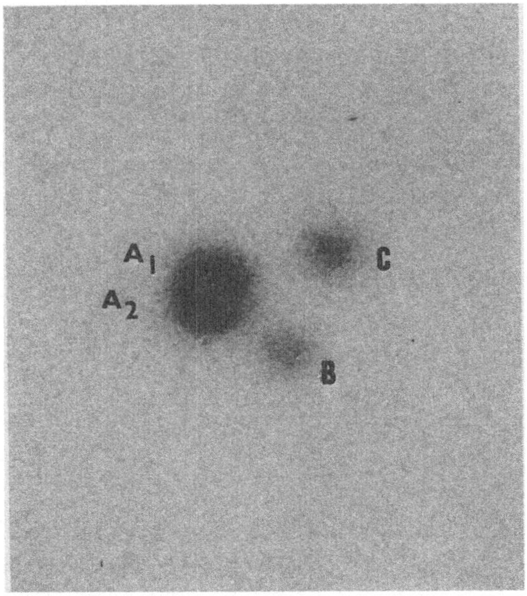

P.G. 1115 + 080 : Electronographic plate obtained by G. Wlérick at the C.F.H. telescope ; seeing = 0.75" (F.W.H.M.).

(Weyman and Foltz, 1983). Rather large optical variability has been detected (Vanderriest et al., 1983), but the images seem to vary almost in phase. This is probably because the condition $\tau << \Delta T_{i,j}$ is not fulfilled, the time delays being of the order of a few days or weeks only (Narasimha et al., 1982).

C - 2345 + 007 :

Again, the third gravitational mirage was found unexpectedly, during a grens survey aimed at finding new quasars (Weedman et al., 1982). 2 images are seen, with the largest separation yet recorded (7.15"). It is thus very probable that cluster plays a dominant role in the phenomenon. Such a cluster is certainly at $Z>0.5$. Between the two images, a very faint object is marginally detected. Its magnitude $V = 24.9$ (Sol et al., 1984) is among the largest ever measured. This stress the difficulty to observe gravitational mirages from the ground.

Finally, a common absorption system has been detected also in the spectra of the 2 images (Foltz et al., 1984).

D - The latest cases :

Recently, 2 new possible cases of gravitational mirages have been brought to light (see, for example, Sky and Telescope, April 1984 issue)

2016 + 112 is a very faint system (2 images at $r \approx 22.5$ and a third component at $r \simeq 23$). The third component is diffuse and can be tentatively assimilated to the main deflecting galaxy. The 2 others images show identical spectra with *narrow* emission lines. Maybe are we observing the first gravitational mirage, not on a quasar but on a very distant Sy II nucleus ?

1635 + 267 is the possible fifth gravitational mirage. Only 2 point-like images are seen.

IV) <u>Conclusion</u> :

A look at the following table shows that many data are still missing for a complete study of the detected gravitational mirages. The faint third or fifth images are systematically undetected ; the main deflecting galaxy is well studied only for 0957 + 561.

Summary of known gravitational mirages :

quasar :	θ(") :	Z_Q :	radio emission:	Z_G :	m_G :	missing :
0957 + 561	6.17	1.41	Yes	0.36±0.05	V=18.9	Image 3
P.G.1115+080	{2.28	1.72	no	≧ 0.7	V>22	Galaxy G
	1.77					Image 5
2345 + 007	7.15	2.15	no	≧ 0.5-0.8	V>23	Galaxy G
						Image 3
2016 +112	3.4	3.27	yes	≃ 0.7	r=23.2	image 3
1635 + 267	3.8	1.96	no	≧ 0.5	r>23.5	Galaxy G
						image 3

Also, the measured separations do not reflect the expected distribution in the hypothesis of a single galaxy deflector. Besides of (probably important) selection effects, the clustering of galaxies is certainly responsable for it. This would complicate the use of gravitational mirages for finding H_o.

Things to do in the coming years are obvious : we have to study more precisely the known cases (the Space Telescope will be of crucial help), and we have to find new cases, perhaps more easy to analyse.

We feel that the continuous monitoring of the variations of 0957+561 remains important, even in a complex imaging context.

Finally, the gravitational mirage phenomenon, observed for the first time 60 years after its prediction, as focussed attention, not only on the importance of extragalactic gravitational mirages, but also on possible stellar mirages (those searched for by Eddington), and on the more common effects of *gravitational lensing*.

It can be guessed that, from now on, the FRW reference Universe will be considered, for many research topics, only as a rather crude first approximation of the real Universe in which we live.

REFERENCES :

- Bahcall J., Milgrom M., 1980, Astrophys. J., 236 ,24.
- Barnothy J., 1965, Astron. J., 70, 666.
- Bourassa R., Kantowski R., 1975, Astrophys. J., 195, 13.
- Einstein A., 1936, Science, 84,506.
- Etherington I., 1933, Phil. Mag.,Ser. 7, 15, 761.
- Foltz C.et al., 1984, Astrophys. J., 281, L.1.
- Gondhalekar P., Wilson R., 1980, Nature, 285, 461.
- Gorenstein M. et al., 1984, Astrophys. J., (in press).
- Gott J.R., 1981, Astrophys. J., 243, 140.
- Hege E., Hubbard E., Strittmatter P., Worden S., 1981, Astrophys. J., 248, L.1.
- Lacroix G., schneider J., 1982, Astron. Astroph., 115, 54.
- Lebofsky M. et al., 1980, Nature, 285, 385.
- Link F., 1937, Bull. Astron., 10, 73.
- Narasimha D., Subramanian K., Chitre S., 1982, Mon. Not. Roy. Astr. Soc., 200, 941.
- Nottale L., 1982a, Astron. Astroph., 110, 9.
- Nottale L., 1982b, Astron. Astroph., 114, 261.
- Pooley P. et al., 1979, Nature, 280, 461.
- Refsdal S., 1964a, Mon. Not. Roy. Astr. Soc., 128, 295.
- Refsdal S., 1964b, Mon. Not. Roy. Astr. Soc., 128, 307.
- Refsdal S., 1966, Mon. Not. Roy. Astr. Soc., 132, 101.
- Sachs R., 1961, Proc. Roy. Soc., 264, 309.
- Shaver P., 1983 in: "Quasars and gravitational lenses" , 24th Liège Astrophysical Colloquium, p. 289.
- Sol et al., 1984, Astron. Astroph., 132, 105.
- Stockton A., 1980, Astrophys. J., 242, L. 141.
- Vanderriest C., 1982, Astron. Astroph., 106, L. 1.
- Vanderriest C. et al., 1982, Astron. Astroph., 110, L. 11.
- Vanderriest C. et al., 1983 in: "Quasars and gravitational lenses" , 24th Liège Astrophysical Colloquium, p. 182.
- Walsh D., Carswell R., Weymann R., 1979, Nature, 279, 381.
- Walsh D., 1983 in: "Quasars and gravitational lenses" , 24th Liège Astrophysical Colloquium, p. 106.
- Weedman D. et al., 1982, Astrophys. J., 255, L. 5.
- Weymann R. et al., 1980, Nature, 285, 641.
- Weymann R., Foltz C., 1983, Astrophys. J., 272, L. 1.
- Young P. et al., 1980, Astrophys. J., 241, 507.
- Young P. et al., 1981a, Astrophys. J., 244, 736.
- Young P. et al., 1981b, Astrophys. J., 244, 756.
- Zeldovitch Y., 1964, Soviet Astron. (A. J.), 8, 13.
- Zwicky F., 1937, Phys. Rev. Ser. II, 51, 290 and 679.

AMPLIFICATION OF LIGHT BY GRAVITATIONAL LENS : DYNAMICS AND THICK LENS EFFECTS

F. HAMMER
Laboratoire d'Astrophysique
DAPHE, Observatoire de Meudon
F-92195, Meudon Principal Cedex, France

The specific effects of dynamics and extension of a gravitational lens on the amplification of light are studied in the frame of a two-step vacuole model.

I - HYPOTHESES AND THE MODEL

To compute amplification of light, we use an exact solution of relativistic optics equations, and relinquish any simplifying hypotheses (except the geometric optics hypothesis), which had been used by many previous studies (Bourassa and Kantowski, 1975; Young et al, 1980, 1981 ; Young, 1981).

We consider a radial beam passing through a two-step vacuole model which is an exact solution of Einstein field equations.

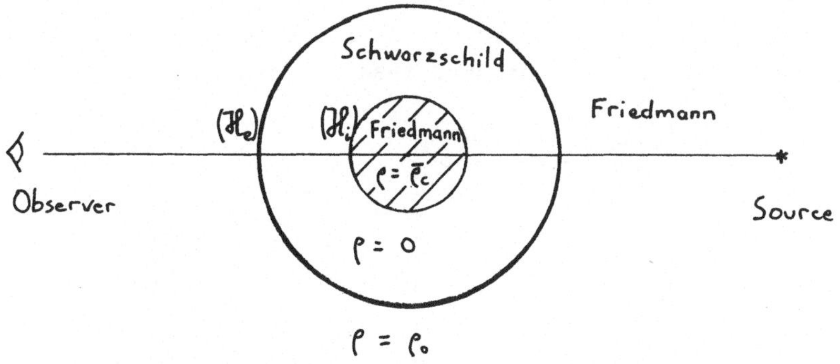

Figure 1 : Space-like section of the model.

The two-step vacuole model is an inhomogeneous model, where the lens (excess density region) is described by a high density Friedmann solution, separated from the background universe (a low density Friedmann solution) by an empty Schwarzschild solution. To solve exactly the Einstein field equations, the density profile must not perturbate the background universe, which implies a "null apparent mass" (see Eisenstaedt, 1977, for a general demonstration).

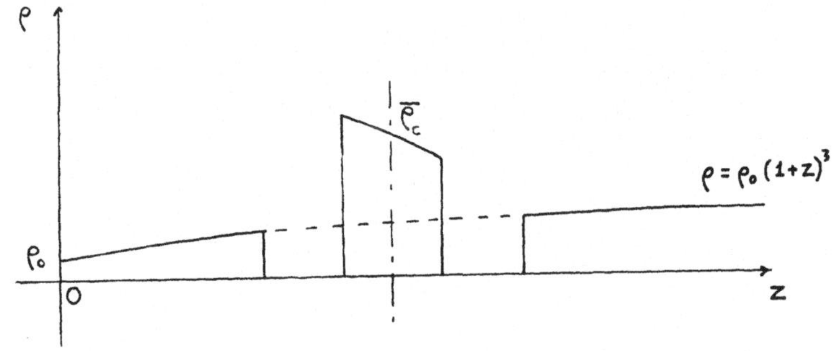

Figure 2 : Representation of the density distribution crossed by the radial beam.

For such a relativistic model, the propagation of a beam is given by the Optical Scalar Equation (Sachs, 1961). For a radial beam, it can write in the Kantowski (1969) form :

$$\frac{d^2\sqrt{A}}{d\omega^2} = \frac{1}{2} R_{ab} \; k^a k^b$$

where A is the cross sectional area of the beam, ω is an affine parameter, R_{ab} is the Ricci tensor and k^a is the wave vector. The beam equations will be successively written in the five regions encountered, and $\sqrt{A}$ and $d\sqrt{A}/d\omega$ matched at the limiting hypersurfaces (see Nottale, 1982b, Nottale and Hammer, 1984 and Hammer, 1984).

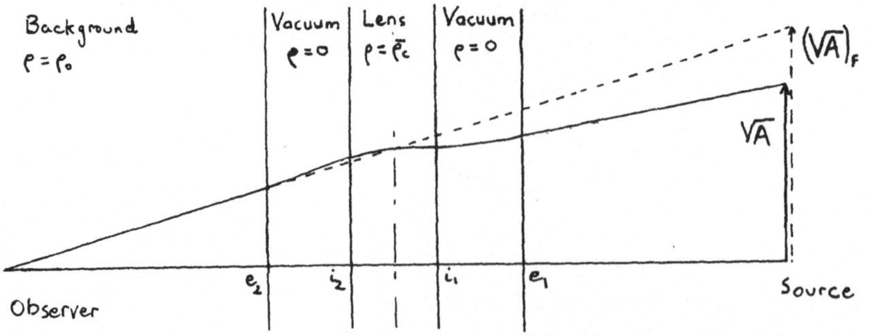

Figure 3 : Propagation of the beam diameter $\sqrt{A}$ from the source to the observer ; dotted line describes the propagation of the beam diameter $(\sqrt{A})_F$ in an homogeneous Friedmann model.

We obtain an exact relation between quantities ($\sqrt{A}$ and $d\sqrt{A}/d\omega$) just behind the hole in function of these just in front of the hole (see Hammer, 1984, for computation and result).

II - AMPLIFICATION

To compute amplification, we have to compare the "diameter" ($\sqrt{A}$) of the beam passing through the lens, with the "diameter" ($(\sqrt{A})_F$) of the beam in an homogeneous Friedmann model, both beams starting from the source at the same epoch. We obtain the result in terms of power series expansion in $H_o r_i/c$ (r_i, radius of the excess density region ; $H_o r_i/c \sim 10^{-3}$ for a rich cluster of galaxies, and $\sim 10^{-5}$ for a galaxy) :

$$\frac{\sqrt{A}}{\sqrt{A_F}} = 1 - \frac{4\pi G}{c^2} \int_L \partial\rho \, dr \, \frac{|D_{\hat{c}} \, |D_{\hat{c}s}}{|D_s}$$

with

$$\frac{4\pi G}{c^2} \int_L \partial\rho \, dr = 6\, q_o \left(\frac{\bar{\rho}_c}{\rho_o} - \left(\frac{\bar{\rho}_c}{\rho_o}\right)^{1/3} (1+z_c)^2\right) \frac{H_o r_i}{c} \left(1 - \frac{\bar{H}_c r_i}{c}\right)$$

which is the integral made over the density disgression with respect to the mean cosmological density (Nottale and Hammer, 1984).

q_o, ρ_o : deceleration parameter and density of the background. $\bar{\rho}_c$: density of the lens.

z_c : redshift of the lens center c.

$\bar{H}_c$: Hubble constant of the excess density region.

$|D_{\hat{c}}$, $|D_s$ and $|D_{\hat{c}s}$: respectively the angular diameter distances of the point $\hat{c}$, s (source) and between $\hat{c}$ and s. Point $\hat{c}$ have the same spatial comoving coordinates as the lens center c, but its temporal coordinates are defined by the epoch when the photon of a radial beam (which would reach the observer at the same time as the "inhomogeneous beam"), in an homogeneous model, reaches the center of the lens (see Nottale, 1982b).

Finally the amplification is given by :

$$\text{Amp} = \left(\frac{\sqrt{A_F}}{\sqrt{A}}\right)$$

in terms of magnitude :

$\partial m = 2.5 \log (Amp)$.

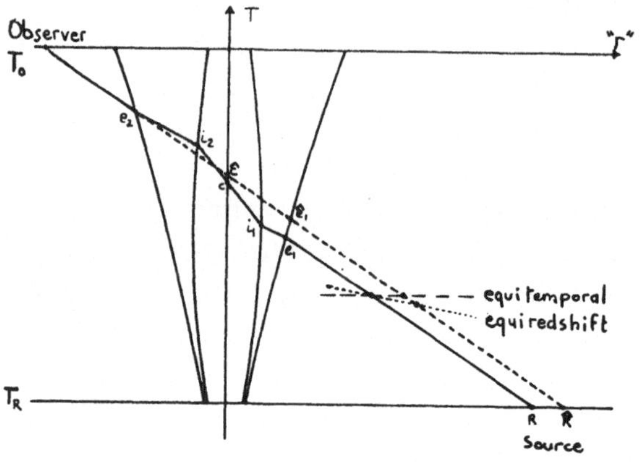

Figure 4 : Space-time representation of the model.

III - DISCUSSION AND CONCLUSION

Many previous works are based on Bourassa and Kantowski (1975), who obtained :

$$\sqrt{A}/\sqrt{A_F} = 1 - \frac{4\pi G}{c^2} \int_L \rho \; dr \; \frac{!D_c \; !D_{cs}}{!D_s}$$

They considered many hypotheses (linearized Einstein equations; vanishing potential $\nabla\Phi$ at infinity ; thin lens approximation ; bounded and stationary mass), and all of them are relinquished in this paper. Relinquishment of the bounded and stationary mass hypothesis allows us to show the effect of dynamics, which gives a relative difference in magnitude :

$$\frac{\Delta(\partial m)}{\partial m} = \frac{2(\sqrt{Amp} - 1)}{\text{Log Amp}} \; \frac{\overline{H}_c}{c} \; r_i$$

For very strong and observable amplification (which is typically the case of double quasar Q 0957 + 561 A, B, see Young et al, (1980)) we obtain $\Delta(\partial m)/\partial m \sim 0.2$.

From a theoretical viewpoint, relinquishment of the thin lens approximation leads us to write the amplification not in terms of the observed redshift of the lens center (z_c), but in terms of ($z_{\hat{c}}$).

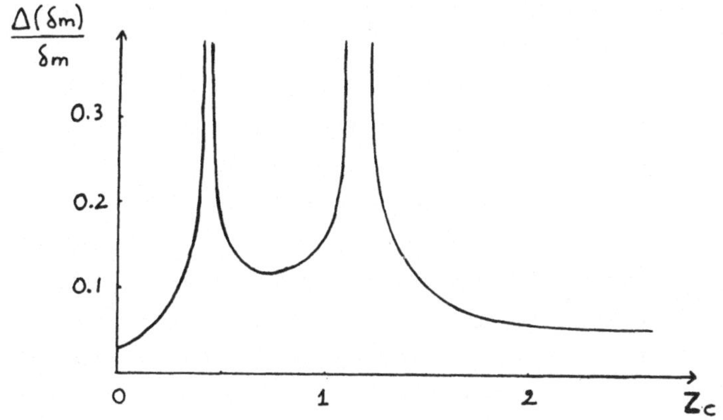

Figure 5 : Dynamics effects in term of magnitude for a lensing rich cluster of galaxies ($q_o\gamma$ = 300, u = 0.001). $\Delta(\partial m)/\partial m$ is the relative difference with respect to the classical formula.

This computation allows us to consider eventual violent dynamics effects in the lens (matter in collapse or collision between galaxies). We generalize this to the non-radial case (Hammer, 1985) which leads us to the crucial point of energy conservation (Hammer and Nottale, 1985), and allows us to apply this theory to actual astrophysical situations implying various cases of lenses (supermassive objects, galaxies, clusters of galaxies etc...).

REFERENCES

Bourassa, R.R. and Kantowski, R., Astrophys. J. 195, 13 (1975).
Eisenstaedt, J., Phys. Rev. D16, 927 (1977)
Hammer, F., submitted (1984).
Hammer, F., in preparation (1985).
Hammer, F., Nottale, L., in preparation (1985).
Nottale, L., Astron. Astrophys. 114, 261 (1982b).
Nottale, L., Hammer, F., Astron. Astrophys., accepted (1984).
Sachs, R.K., Proc. Roy. Soc. London A264, 309 (1961).
Young, P., Gunn, J.E., Kristian, J., Oke, J.B. and Westphal, J.A., Astrophys. J. 241, 507 (1980).
Young, P., Gunn, J.E., Kristian, J., Oke, J.B. and Westphal, J.A., Astrophys. J. 244, 736 (1981).
Young, P., Astrophys. J. 244, 756 (1981).

THERMODYNAMICAL FLUCTUATIONS OF MASSIVE BLACK HOLES

D. Pavón and J.M. Rubí
Departamento de Termología
Universidad Autónoma de Barcelona
Bellaterra (Barcelona) Spain

After Hawking discovery of black hole radiance |1| it has became apparent that these collapsed objects obey the thermodynamical laws |2,3|. Despite of this it is not clear at all which is the meaning of black hole entropy, S_{bh}, -which has been associated to the area of event horizon- nor even wheter or not that concept may be defined in an unambiguous way. So far the more popular interpretation of S_{bh} is the one provided by Bekenstein |4|. According to this author, that quantity should measure the number N of black hole interior configurations compatible with the given black hole state through the formula

$$S_{bh} = (1/8\pi) \ln N , \tag{1}$$

(we use units in which $G = c = \hbar = 1$ and $k_B = 1/8\pi$ here and below). The main problem with this interpretation lies in the singularity, located in the hole interior, which will probably prevent to count the above number. Nevertheless works by Zurek |5| and the authors |6,7| have recently proportioned some support to Bekenstein conjecture.

Our objective here is to analyse classically the thermodynamical fluctuations around equilibrium of massive Schwarzschild and Kerr black holes by resorting to Bekenstein conjecture in conjuction with Einstein-Boltzmann expression

$$\Pr \propto \exp(8\pi \delta S_t) \tag{2}$$

for the probability of spontaneous thermodynamical fluctuations around equilibrium. In (2) S_t stands for the total entropy of our system: a black hole surrounded by thermal radiation in equilibrium with it and enclosed in a rigid box of perfectly reflecting walls. We consider the black hole massive enough to limit seriously the emission of charged particles and the box large enough to make the flat space-time limit approximation valid. We shall use the microcanonical ensemble as the canonical one is inapplicable to this situation |8|.

Starting from (2) besides state equations of thermal radiation and the Schwarzschild entropy formula $S_{bh} = M^2/2$, it is straightforward to derive the second moments for the fluctuations around equilibrium between the Schwarzschild hole and thermal radiation $\langle \delta M \, \delta M \rangle = \langle \delta E_{rad} \, \delta E_{rad} \rangle = - \langle \delta M \, \delta E_{rad} \rangle = T^2 C_{sch}/8\pi$. Here and below $\langle \ldots \rangle$ means microcanonical average whereas T denotes the equilibrium temperature $T = T_{rad} = T_{bh} = 1/M$, and heat capacity C_{sch} is given by $4aVT^3/(1-4aVT^5)$. Since $\langle\delta M\ \delta M\rangle$ as well as $\langle\delta E_{rad}\ \delta E_{rad}\rangle$ cannot be negative quantities we automatically get the restriction $C_{sch} > 0$, which is nothing but the stability criterion $E_{rad} < M/4$ first derived by Hawking |8|. As it is well known heat capacity of a Schwarzschild hole is negative and accordingly some second moments have to be negative. Thus, for instance, if the hole absorbs a tiny quantity of energy from the radiation, both δM and δS_{bh} result positive quantities, while δE_{rad}, δS_{rad}, δT_{bh} and δT_{rad} negative. It can be immediately realized |6| that the second moments $\langle\delta M\ \delta T_{bh}\rangle$, $\langle\delta M\ \delta T_{rad}\rangle$, $\langle\delta S_{bh}\ \delta S_{rad}\rangle$, $\langle\delta S_{bh}\ \delta T_{rad}\rangle$, $\langle\delta M\ \delta S_{rad}\rangle$ and $\langle\delta S_{bh}\ \delta E_{rad}\rangle$ must be negative whereas the remaining ones must be positive. This issue reveals that (2) works out well when applied to a thermodynamical system in equilibrium containing a Schwarzschild black hole.

It is straightforward to extend our analysis to a Reisnerr-Nordström black hole whereby we will omit it. However for a Kerr hole the situation is more involved as thermodynamical equilibrium between a Kerr hole and surrounding radiation requires not only equality between temperatures but also that the radiation corotates along with the hole at its same angular velocity Ω. Otherwise the black hole would emit particle preferently with orbital angular momentum axial component parallel rather than antiparallel to the spin vector, $\underset{\sim}{J}_{bh}$, of the hole. In this way $\underset{\sim}{J}_{bh}$ would diminish and consequently thermodynamical equilibrium would be impossible |2|. To avoid viscous dissipation through shear stresses it is needed the rotation of the radiation to be uniform. This besides the large volume V of the box could give rise to superluminal speeds. However this difficulty dissapears if we assume the ratio J_{bh}/M^2 much smaller than unit. Because of the uniform rotation Gibbs equation, for thermal radiation must be generalised to read $T_{rad}\, dS_{rad} = dE_{rad} - \omega_{rad}\, dJ_{rad}$, and Gibbs equation for Kerr hole adopts the form $T_{bh}\, dS_{bh} = dM - \Omega\, dJ_{bh}$.

Before calculating the second moments of the fluctuations it is convenient to write down Kerr entropy as a function of M and J_{bh}, $S_{bh} = (M^2/4)\left\{1+\left[1-(J_{bh}^2/M^4)\right]^{1/2}\right\}$, and to define the following quantities: $C_{Jbh} = (\partial M/\partial T_{bh})_{J_{bh}}$, $C_{Jrad} = (\partial E_{rad}/\partial T_{rad})_{J_{rad}}$, $\Lambda_{bh}^{-1} = -T^2(\partial^2 S_{bh}/\partial J_{bh}^2)$, and $\Lambda_{rad}^{-1} = -T_{rad}^2(\partial^2 S_{rad}/\partial J_{rad}^2)$. By C_{Jbh} and C_{Jrad} we indicate the heat capacities at constant angular momentum for the black hole and for the radiation respectively. The second one is always positive but not so the first one which is negative for low values of the ratio J_{bh}/M^2. In turn the above Λ parameters are both positive. The former because of the cosmic censorship hypothesis |9| and the latter one since J_{rad} must be an increasing function of ω_{rad}. In view of their definitions Λ_{bh} and Λ_{rad} can be understood as a sort of rotational heat capacities.

Starting from (2) we immediately have $\langle\delta M\ \delta M\rangle = \langle\delta E_{rad}\ \delta E_{rad}\rangle = T^2 C_k/8\pi$ and $\langle\delta J_{bh}\ \delta J_{bh}\rangle = \langle\delta J_{rad}\ \delta J_{rad}\rangle = T^2\Lambda/8\pi$, with $C_k = C_{Jbh}\ C_{Jrad}/(C_{Jbh}+C_{Jrad})$ and $\Lambda = \Lambda_{bh}\ \Lambda_{rad}/(\Lambda_{bh}+\Lambda_{rad})$. Crossed second moments like $\langle\delta M\ \delta J_{bh}\rangle$ etc.. vanish which indicates that probabilities for black hole emission in different modes are independent. The stability criterion, $C_k > 0$, is obtained again as a by-product of the positive character of both $\langle\delta M\ \delta M\rangle$ and $\langle\delta E_{rad}\ \delta E_{rad}\rangle$. Of course, for vanishing J_{bh} it reduces to $C_{sch} > 0$.

Christodoulou formula $M^2 = 2S_{bh} + J_{bh}^2/8S_{bh}$ relates the square of irreducible mass, $M_{irr}^2 = 2S_{bh}$, and the square of rotational mass, $M_{rot}^2 = J_{bh}^2/8S_{bh}$ with the total energy of the Kerr hole. Hence, it is obvious that the equilibrium fluctuations of these quantities have to be correlated. A straighthforward calculation shows $\langle\delta M\ \delta M_{rot}\rangle < 0$, and $\langle\delta M_{irr}\ \delta M_{rot}\rangle < 0$. The former implies that,on the average, when a Kerr black hole absorbs from the radiation a particle carrying a small quantity of energy, $m = \Delta M$, and angular momentum, j, the increase of black hole angular momentum, $\Delta J_{bh} = j$, is lower than the increase of entropy, ΔS_{bh}, experienced by the hole. Both increments are such that the ratio $J_{bh}/(8S_{bh})^{1/2}$ is lower after than before the particle is swallowed up. That is to say, an increase of irreducible mass means -on the average- an increase of total mass and a decrease of rotational mass. This result is related, in some way, to the second gendaken experiment proposed by Wald |10| supporting cosmic censorship hypothesis. Effectively, the relation $\langle\delta M\ \delta M_{rot}\rangle < 0$ shows that the black hole absorption of particles with arbitrary high value of the ratio j/m is strongly inhibited. Likewise the relation $\langle\delta M_{irr}\ \delta M_{rot}\rangle < 0$ is a direct consequence of S_{bh} being a decreasing function of J_{bh}.

The reader interested in another second moments is referred to |7|.

Because of the calculated second moments bear the correct sign we can say our results support Bekenstein interpretation of black hole entropy. However it remains to be given a conclusive proof of equation (1). The main difficulty about this is the tacitly assumed black hole interior singularity. However it may be possible the singularity does not exist at all as it is obtained as an extrapolation of gravitational collapse when quantum effects are not properly taken into account.

It can be argued that in reality black holes are not surrounded by thermal radiation but by Hawking radiation. However the latter one behaves like thermal radiation in flat space-time limit approximation which has been used here. Moreover according to Pavón and Israel |11| thermal stability between a black hole and its own Hawking radiation is possible even for Planck-mass black holes and box sizes comparable with the black hole radius.

ACKNOWLEDGEMENTS

This work has been partially supported by the Comisión Asesora de Investigación Científica y Técnica of the Spanish Government.

REFERENCES

|1| S W Hawking, Commun. Math. Phys., 43 (1975) 199.
|2| P C W Davies, Rep. Progr. Phys., 14 (1978) 1313.
|3| W Israel, Sci. Progress, 68 (1983) 333.
|4| J D Bekenstein, Phys. Rev., D12 (1975) 3077.
|5| W H Zurek, Phys. Rev. Lett., 49 (1982) 1683.
|6| D Pavón and J M Rubí, Phys. Lett., 99A (1983) 214.
|7| D Pavón and J M Rubí, preprint(1984).
|8| S W Hawking, Phys. Rev., D13 (1976) 191.
|9| R Penrose, Rivista del Nuovo Cimento, 1 (1969) 252.
|10| R Wald, Ann. Phys. (N.Y.), 83 (1974) 548.
|11| D Pavón and W Israel, Gen. Rel Grav., (1984) to appear.

NEWTONIAN AND RELATIVISTIC BIANCHI I MODELS OF THE UNIVERSE

B. Barberis and D. Galletto
Istituto di Fisica Matematica "J.-Louis Lagrange"
Università di Torino
Via Carlo Alberto 10
10123 Torino (ITALY)

1. - Among the most simple and interesting homogeneous and anisotropic models of the Universe there are the well-known Bianchi I models, which are characterized by the metric

$$ds^2 = c^2 dt^2 - \sum_{1}^{3} {}_i R_i^2(t)\,(dy^i)^2 . \tag{1}$$

They include, as a particular case, the Einstein-de Sitter model whose metric is the Robertson-Walker metric in the case when the sections $t = \text{const.}$ are the ordinary Euclidean space.

The models of the Universe correspondent to the metric (1) have been studied by Heckmann & Schücking ([17], [13]), Saunders ([16]), Ryan & Shepley ([15]), etc. In particular Heckmann & Schücking obtained the solutions of the Einstein equations for this metric, without the cosmological term and without pressure, that is when the scheme used for the description of the Universe is the usual incoherent matter scheme. These models correspond to the case in which there is shear but not rotation.

The study of the homogeneous and anisotropic models of the Universe within the framework of Newtonian mechanics was begun by Heckmann & Schücking in the years 1955-'56 ([10, [11]) who took the Poisson equation as their starting point.

A similar study of these models was made by Davidson & Evans in the years 1971-'73 ([2], [3]) starting, not from the Poisson equation, but from the very much stronger consideration that the homogeneity hypothesis implies that at any instant the distribution of matter for the pressureless cosmological fluid with which the Universe is described (and which in the following will be indicated by $\mathcal{U}$) is spherically symmetric with respect to any one of its elements.

Other studies in the Newtonian framework related to the Heckmann & Schücking studies were made by Narlikar ([14]), Zel'dovich ([19]) and Shikin ([18], etc.). Other general studies in the Newtonian framework for the homogeneous and anisotropic models of the Universe have been made by Galletto & Barberis in [7], [8].

In 1977 Davidson & Evans ([4]) studied in detail, within the Newtonian framework, the case which corresponds within the relativistic framework to the case of the metric (1).

In the present paper, on the basis of the studies made by Galletto & Barberis in [7], the case studied by Davidson & Evans in [4] is briefly re-examined and, analogously to what happens in the homogeneous and isotropic case (see [6]), we outline that it leads smoothly to the relativistic case described by metric (1).

2. - The absence of rotation of the Universe implies that the symmetry axes orthogonal to the planes of symmetry of spiral galaxies have invariable directions one with respect to the other. It follows that it is possible to connect to any galaxy (and hence to any element of the fluid $\mathcal{U}$) a frame of reference determined by three axes parallel to any three symmetry axes among the ones introduced above. We shall call these frames natural frames of reference. They are in translatory motion one with respect to the other.

Let O be any galaxy, that is any element of $\mathcal{U}$, and let $\mathcal{R}_0$ be the natural frame of reference which has its origin in O and which is determined by means of a system of orthogonal Cartesian coordinates x^i with origin at O. With a suitable choice of the coordinates x^i, the application of the cosmological principle implies that in the present case (absence of rotation) for any element P of $\mathcal{U}$ we have

$$\dot{x}^i = h_i(t)\, x^i , \tag{2}$$

which are the extension of the Hubble law to the present case and where we do not have to sum with respect to the index i. Obviously these expressions are independent of the choice of the natural frame of reference $\mathcal{R}_0$. From (2) it follows that

$$\ddot{x}^i = (\dot{h}_i + h_i^2)\, x^i . \tag{3}$$

The application of the principle of superposition of simultaneous forces (see [5],[7], [9]) and the spherical symmetry for the distribution of matter of $\mathcal{U}$ with respect to O imply that from (3) it follows that

$$\frac{\ddot{x}^i}{x^i} = -\frac{4}{3}\pi k \mu \tag{4}$$

(where μ is the density of $\mathcal{U}$), without resorting at all to Newton's theory of gravitation. These equations preserve their form unchanged in every natural frame of reference.

If the instant t_0 is fixed once and for all, and if we define

$$R_i(t) = \exp \int_{t_0}^{t} h_i(t)\, dt , \tag{5}$$

equations (4) assume the following form

$$\frac{\ddot{R}_i}{R_i} = -\frac{4}{3}\pi k \mu \tag{6}$$

and coincide with the equations obtained by Davidson & Evans in [4]. It must be underlined that Davidson & Evans obtained equations (6) by making recourse to Newton's theory of gravitation. The result expressed by equations (6) shows that the statement made by Zel'dovich in [19], and subsequently in [20], Chap. 19, that the study of the homogeneous and anisotropic Newtonian models of the Universe is indeterminate, and even more that it has a high degree of indetermination, is not true. This indetermination appears also in the study of these models made by Heckmann & Schücking in [10] and [11] and reconsidered in [12]. This indetermination does not subsist and, as we have seen, to prove this it is not necessary to make recourse to Newton's theory of gravitation.

3. - Remembering definition (5), let us introduce now, together with the Cartesian coordinates x^i, the following coordinates (comoving coordinates):

$$y^i = \frac{x^i}{R_i(t)} \quad . \tag{7}$$

If we take into account the property of the velocity of light revealed by the Michelson-Morley experiment, we have that the local velocity of light is the same with respect to any natural frame of reference. From this result, making use of the comoving coordinates (7) and making recourse to the Galilean law of addition of velocities, it follows that the metric of the space-time manifold is necessarily expressed by metric (1) which in the relativistic context characterizes the Bianchi I models.

At this point it is possible to develop all the relativistic considerations concerning the red-shift, the visual horizons, etc. These considerations become a consequence of the constancy of the local velocity of light and of the Galilean law of addition of velocities.

Finally, if we require equations (6) of the Newtonian case, with a suitable modification (if it is necessary), to be compatible with metric (1), that is if we look for an intrinsic form of equations (6) in the context of this metric, it follows that this intrinsic form is precisely expressed by the Einstein gravitational equations of the general theory of relativity and furthermore that the modified form of equations (6) so obtained is precisely given by the explicit form assumed by the Einstein gravitational equations in the case of metric (1).

Analogously to the isotropic case, in which the considerations made for the case of the Einstein-de Sitter model (see [6]) can be extended to the general case of the homogeneous models (see [1]), the considerations and the results outlined in this paper can be extended to the other Bianchi models with suitable adjustements.

All these results are presented in detail in [9] and in forthcoming papers.

REFERENCES

[1] BARBERIS B., GALLETTO D., Atti del 7° Congresso Nazionale A.I.M.E.T.A., Trieste, 2-5 ottobre 1984 (in print).

[2] DAVIDSON W., EVANS A.B., Nature Phys. Sci., **232**, pp. 29-31 (1971).

[3] DAVIDSON W., EVANS A.B., Int. J. Theor. Phys., **7**, pp. 353-378 (1973).

[4] DAVIDSON W., EVANS A.B., Comm. Roy. Soc. Edinburgh, **10**, pp. 123-145 (1977).

[5] GALLETTO D., Atti del 3° Convegno Nazionale di Relatività Generale e Fisica della Gravitazione, Torino, 18-21 settembre 1978; Accademia delle Scienze, Torino, 1981; pp. 111-157.

[6] GALLETTO D., Atti del Convegno Internazionale "Aspetti matematici della teoria della relatività", Roma, 5-6 giugno 1980; Accademia Nazionale dei Lincei, Roma, 1983; pp. 59-83.

[7] GALLETTO D., BARBERIS B., Atti del Convegno su "Problemi attuali di fisica teorica", Torino, 12-13 dicembre 1980; Accademia delle Scienze, Torino, 1981; pp. 211-224.

[8] GALLETTO D., BARBERIS B., Proceedings of "Journees Relativistes 1983", Torino, May 5-8 1983; Accademia delle Scienze, Torino (in print).

[9] GALLETTO D., BARBERIS B., Boll. Un. Mat. Ital., Suppl. (1984) (in print).

[10] HECKMANN O., SCHÜCKING E., Z. Astrophysik, **38**, pp. 95-109 (1955).

[11] HECKMANN O., SCHÜCKING E., Z. Astrophysik, **40**, pp. 81-92 (1956).

[12] HECKMANN O., SCHÜCKING E., Handbuch der Physik, LIII, pp. 489-519 (1959).

[13] HECKMANN O., SCHÜCKING E., in: Witten L. (ed.), "Gravitation: an Introduction to Current Research", Wiley, New York, 1962; Chap. 11, pp. 438-469.

[14] NARLIKAR J.V., Mon. Not. Roy. Astronom. Soc., **126**, pp. 203-208 (1963).

[15] RYAN M.P. Jr., SHEPLEY L.C., **Homogeneous Relativistic Cosmology**, Princeton Univ. Press, 1975.

[16] SAUNDERS P.T., Mon. Not. Roy. Astron. Soc., **142**, pp. 213-227 (1969).

[17] SCHÜCKING E., HECKMANN O., Actes du 11e Conseil de Physique Solvay, Bruxelles, 1958; pp. 149-159.

[18] SHIKIN I.S., Soviet Phys. JETP, **32**, pp. 101-107 (1971).

[19] ZEL'DOVICH Ya.B., Soviet Astron. AJ, **8**, pp. 700-707 (1965).

[20] ZEL'DOVICH Ya.B., NOVIKOV I.D., **The Structure and Evolution of the Universe**, Univ. Chicago Press, 1983.

THE COSMOLOGICAL CONSTANT

A.Blanchard[1] and F.X.Désert[2]

1 : Groupe d'Astrophysique Relativiste, Observatoire de Paris-Meudon L.A.M., 92195 MEUDON Cedex

2 : Equipe de Radioastronomie,Laboratoire de l'Ecole Normale Supérieure, 24 rue Lhomond, 75005 PARIS

Abstract : In this paper we recall briefly the "zoology" of dust-filled universes wih a positive cosmological constant Λ. We analyse present observational limits and we point out the absence of any convincing argument against a non-vanishing Λ along with important consequences for cosmology. We present some possible observational tests able to improve existing limits. Finally we mention interesting implications of a positive Λ-term.

I) Introduction

The Λ constant was introduced by Einstein to allow for static universes which were thought to be the only acceptable ones. Since the cosmological constant has always suffered from strong prejudices. However it is interesting to notice that Λ is the only free parameter in general relativity and that observational constraints have not been improved, unlike a large number of theories competing with general relativity[1].

Einstein's equations with Λ read:

$$R_{ab} -(1/2R+\Lambda)\, g_{ab} = 8\pi G\, T_{ab}$$

The term Λg_{ab} could be interpreted as a geometrical term,however it can be looked as a term of the energy momentum tensor T_{ab}, the medium described by such a term is the vacuum and the equation of state is :

$$\rho = - p = \Lambda/8\pi G$$

II) The "zoology" of cosmological models with Λ≠0

We are interested here in homogeneous,isotropic dust-filled universes with positive Λ (the possibility of a negative Λ is not excluded but consequences for cosmology are less interesting). A more complete description could be found elsewhere[2,3]. The Eintein's equations are now :

$$\begin{cases} 2\dfrac{\ddot{R}}{R} + \dfrac{\dot{R}^2}{R^2} = -\dfrac{kc^2}{R^2} + \Lambda \\ \dfrac{\dot{R}^2}{R^2} - \dfrac{8\pi G\rho}{3} = -\dfrac{kc^2}{R^2} + \Lambda/3 \end{cases}$$

we adopt the following notations :

$$\Omega_0 = \frac{8\pi G\rho_0}{3H_0^2} \qquad K_0 = \Omega_0 + \lambda_0 - 1$$

$$\lambda_0 = \Lambda/3H_0^2 \qquad \xi = 1 + z = R_0/R$$

with H_0 the today Hubble constant. From equations (1) one obtains a dynamical constant :

$$K_0 = k/H_0^2R_0^2 = \Omega_0\xi + \lambda_0/\xi^2 - \dot{R}^2/H_0^2R_0^2$$

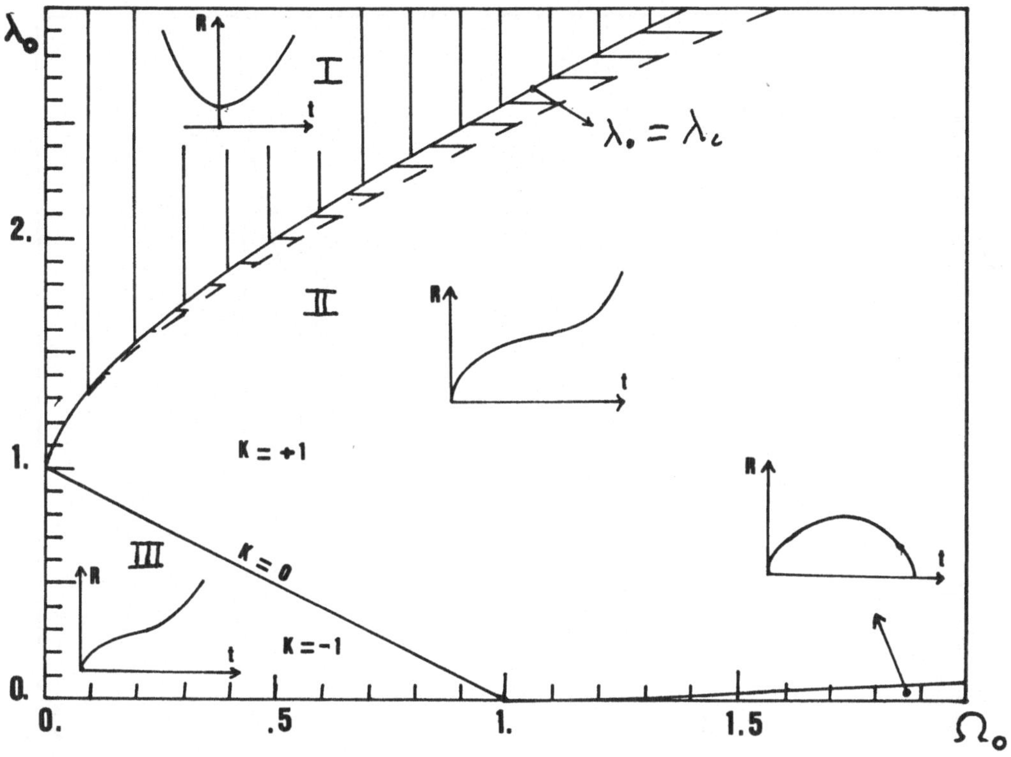

Figure 1: Diagramm of the evolution of the scale factor R with time for different regions in the (Ω_o,λ_o) plane. The region I vertically hachured (without initial singularity) is excluded by the existence of high redshift objects. The small region (horizontally hachured) is excluded by others observations (see the text).

This gives the Hubble Radius : $R_o = 1/H_o \, |K_o|^{1/2}$

and a useful expression of $\dot{R}^2$:

$$\dot{R}^2 = \frac{1}{|K_o|} \left(\Omega_o \xi + \frac{\lambda_o}{\xi^2} - K_o \right)$$

The qualitative feature of the scale factor evolution with time for different regions of the plane Ω_o,λ_o are plotted in figure 1. One can notice that there are many behaviours quite different than in models without cosmological constant. The region I and II are separated by the curve $\lambda_o = \lambda_c$ corresponding to the existence or absence of initial singularity. This curve can be parametrised by :

$$\Omega_o = \frac{2}{(\alpha-1)^2(\alpha+2)} \quad \text{and} \quad \lambda_c = \frac{\alpha^3}{(\alpha-1)^2(\alpha+2)}$$

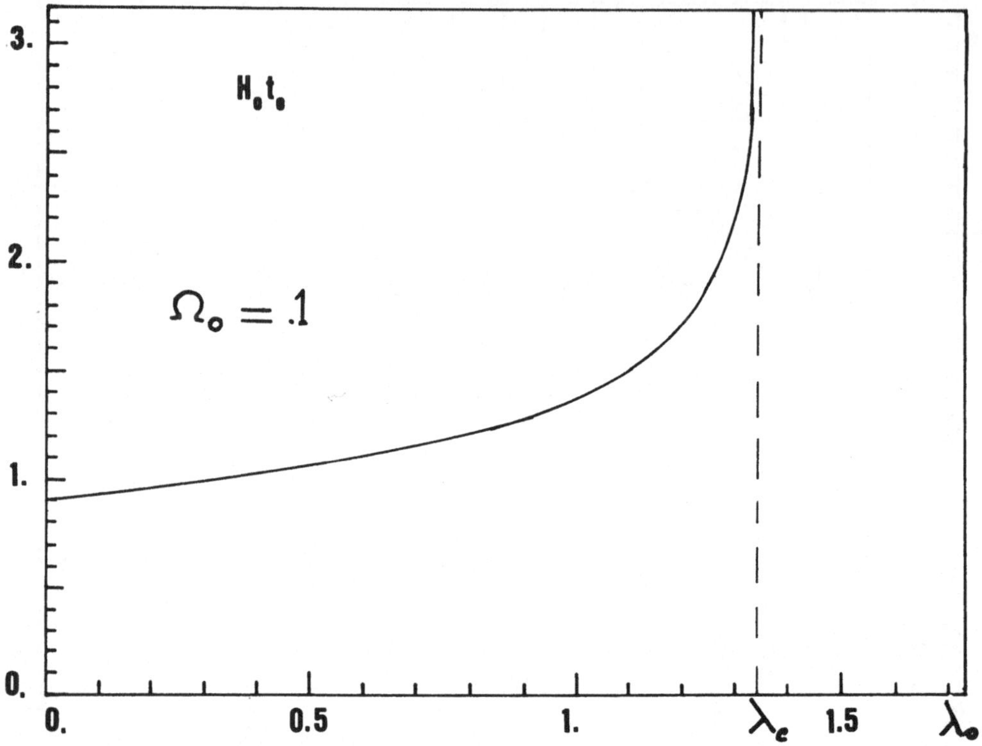

Figure 2: The age of the universe in unit of $1/H_o$ as a function of λ_o (for Ω_o=.1).

A crucial difference holds in the age of the universe. From the expression (2),it follows :

$$H_o t_o = \int_x^1 \frac{dx}{\sqrt{\Omega_o x^{-1} + \lambda_o x^2 - k_o}} \quad \text{with } x = 1/\xi$$

(this holds only if $\lambda_o < \lambda_c$,else the age is not finite). We present the age of the universe with Ω_o = .1 for the different value of λ_o. The age is a growing fonction of λ_o.

III) Observational constraints

It is generally thought that the cosmological constant is very well constrained. Actually we will see that this is quite wrong. The dynamics of the solar system give a first upper limit[4] :

$$|\Lambda| < 10^{-42} \quad \text{i.e. } \lambda_o < 10^{12}$$

But this limit is well above that obtained from cosmological observations. A strong constraint is derived from the existence of high redshift objects ($Z>3.78$ at this day). Models with a cosmological constant $\lambda_o > \lambda_c$ have no initial singularity (see fig.1), blueshifts are possible and there is a maximal redshift Z_m corresponding to the minimal possible value for the scale factor R. from this, assuming $\Omega_o>.05$, the region I is excluded. Therefore, it is interesting to note that the observation of quasars gives a strong limit on :

$$\lambda_o<\lambda_c \qquad \text{(for a given value of } \Omega_o\text{)}$$

However even with $\Omega_o<.05$, one obtains that λ_o could not be much larger than 1. Classical tests as the Hubble diagram for galaxies or the angular-diameter relation are not able to give an interesting constraint because too many biases can affect these relations. One of the most uncontrolable of them certainly is the evolution effect. Therefore we will assume that the following limits (from these tests) are reliable:

$$0< \Omega_o <5$$
$$-5< q_o <5$$

An important way to know if a $\lambda_o\neq 0$ is relevant for cosmology stands in the age of the universe t_o . The age of the oldest stellar systems are generally estimated to be in the range 12-16 Gyr and it would be difficult to consider that $t_o<$ 10 Gyr. A Hubble constant of 50 km/s/Mpc is still compatible with an age up to 20 Gyr (and $\lambda_o=0$), while on the other hand a value of H_o =100 km/s/Mpc leads to $t_o<9.8$ Gyr if one keeps $\lambda_o=0$. This is a classical strong argument favoring a non-vanishing Λ[17] (or a Hubble constant of 50 km/s/Mpc). However confidence in models and observations needs to be increased to conclude firmly.

A recent upper limit on λ_o has been obtained by Tytler[5] by counts of absorption lines in quasars spectra. For a given set of parameters (Ω_o,λ_o), one can compute the value of ε :

$$1+\varepsilon = 27\lambda_o\Omega_o^2/4/K_o^3$$

The limitation obtained by Tytler reads:

$$\varepsilon > .1 \qquad \text{(for } \Omega_o<.2\text{)}$$

The region excluded by this condition has been indicated on fig.1. As one can see the new region rejected is very small. We will conclude this chapter by saying one time again that the value H_o = 100 km/s/Mpc is clearly incompatible with classical age of the oldest stellar systems. This suggests the possibility of a positive cosmological constant. The second point one should keep in mind is that Λ is a cosmological parameter for which observational limits give a very wide range of values. It is clear that limits on λ_o are worse than those on Ω_o (just because $q_o = \Omega_o/2 - \lambda_o$ is much more difficult to estimate than the density parameter Ω_o). For example, Fliche, Souriau, Triay[6] have developped Lemaitre's model ($K_o > 0$ and $\lambda_o > 0$) which presents an empty region in quasars distribution separating the universe into two equal halves leading to $\lambda_o = 1.17$ for $\Omega_o = .1$. This model does not suffer of any incompatibility with present observations.

IV) Possible improvements

At first, the next improvement one is waiting for a better estimate of H_o than the one presently available. With the Space Telescope, we can expect that in the near future the uncertainty will be much less than now. But direct improvements of limit on λ_o by the use of classical tests, even with better observational data, seems to be problematic as it will be very difficult to eliminate evolution effect.

Therefore it appears very interesting to find new tests which are insensitive to evolution effects. Alcock and Pacsynski[7] have proposed such a test. For a comoving sphere, the value of the quantity :

$$Z.\Delta\theta/\Delta Z$$

(where $\Delta\theta$ is the angular diameter, and ΔZ is the redshift thickness of the sphere) is insensitive to Ω_o and to confusion effect but depends more on λ_o. However such a test is difficult to use in practice. At first it is difficult to "see" a comoving sphere. A cluster has ist own dynamics, and a supercluster is generally highly non spherical. A possible way to use this test lies in chosing voids as comoving spheres. It seems reasonable to think that voids are quite spherical, and are approximatively comoving. But in any case this needs high quality observations.

Another way to obtain a good improvement in limits on λ_o, lies in the use of the Cosmic Microwave Background fluctuations[8]. Fluctuations of C.M.B. have been computed by several authors. The minimal level one can wait is unclear and depends on the scenario. However small scale angular fluctuations appear to be quite similar in all models without late reheating. Because of the finite thickness of the recombination, fluctuations smaller than the the scale θ_o of this thickness are blurred. Typically, we have:

$$\theta_o \sim 3' - 10'\sqrt{\Omega_o}$$

However, because the angular diameter distance is different if $\Lambda \neq 0$ the scale θ_o is increased:

$$(\theta_o)_\lambda = K_\lambda . \theta_o$$

K_λ is the ratio of this scale when $\lambda_o \neq 0$ to the scale given by the standard model (for a given Ω_o).

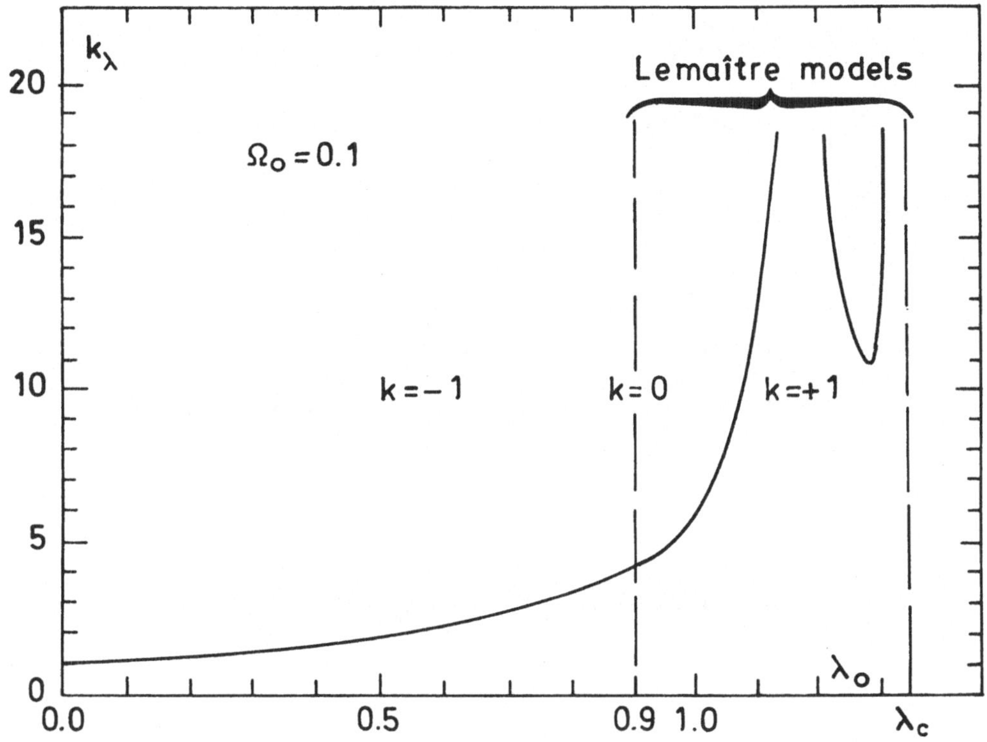

Figure 3: Dilatation coefficient K_λ vs. λ_o with Ω_o=0.1.

K_λ is shown in fig. 3 with $\Omega_o \sim 0.1$. As one can see, as soon as $\lambda_o > 1$ the value of θ_o is greater by one order of magnitude. Therefore it appears that the observation of this angular scale θ_o could determine crucially whether $\lambda_o > 1$, a value needed to solve the inconsistency of $H_o \sim 100$ km/s/Mpc with $t_o \sim 14$Gyr. This is a sensitive test because we use angular distance with the more distant object we can see (the CMB is located at $Z \sim 1000$). The only trouble is that a later reheating leads to the same effect[9] (see fig. 4). However one can hope to get an improved upper limit on λ_o by this mean.

V) Theoretical Aspects

We have already mentioned that a cosmological constant can solve the problem (if any) of the age of the universe. In this discussion we have implicitely assumed $\Omega_o \sim .1$. However the possible existence of a homogeneous invisible component of matter has been examined and would be probably difficult to detect. This leads to a value of Ω_o greater than one. But in such a case the problem becomes more stringent since $\Omega_o > 1$. gives $t_o < 6.6/h$ Gyr with H_o = 100.h km/s/Mpc (remember that the age of solar system is about 5 Gyr!). The necessity of $\Lambda \neq 0$ is then very strong[10,11].

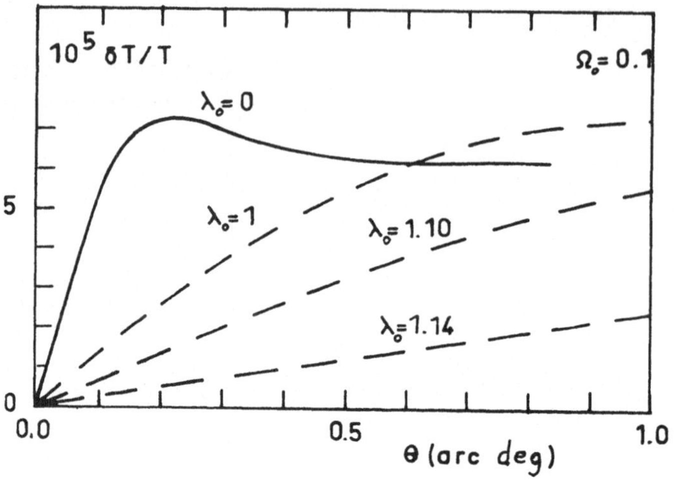

Figure 4: Typical fluctuations $\delta T/T$ vs. θ with $\lambda_o=0$, $\Omega_o=0.1$ and the same quantity with $\lambda_o=1$ ($K_\lambda=5$), $\lambda_o=1.1$ ($K_\lambda=10$) and $\lambda_o=1.14$ ($K_\lambda=30$).

This is not the only point of cosmology for which the introduction of a Λ-term could be useful. Recent theoretical models in particle physics introduce a very large cosmological constant in the early universe. Therefore it is not unconceivable that a remnant term still remains.For example , Zeldovich[12] has shown that the Λ-term arises naturally when dealing with zero-point energy oscillations of all particles with the equation of state: $\varrho=-p$ but the numerical value obtained is exceedingly large. In addition, inflation scenarios predict that the curvature term must be zero i.e. $K_o=\Omega_o+\lambda_o-1$ is closed to zero with a very high accuracy. On the other hand, the matter which appears clumpy leads to the value of density within galaxies of $\Omega_o \sim .1$. As noted by Peebles[13] these two points could be considered as an indication of a non-vanishing Λ.

Another point where λ_o could help is in the problem of the level of the CMB we have discussed in the last chapter[8]. Another reason lies in the fact that gravitational linear perturbations of density grow faster in model with $\Lambda>0$[14,15]. By the way, we mention that the horizon problem can be solved by the induction of Λ. However this solution semms unlikely since it needs fine-tuning of the value of λ_o.

VI) Conclusion

Modern developements in particle physics leads to the "cherished" inflation. These are based on a large cosmological constant in the the early universe (which is vacuum energy). The this term is presently so smaller is deeply problematic. A remaining term is therefore quite likely, and challenges the observations. Presently available observations are still compatible with a value of Λ which can be important for cosmology ($\lambda_0 \sim 1.$). New theoritical developments are possible[16] and the richness of this extension of standard cosmology has certainly been underestimated. In addition, further improvements of existing limits are within reach, in a near future.

Aknowledgments : One of us (A.B.) thanks J. Eisenstaedt for his interesting remark during the conference about the status of general relativity. We thank Guy Lacroix for having brushed up the English of this paper.

<u>References</u> :

[1] T. Damour :"Les Houches" Rayonnement Gravitationnel (1984)

[2] R. Stabell and S. Refsdal : M.N.R.A.S. <u>132</u> 379 (1966)

[3] S. Refsdal,R. Stabell and de Lange : Mem. Astron. Soc. <u>71</u> 143

[4] J.N. Islam : Physics Letters <u>97A</u> 239 (1983)

[5] D. Tytler : Nature <u>291</u> 289 (1981)

[6] H.H. Fliche,J.M. Souriau and R. Triay Astron. Astrophys. <u>108</u> 256 (1982)

[7] C. Alcock and B. Paczynski : Nature <u>281</u> 358 (1979)

[8] A. Blanchard : Astron. Astrophys. <u>132</u> 359 (1984)

[9] M. Davis : Physica Scripta <u>21</u> 717 (1980)

[10] J.P. Luminet and J. Schneider : Astron. Astrophys. <u>98</u> 412 (1981)

[11] Ya.B. Zeldovich and R.A. Sunyaev : Pis'ma Astr. Zh. <u>6</u> 451 (1980)

[12] Ya.B. Zeldovich : Soviet Physics Uspeckhi <u>11</u> 381 (1968)

[13] P.J.E. Peebles : to be published.

[14] K.Brecher and J. Silk : Ap.J. <u>158</u> (1968)

[15] F. Occhionero,N. Vittorio,P.Carnevali and P. Santangelo : Astron. Astrophys. <u>86</u> 212

[16] S.A Bludman:Nature <u>308</u> 319 (1984)

THE INFLATIONARY UNIVERSE : A PRIMER

Rémi HAKIM
LA 173 - DAF
Observatoire de Paris-Meudon
92195 - Meudon Principal
(France)

I. INTRODUCTION

These last years an abundant literature on the so-called inflationary universe[1] has been published, essentially since the interesting idea of Guth[1]. The notion of an inflationary era in the very early universe has completely changed our views on cosmology before, say, one second, and cannot be ignored eventhough the technical details are necessarily in constant evolution. Accordingly, in this talk, only the "substantifique moelle" of the basic ideas is given without entering into models subject to revision : our purpose is only pedagogical.

The idea of inflation - or, exponential growth of the scale factor of the expanding universe - has its origin in the study of Grand Unified Theories[2] (GUT's) and, more particularly, in the study of the phase transitions to which it gives rise (through spontaneously broken symmetries [3]) in the so-called GUT era ($t \sim 10^{-35}$ sec., $T \sim 10^{15}$ GeV).

The concept of Grand Unified Theory is quite natural : since the electro-weak interactions have a $SU_L(2) x U(I)$ gauge invariance[4] and since the strong interactions have an SU(3) gauge invariance (Quantum Chromodynamics (QCD)[5]), all these interactions should be unified by imposing an invariance under a more general gauge group G such that :

$$G \supset SU(3) \times SU_L(2) \times U(I) ;$$

this view is supported (at a theoretical level) by the fact that the effective coupling constants [6] of these various interactions do coincide at high energy ($\sim 10^{15}$GeV, an energy that determines the scale of GUT's), at least within one order of magnitude.

There is, however, a number of problems connected with GUT's which make their application to cosmology rather speculative, so that no definite conclusions can be drawn.

(1) On the experimental side first, it is clear that energies of the order of 10^{15} GeV are far beyond our present possibilities (with our technology, this would require accelerators of the size of the Galaxy !) and only some low energy clues could confirm the GUT's. There exist essentially three such possible clues, which are under active experimental investigations : (i) the lifetime of the protons is predicted to be $\gtrsim 10^{31}$ years [7], (ii) the detection of monopoles [8] (topological defects [9] at the frontier of different spatial regions of spontaneously broken symmetries leaving a U(1) invariance) is also predicted as well as (iii) the existence of an electric dipole moment for the neutron.

(2) There exists a large number of possibilities for the choice of the gauge group G, SU(5) being the "smallest" possible one (although it seems to be eliminated by present data on the lifetime of the proton, it can be considered as a prototype of GUT's group and hence, possesses heuristic virtues). However, besides the choice of gauge group, one has also to choose the multiplets in which the known particles are to be accomodated although one generally chooses the fundamental representation of G. Finally, a large number of unknown constants (masses, couplings, etc) have to be determined (how ?).

(3) On the other hand, while unifying different types of interactions, one also gathers their unsolved problems (e.g. the difficulty to master the inherent linearity of non-abelian gauge theories, the quark confinement ; the unknown Higg's sector etc.) and adds some new ones (e.g. the hiérarchy problem, solved by the supersymmetric theories which, in turn, bring their own set of new problems). Furthermore, one has to rely on the "desert" hypothesis (essentially, everything is known between $\sim 10^{15}$ GeV and $\sim 10^{2}$ GeV).

(4) Finally, the calculations performed when dealing with dense hot matter in the GUT's era rest on perturbative (or one loop) calculations and often on simplifying assumptions. Unfortunately, it is difficult to control the validity of the approximations effected and of the various assumptions made. Also,

it should be remarked that the smallness of the effective coupling constant ($\sim 1/50$) at 10^{15}GeV does not guarantee the adhequacy of perturbative calculations : think of the smallness of $e^2 = 1/137 \ll 1$ and of the phenomenon of superconductivity (which is non perturbative) ; perturbative calculations could never have predicted this phenomenon. Let us also add that, at high density, Hartree-like (or other non-perturbative) calculations might prove more useful..

From this brief discussion, it is clear that Grand Unified Theories are not yet in a state where credible conclusions can be drawn as to the early universe. Nevertheless, the idea of inflation is very interesting in so far as it provides a solution to a number of conceptual problems, to which we come back below. The only problem (perhaps) of observational importance solved by inflation is the one of the spectrum of initial perturbations although it might be argued that this spectrum could form in the course of many other phenomena and at later (or earlier !) times. For instance, one might argue that this spectrum could form via some non linear phenomena occuring in the cosmic plasma after ~ 1 sec (i.e. in a period where temperatures and densities are those occuring in known physics).

Also, it should be born in mind that an inflationary era can result from other kinds of phase transitions (i.e. not only the one(s) occuring because of the spontaneous breaking of the symmetry of the state of the Universe under G) or from different dynamical approaches to highly dense matter. We come back to this question in Sec. 4. In fact, the presently observable properties of our Universe (expansion, fluctuations of the thermal radiation at 3°K, abundance of light elements, statistical properties of the cosmic "gas" of galaxies, etc) are most likely determined - eventhough we don't know (as yet) how - by physical phenomena occuring after ~ 1 sec. where conventional theories can be used. Therefore, there are two possible extreme attitudes toward the primordial universe ($t \lesssim 1$ sec.) : either one adopts a strict operationalism and any investigation on this period is to be considered merely as an irrelevant speculation ; or the solution of some difficulties, essentially conceptual, is rejected to inaccessible epochs and/or energies where no experimental data are available and where the Theory is not well established. Anyhow almost everything concerning the very early universe is, for the time being, only speculative eventhough quite fascinating. However, the main virtue of such speculations lies in their heuristic value : they suggest new phenomena new problems and new ways to be investigated. Furthermore, it is often thought that the Universe could be a "high energy laboratory" able to select acceptable theories of elementary particules among all possible ones... This would certainly be correct if present observational data would be accurate enough ; unfortunately, these data are plagued with a lot of uncontrolled errors and it is quite difficult to get trustworthy conclusions even on the astrophysics side.

Accordingly, we are led to speculate on the very early universe (see the excellent proceedings of the Nuffield Workshop[10]) feedings our views and speculations with numerous implicit philosophical prejudices. Let us also add that there exist several degrees in speculation : roughly, they increase as we go back in time, beginning with reasonable ones (e.g. the quark / hadron transition and / or the Weinberg-Salam one), passing by much less reasonable ones (e.g. the GUT's transition(s)) and ending with highly fascinating speculations although not reasonable at all (supersymmetric GUT's, super-gravity / quantum gravity, etc.).

In this talk our own philosophy as to the inflationary universe is summarized in the following questions to which we limit ourselves : (i) what are the problems that can possibly be solved by inflation ? (ii) what kind of physical phenomena can give rise to an inflationary behavior ? (iii) what are the new ways suggested by inflation and (iv) what can be extracted from inflation that is <u>independant</u> of the models from which it comes?

This paper is organized as follows. In Sec. 2 the role of the vacuum is emphasized while Sec. 3 is devoted to the problems solved by inflation. In Sec. 4, we briefly outline how to realize inflation. In Sec. 5 some conclusions are drawn.

2. THE ROLE OF THE VACUUM

During the last decade the physical character of the vacuum has been more and more emphasized leading to the consideration of a true "material medium" (see e.g. the book by T.D. Lee[11]) possessing symmetries, energy, etc. Hence, it is quite natural that cosmological consequences have to be investigated[12]. In particular, conventional models of inflation (either the original one, or the so-called "new inflation[13]) are essentially based - as we see below - on its properties and on the fact that it might dominate the evolution of the early universe, instead of radiation.

Before looking at the effects of the vacuum in early cosmology, let us briefly summarize the basic equations of standard cosmology [14]. The assumed homogeneity and isotropy of spacetime joined to the assumption of the existence of a universal time leads to the Robertson-Walker metric, supposed to describe our Universe

$$ds^2 = dt^2 - R^2(t)\left\{\frac{dr^2}{1 - kr^2} + r^2\, d\Omega^2\right\} \tag{2.1}$$

(k = normalized spatial curvature = 0, $\pm$ 1).

where the scale factor $R(t)$ is determined as a solution of Einstein's equations :

$$R_{\mu\nu} - \frac{1}{2} R\, g_{\mu\nu} = 8\pi G\, T_{\mu\nu} + \Lambda g_{\mu\nu} \tag{2.2}$$

where G is the gravitational constant ; where Λ is the cosmological constant (to which we come back below) and where $T_{\mu\nu}$ is the momentum-energy tensor describing the state of the matter and of the vacuum in the Universe, assuming the absence of dissipative effects,

$$T^{\mu\nu} = (P + \rho)\, U^\mu U^\nu - P\, g^{\mu\nu} \tag{2.3}$$

(P being the pressure and ρ the energy density while U^μ is the average four-velocity of matter).

Einstein's equations and momentum-energy conservation lead to

$$\left(\frac{\dot{R}}{R}\right)^2 = -\frac{k}{R^2} + \frac{8\pi G}{3}\rho + \Lambda \tag{2.4.a}$$

$$P\, d R^3 + d(\rho R^3) = 0 \tag{2.4.b}$$

to which an equation of state $P = P(\rho)$ must be joined.

The momentum-energy tensor can be splitted into a matter part and a vacuum part :

$$T^{\mu\nu}_{vac} = - P_{vac} \, g^{\mu\nu} \tag{2.5}$$

so that the equation of state of the vacuum reads :

$$P_{vac} + \rho_{vac} = 0, \tag{2.6}$$

and :

$$T^{\mu\nu} = (\rho_{mat} + P_{mat}) \, U^{\mu} \, U^{\nu} - (P_{mat} + P_{vac}) \, g^{\mu\nu} \tag{2.7}$$

with $P_{mat} \sim \frac{1}{3} \rho_{mat}$ (in the very early radiation dominated universe).

Let us now come back to the possible role of the vacuum. The Friedman equations (2.4) immediately yield :

$$\ddot{R} = -\frac{4\pi G}{3} [\rho + 3P] \, R \tag{2.8}$$

In general, the quantity $(\rho + 3P)$ is positive and a singularity i.e. $R = 0$, is unavoidable. Let us however inspect the quantity $(\rho + 3P)$ more closely. For ultra relativistic matter, it reads

$$(\rho + 3P) \sim 2 \times (\rho_{mat} - \rho_{vac}) \tag{2.9}$$

which can be positive or negative, leading there by (see fig.1) to a singularity or, possibly, to a non singular behavior of the cosmological spacetime manifold. When the vacuum energy dominates the matter energy, i.e. when

$$\rho_{vac} >> \rho_{mat}, \tag{2.10}$$

then $(\rho + 3P)$ is negative and hence $\ddot{R}$ is positive, leading to the behavior depicted on fig. (1.b) for $R(t)$.

Let us now assume that, during some interval of time Δt, the vacuum energy is dominant (condition (2.10)). Then the first Friedman equation (2.4.a) - with $\Lambda = 0$ (but this is a non essential simplification ; see below a brief discussion of the cosmo-

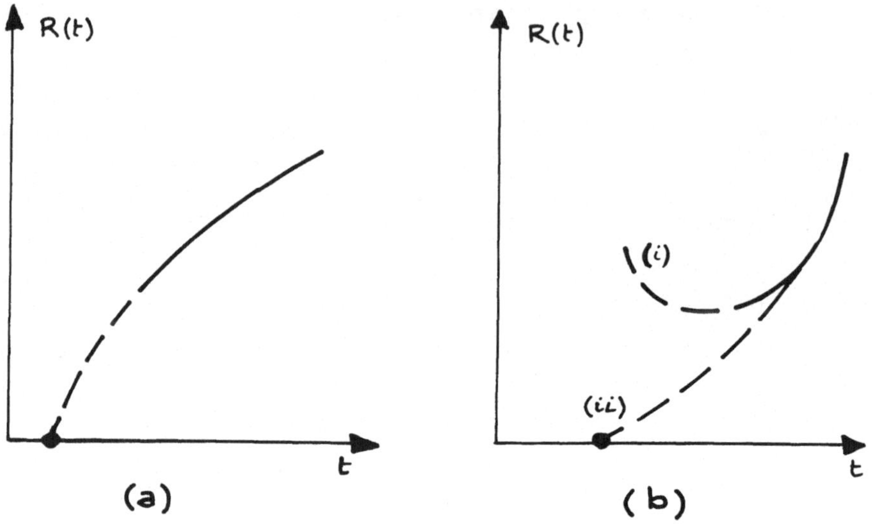

Fig. 1 : Two possible configurations of the scale factor as a function of time. In (a) $(\rho+3P)$ is positive and hence $\ddot{R} < 0$: the singularity cannot be avoided. In (b) the dominance of the vacuum yields $(\rho+3P) < 0$ so that $\ddot{R} > 0$; in this case the singularity may be avoided (case (i)) or not (case (ii)).

logical constant) - reads

$$\left(\frac{\dot{R}}{R}\right)^2 = -\frac{k}{R^2} + \frac{8\pi G}{3}\rho_{vac} \qquad (2.11)$$

whose solutions are of the following form

$$R(t) \sim R(t_o).\ \exp.\ [H_o(t - t_o)] \qquad (2.12)$$

after several Hubble times H_o^{-1}[i.e. when $(t - t_o) >> H_o^{-1}$] where

$$H_o^{-1} = \{3/8\pi G\ \rho_{vac}\}^{1/2} \qquad (2.13)$$

(the solution (2.12) is exact for $k = 0$, otherwise it represents only an asymptotic form). Notice that ρ_{vac} is implicitly assumed to be constant, a natural hypothesis although not necessarily true (for instance, the vacuum could have a periodic behavior; see e.g. Ref. 15).

Several remarks are now in order. First of all, this exponential behavior of the scale factor R(t) is the one of the so-called inflationary universe. Although this possible behavior was known [16] before inflation was discovered, its use as a way to solve various problems was realized by Guth[1]. A second remark is that, in some sense to be specified more precisely in the next section, the "initial conditions" are forgotten after a few Hubble times H_o^{-1} : indeed, Eq. (2.12) does not involve the spatial curvature index k ; everything happens as if $k = 0$ for $t >> H_o^{-1}$ and the original k has been forgotten. This is a kind of "H - theorem" or, in Hawking words[17], a "no - hair" theorem.

This last remark is quite important since it leads to the important prediction (of the inflationary models) that the present density ρ_o in the universe is equal to the critical density ρ_c

$$\rho_c = 3H^2 / 8\pi G \sim 2 \times 10^{-29}\ g/cm^3 \qquad (2.14)$$

(Here, H is the present Hubble constant)

up to negligible exponentially small terms. Therefore, inflation does predict that the dimensionless ratio $\Omega \equiv \rho_o/\rho_c$ is equal to one. What about the "observed" value of Ω? It is relatively safe to assume that Ω is such that

$$.02 \lesssim \Omega \lesssim 2.3 \qquad (2.15)$$

where the upper limit[18] comes from the age evaluated for the oldest astrophysical objects, i.e. globular clusters, while the lowest limit comes nucleosynthesis data ; in fact, it is probably of the order of .1. These data are thus consistent with the value predicted by inflation although the favoured figure is rather smaller than 1, perhaps .2 or .4. Therefore, in order to reach the critical value $\Omega = 1$, it is necessary to rely on the possible existence of "dark matter" (massive neutrinos, photinos gravitinos, etc) whose existence as a result of experiments or observations (e.g. black holes, cold invisible stars, etc) is not (yet ?) established. This dark matter[19] seems to exist around spiral galaxies as a result of the study of their rotation curves, although one can invoke other explanations for their observed shapes [20]. However, even <u>this</u> dark matter in halos of galaxies is not sufficient to yield $\Omega = 1$. If it would appear that (i) there are no "inos" or enough black holes or dead stars to provide $\Omega = 1$ and (ii) finally $\Omega < 1$, the situation could still be saved. Indeed, if the cosmological constant Λ is non vanishing the quantity that replaces Ω is :

$$\Omega \rightarrow \Omega + \Lambda / 3H^2 \equiv \tilde{\Omega} \qquad (2.16)$$

(H being the present value of the Hubble constant)
so that it is actually $\tilde{\Omega}$ which should be equal to one[21]. In fact, the observational data on Λ are quite uncertain. Nevertheless they are consistent with $\Lambda/3H^2 \sim O(1)$.

These last remarks point out the possible role of the cosmological constant ; it appears in Einstein's equations (2.2) as a <u>vacuum term</u>[12], with

$$T^{\mu\nu}_{vac} = \frac{\Lambda}{8\pi G} g^{\mu\nu} \qquad (2.17)$$

However, we should have :

$$\Lambda / 8\pi G \sim m^4 \qquad (2.18)$$

where m is a typical energy scale ; if m is taken to be the proton mass, then $\rho_{vac} = \Lambda / 8\pi G \sim 10^{14} g/cm^3$ whereas astronomical data provide $\rho_{vac} \lesssim 10^{-29} g/cm^3$ so that this discrepancy of 43 orders of magnitude should be explained (even if one would assume a neutrino mass of $\sim$ 10eV, there would remain a discrepancy of 11 orders of magnitude to be explained). This is one of the basic (conceptual) problem of cosmology. It is not explained at all by inflation and the smallness of the cosmological constant is rather to be considered as an empirical fact.

3. THE CONCEPTUAL PROBLEMS OF THE EARLY UNIVERSE

We now focuse our attention to those problems that can be solved by the exponential behavior of the scale factor. All these problems are essentially conceptual questions with no direct observational bearing. Most of them could be discarded as real problems and considered as being part of the "initial conditions" of our universe : whether they are solved or not is immaterial. For instance, whether the Universe started with a symmetry between matter and antimatter, or not, is a matter of taste and philosophy (this problem - explaining the baryonic asymmetry of the observed universe[22] while the early universe was symmetric - is, of course, not explained by inflation ; its solution[23] (at least at a qualitative level) is one of the great successes of the marriage of GUT's and cosmology)). There is however a notable exception to what has been said above. It concerns the spectrum and the amplitude of those primordial fluctuations whose growth is supposed to give rise to the observed structures (galaxies or clusters, etc...) ; yet it is highly model dependant and there are still much improvments to be effected before the problem can be considered as solved (see e.g. Ref. 10). Accordingly in this talk we shall not consider this question.

Let us also add that the problems considered below (horizon, homogeneity, isotropy, rotation), are, in fact, involved all together and they have not the same importance.

The horizon problem

The cosmological background radiation is known to be isotropic with a high degree of precision, about one part in thousands[24] : two microwaves - or infrared - antennas pointed in opposite directions in the sky do collect thermal radiation with $\Delta T/T \lesssim 10^{-3}$, T being the black body temperature. In the conventional big bang model[14] this is quite puzzling since the two emissive regions could never have been in causal connection (see fig. 2) ; the radiation received is thought to have been emitted roughly at the recombination of hydrogen ($t_R \sim 10^5$yr, $z_R \sim 3000$). This problem of causality is thus connected to the uniformity of the cosmic radiation background.

Let us specify this question a little further. Let us call t_o our age since the big bang ($t_o \sim 10^{10}$yr) and let

$$R(t) = \begin{cases} R_{mat}(t) = a\, t^{2/3} & t \gtrsim t_R \qquad (3.1.a) \\ R_{rad}(t) = b\, t^{1/2} & t \lesssim t_R \qquad (3.1.b) \end{cases}$$

be the scale factor when matter or radiation dominates the evolution of the universe (note that, roughly, $R_{mat}(t_R) = R_{rad}(t_R)$ so that $b/a \sim t_R^{1/6} \sim 10^{5/6}$). The coordinate radius of our horizon (i.e. the coordinate distance travelled by a light signal since the big bang (or, equivalently, since t_R , because $t_R \ll t_o$) ; it is given by the integration of $ds^2 = 0$) is

$$\Delta r(t_o,0) \,\#\, \Delta r(t_o,t_R) = 2\int_{t_R}^{t_o} \frac{dt}{R_{mat}(t)} \qquad (3.2)$$

while the coordinate radius at time t_R of a causally connected region (i.e. the locus of all those points inside a typical forward light cone at t_R and whose origin it at $t = 0$) is

$$\Delta r(0,t_R) = 2\int_{0}^{t_R} \frac{dt}{R_{rad}(t)} \qquad (3.3)$$

so that there are

$$N \sim \Delta r(t_o,t_R) \;/\; \Delta r(0,t_R) = \frac{3}{2} \times \frac{b}{a} \times t_R^{1/6} \sim 70 \qquad (3.4)$$

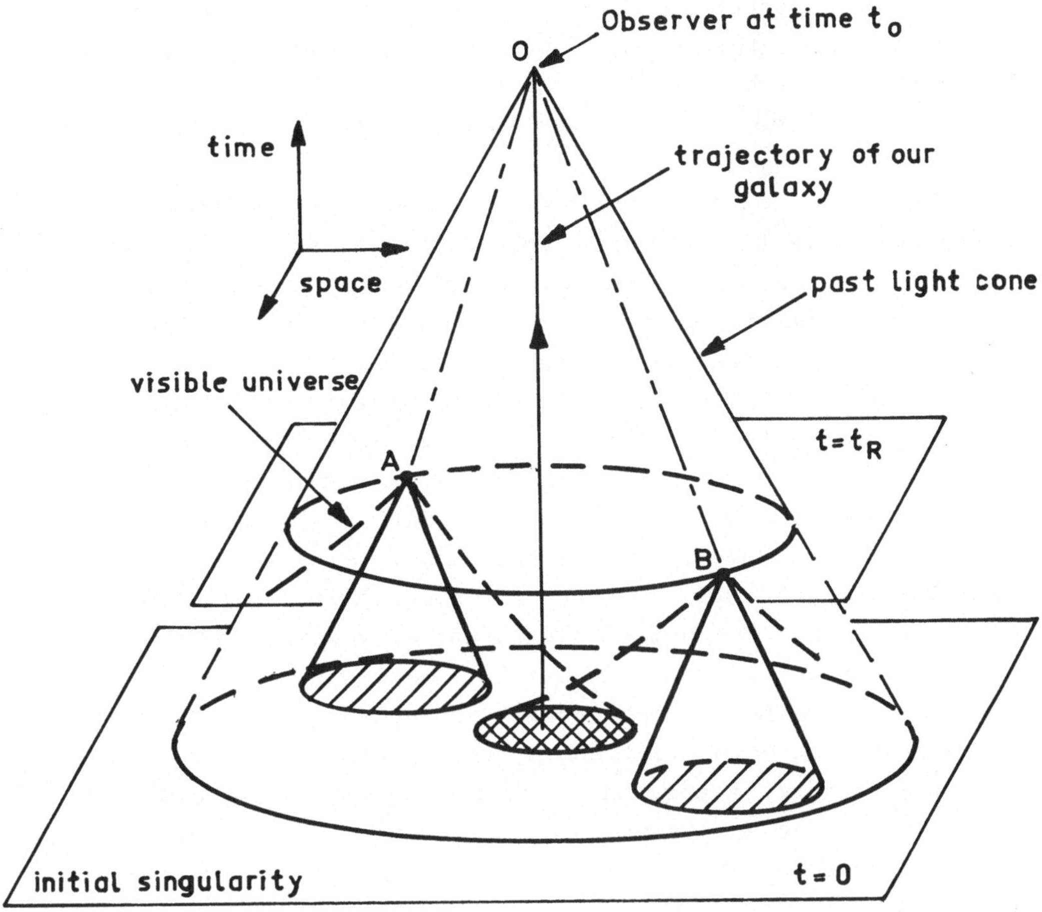

Fig. 2 : The horizon problem. In principle a galactic observer O can observe up to the recombination time t_R. Two regions A and B at t_R - in opposite directions in the sky - appear quite similar although in the standard big bang model they are unconnected (the pass light cones originating from A and B do not intersect). An inflationary era, by "enlarging" light cones (dotted lines) make them intersect so that A and B have been in causal connection.

non causally connected regions that separate two extreme emissive regions at the recombination time (therefore, there are $\sim N^3$ such regions in the visible universe and $\sim N^2$ apparent ones). At times earlier than t_R. N is even larger so that the problem still persists.

Let us now examine how an exponentially increasing scale factor could solve the horizon problem. This is shown qualitatively on fig. 2 : the essential feature of an inflationary period is that the (forward and backward) light cone is much larger than the one in radiative (or, matter dominated) era ; as a consequence backward "inflationary light cones" do intersect (see fig. 2) leading thereby to a possible connection which opens the door to a causal explanation of the similar observational properties of opposite regions in the sky. Analytically, the solution of the horizon problem is obtained with the same kind of calculations that lead to Eq. (3.4) with due account of an inflationary period Δt and by imposing that the number of non causally connected regions N, at the time under consideration t_1 , be less than one, i.e. our horizon is engulfed in one causaly connected region. This yields a constraint on both Δt, t_1 and H_o (Eq. (2.13)).

Accordingly, a possible inflationary era - whatever its origin - provides an elegant solution to this conceptual problem. However, this solution is not necessarily the only possible one the causal structure of our universe in the quantum (gravitational) era is not at all elucidated and is certainly quite involved (think e.g. to sums "over topologies"[25]) ; furthermore, if the topology of a k = 0 or -1 universe is the one of a cylinder there is no horizon problem at all since the backward light cone has an infinite number of intersections ; when k = +1, this is also the case provided there does not exist any singularity in our past, i.e. provided the universe has a bounce due to a period of vacuum dominance (e.g. $\Lambda \neq 0$).

The flatness[26], rotation[27] and anisotropy[28] problems

These three problems are of a totally different nature when compared to the preceding one : while the horizon problem is essentially a topological problem, these ones are related to the initial conditions that prevailed in our past. They can be stated as follows. Why is our universe so nearly flat ? Why it is isotropic ? Indeed, among all possible initial conditions the ones that lead to the observed universe are very special and, in fact, we would like to obtain these flatness, slow rotation and isotropy for almost all possible initial conditions ; they should not play any role and we are looking for a kind of "no hair" therem [7] according to which they could be forgotten. Here, only the flatness problem will be stated precisely although the other two will be solved simultaneously with an inflationary era.

We start again with the Friedman equation (2.4.a) with $\Lambda=0$ (this is not essential) which we rewrite as :

$$\Omega(t) = 1 + \frac{k}{H^2(t)R^2(t)} \tag{3.5}$$

with :

$$H(t) \equiv (\dot{R}/R)^2 \tag{3.6}$$

In the usual radiative era where R(t) is given by Eq. (3.1.b), Eq. (3.5) reads :

$$\Omega(t) = 1 + 4\ kt/b^2 \tag{3.7}$$

and it is found that at $t \sim 1$ sec $\Omega(1 \text{ sec}) = 1 \pm O(10^{-15})$. Accordingly, Ω should have been chosen extremely close to one in the past (this fine tuning of the initial Ω is even finer at $t \sim 10^{-35}$ or 10^{-43} sec) : a slightly higher value of Ω (1 sec) would have led to a re-collapse of the k = +1 universe while a slightly lower value would imply that the black body background would have a nearly vanishing temperature to day ! Stated in another way, the flatness problem amounts to considering that the only available length scale is defined by m_p^{-1} (m_p being the Planck mass : $m_p = (hc/G)^{1/2} \sim 10^{19}$ GeV) so that the only "natural" initial condition for the k/R^2 term is $k/R^2 \sim O(m_p^2)$, i.e. the universe would have been curvature dominated very early.

Eq. (3.5) considered during an inflationary era shows how the problem is solved : H(t) is constant (see Eq. (2.13)) and the last term is vanishingly small so that $\Omega(t)$ can be very close to one.

More generally, our universe should be anisotropic owing to possible shear [29] and vorticity [29] of the four-velocity U^μ of matter.

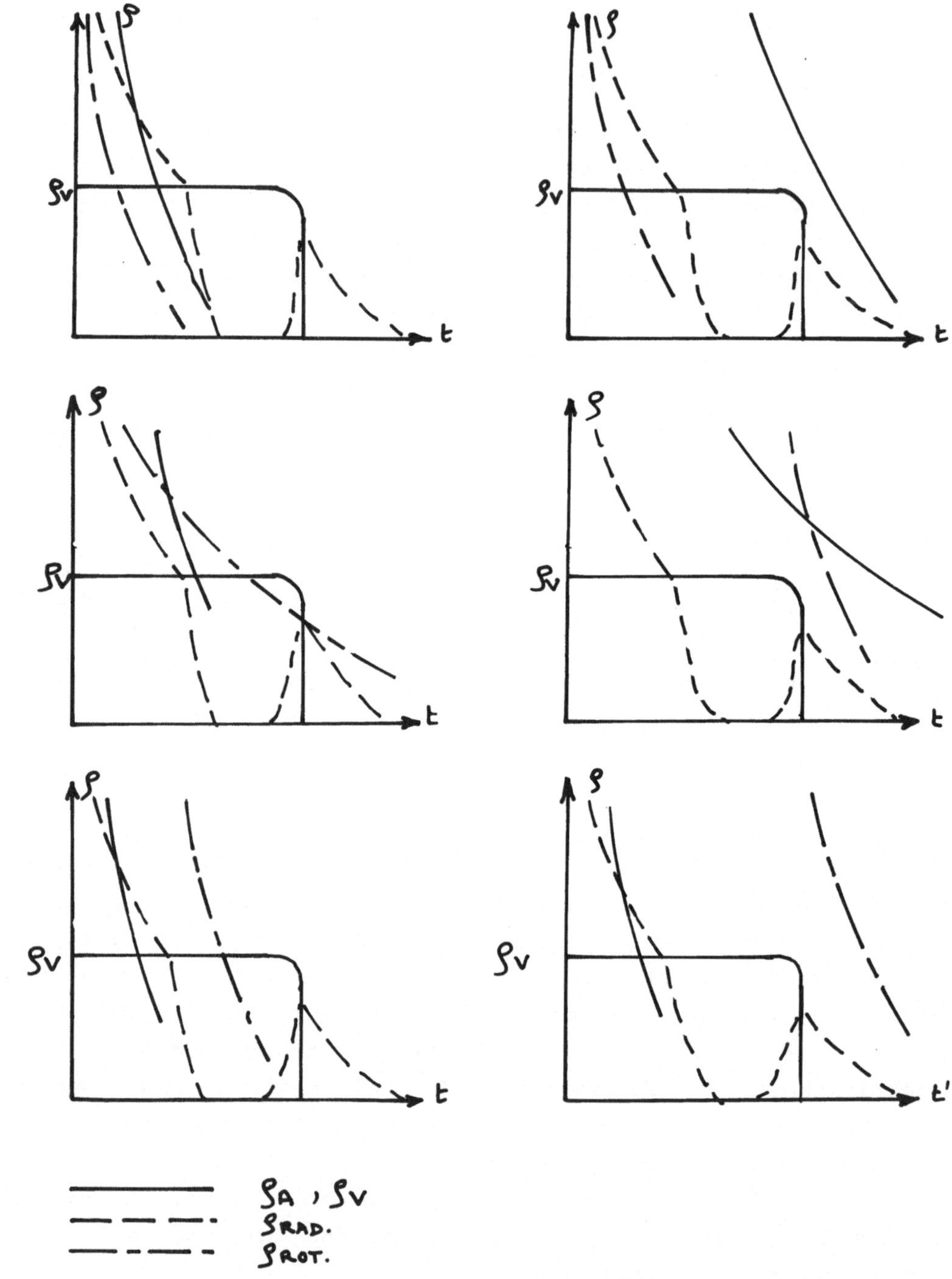

Fig. 3 : Relative importance of various energy densities after Ref. 28 (ρ_V: vacuum ; $\rho_{RAD.}$: radiation ; ρ_A: anisotropy ; ρ_{ROT}: rotation) ; not all possibilities (which are very numerous) have been depicted ; case (i) is the only one where $\rho_V > \rho_{ROT}, \rho_{RAD}, \rho_A$ and hence where there is inflation.

In such a case, instead of the Friedman equation (2.4.a) (still with $\Lambda = 0$), one can write an equation of the following form[29]:

$$\left(\frac{\dot{R}}{R}\right)^2 = \frac{8\pi G\rho}{3} + \frac{k}{R^2} + \frac{\sigma^2}{R^6} + \frac{\Omega^2}{R^m} = 0 \qquad (3.8)$$

obtained from the Raychaudhury's equation (29,30). In Eq. (3.8) σ^2 and Ω^2 are constants related to the shear and rotational anisotropy respectively while m is a constant depending on the equation of state ; for instance, when radiation dominates $m = 2$ while $P \sim \rho/3$ and when $P \simeq 0$ $m = 4$. In the case of a vacuum dominated universe ($P = -\rho$) then $m = 8$. These values can be inferred from the interpretation of the last term of Eq. (3.8) as an <u>energy of rotation</u> (divided by R^2) and from the conservation of angular momentum (or, $\omega M R^2$ = const ; equivalently, $\rho R^5 \omega$ = const. and since $\rho \sim R^{3\gamma-5}$ (from Eq. (2.4.b) with $P = (\gamma-1)\rho$) then $\omega \sim R^{3\gamma-5}$ where ω is the rotation velocity)[27].

Equation (3.8) shows that the "rotation problem" is on the same footing as the flatness problem in our dust universe (m=2) while the anisotropy one is (slightly) less important for a large value of σ. It also shows that if the primordial universe is vacuum dominated and does possess an inflation period, then not only is the flatness puzzle solved but also does the other two problems since they correspond to smaller terms in this equation.

However, as remarked by Barrow and Turner [28](in the case $\Omega=0$) the situation is not so simple and a too large initial anisotropy can prevent the occurence of inflation. Indeed, assuming the dominance of the vacuum during a definite period of the early universe (and still with $\Omega = 0$) Eq. (3.8) can be solved as

$$R(t) = \left\{\frac{\sigma}{H_o} \sinh\,(3H_o t)\right\}^{1/3} \qquad (3.9)$$

which behaves like $\sim t^{1/3}$ when anisotropy dominates (i.e. when $\sigma^2/R^6 \gg 8\pi G\rho_{vac}/3$) and has the expected inflationary behavior in the opposite case. It should be noticed that the situation is not only similar but, as a matter of fact, worse when $\Omega \neq 0$ owing to the higher power of the vorticity term in R^{-1} (than the one of the shear term). The various cases are depicted on fig. 3 drawn after Ref. 28.

In fact, Demianski[31] has given an argument against this remark by Barrow and Turner[28] ; however, his argument (i) is model dependant and (ii) rests on a number of implicit or explicit assumptions. Consequently, the problem cannot be considered as being solved and rather we think that the objection of Barrow and Turner is quite valid. Finally, it should be remarked that these fine tuning problems - although solved by inflation if it takes place during a sufficient period of time - are perhaps not real problems. Guth himself recognizes in his original paper[1] that not every physicist will be convinced by the arguments presented !

The primordial monopole problem

Magnetic monopoles are predicted to be copiously produced in the early universe as a result of the spontaneous breakdown of a GUT symmetry that leaves a U(1) invariance[8,9]. However, no such monopoles have yet been detected (with, perhaps, an exception [32]) : this is the monopole problem. However, it should be realized that if Grand Unified Theories had to be discarded for experimental or theoretical reasons, the problem would be automatically solved ! For the moment, whether one believes or not in GUT's (or supersymmetries, etc) is just a matter of taste and philosophy and so is the case for the monopole problem.

In Sec. 1 we mentionned that monopoles are produced when the GUT gauge symmetry is spontaneously broken while an U(1) symmetry still survives : regions of uncorrelated Higg's fields do exist and their frontiers contain singularities which are topological defects occuring because of the different directions of the average Higg's fields in the various regions[9]. These magnetic monopoles are quite heavy - of the order of $M/\alpha \sim 10^{16}$ GeV (where M is the typical mass scale of the gauge bosons once they have aquired a mass and where α is the fine structure constant at the GUT's energy scale) - and must have been produced abundantly. In fact, topological defects other than monopoles - e.g. strings[9,33] - can also be produced when other subgroups persist when G is spontaneously broken (e.g. when $G \equiv SO(10)$ then strings are produced[9,33])

Also, besides the primordial monopole problem, the spontaneous breakdown of a discrete symmetry (in the case of Higg's field : $\phi \to -\phi$) leads to the so-called domain wall problem : the interface of regions of different symmetry (+ or −) appears to be very massive and would lead to (unobserved) very strong anisotropies[34]. In fact, this problem exists only because the choice of the Higgs' field potential $V(\phi)$ is initially symmetric (i.e. it does not contain terms like $\gamma\phi^3$) : whether or not this choice has actually to be done is (yet) an open question.

Let us now come back to the problem of monopoles. If $d_H(t)$ is the horizon length at time t, i.e.

$$d_H(t) = R(t) \int_0^t \frac{dt'}{R(t')} \tag{3.10}$$

then the correlation length ξ of the Higgs' field must be such that $\xi < d_H(t)$ ($d_H(t) = 2t$ in the standard big bang scenario) and the density of monopoles $n_{mon}(t)$ is roughly given[9] by

$$n_{mon}(t) \sim \xi^{-3} > d_H^{-3}(t) \tag{3.11}$$

Owing to the fact that the ratio $n_{mon}(t) / s(t)$ (s/t) being the entropy density at time t) is conserved during the expansion of the universe (monopole / antimonopole annihilation has been shown to be quite small in the GUT era [8,10]), taking into account the presently observed value of s(t) and the typical mass of a monopole ($\sim 10^{16}$GeV), one would find an energy density larger than 10^{-18} g/cm^3 at the present epoch. Equivalently, this would lead to $\Omega \gtrsim 10^{11}$, which is far outside the observational limits ($\Omega \lesssim 2.3$)

On the other hand, the known estimations of the galactic magnetic field leads to an upper estimate of the present density of monopoles (the "Parker limit"). Indeed, since magnetic fields do accelerate magnetic monopoles (see a review of their unusual electromagnetic properties in Ref. 35), the energy density stored in the field (i.e. $B^2/8\pi$) tends to be dissipated and if one demands that the galactic field have not substantially decayed during a typical regeneration time ($\sim 10^8$ yr) then one obtains an upper limit for the monopole flux : $< n_{mon} V > < 10^{-1} M^{-2} yr^{-1}$.

Let us now examine how an inflationary era can solve this problem (whether real or not) : since $n_{mon}(t) \sim R^{-3}(t)$, it follows that $n_{mon}(t_o)$ must be exponentially small at our epoch t_o. However, this raises a new problem : an inflationary era would also lead to an exponentially small baryonic density if the asymsymmetry between matter and anti-matter would be performed before such an era (still assuming that the universe was originally symmetric ; otherwise, there is no problem at all). In fact, in the presently given inflationary universe scenarios, this asymmetry occurs at the end (or after) the de Sitter phase.

4. HOW TO REALIZE AN INFLATIONARY BEHAVIOR ?

In the preceding sections various problems of the early universe have been mentioned and it has been shown how inflation could solve them. However, the main question is "how to get inflation" ? Although in the following we limit ourselves to phase transitions in the early universe as a way to get inflation, this is not at all the only possibility. For instance, some field theoretical models[36] yield an exponential behavior of the scale factor. Also, it could be realized in the quantum era (see e.g. Ref. 37 and quoted papers). It is amusing to remark that during a possible quark era at zero temperature and described by the M.I.T. bag equation of state[38], one gets a mini-inflation!

In this section, we limit ourselves to the most general scenarios that lead either to the original inflation[1] or to the new one[13], as a result of a first order phase transition. In the original scenario, it comes from the metastability of the vacuum while in the new one its origin is rather a loss of stability. These two possibilities have been depicted on fig. 4. Of course, one might imagine other possibilities.

The original inflation[1]...

On fig. (4.a) a plot of the free energy present in the Universe (including both the contributions of the matter, radiations and vacuum) is given as a function of an order parameter (i.e., in the symmetric phase the order parameter is vanishing by definition and is different from zero otherwise) in the case of only one (thermodynamically) stable state and only one metastable state : of course, the latter corresponds to the higher value of the free energy and the broken symmetry state is the one with a non zero order parameter. The various curves correspond to various values of a macroscopic parameter.

In the framework of GUT's the order parameter is the average value of the Higgs field whereas the macroscopic parameter that labels the free energy curves is the temperature. In fact, we could have the same general shape of the free energy as a function of another thermodynamic variable and labelled by a macroscopic parameter e.g. the baryonic density (in a non symmetric and cold universe...). Furthermore, we must assume that the vacuum dominates over all other forms of (free) energy in order to get the form (2.5) for $T^{\mu\nu}$ and hence an exponential behavior for R(t). In the context of GUT's the lower free energy state is called the "true vacuum" and is stable while the other one is termed the "false vacuum", which is metastable.

At very high temperatures, the Universe is in a symmetric state (curve (i)) and as it expands it cools down and a metastable state is supposed to occur (curve (ii)) untill a critical temperature (or an other parameter) is reached where both vacua have the same energy (curve (iii)). Beyond this critical temperature the true vacuum is the less symmetric state (with a non vanishing order parameters) while the "old" true vacuum (the most symmetric state) becomes metastable (curve (iv)) and thus a "false vacuum".

Beyond the critical temperature there exist several possibilities depending on the height and width of the "potential barrier" that separates the two vacua : either the fluctuations of the system (whether thermal and/or quantum) are important enough as to bring it from the symmetric state to the broken symmetric one or not ; equivalently, either the universe is trapped in the (symmetric) false vacuum and it remains in this metastable state during some time before it undergoes a phase transition to the (less symmetric and stable) true vacuum or it falls directly in the lower (free) energy state. In the first case (i.e. when the universe is trapped in the metastable vacuum) there exists an inflation period which persists as long as (i) matter is dominated by the vacuum and (ii) the vacuum is a false vacuum. At the same time the universe begins to nucleate bubbles of true vacuum (either by quantum tunnelling or because of the presence of impurities) which grow : in order for the Universe to stay in the metastable vacuum this rate of nucleation must be small compared with the typical expansion time H_o^{-1} (see Eq. (2.13)).

After this supercooling of the universe the transition - through bubbles formation - is completed and the latent heat of

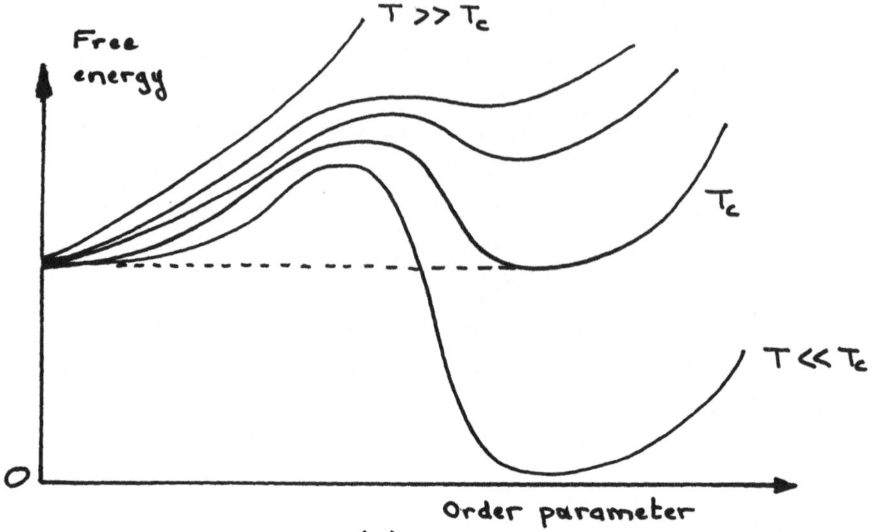

(a)

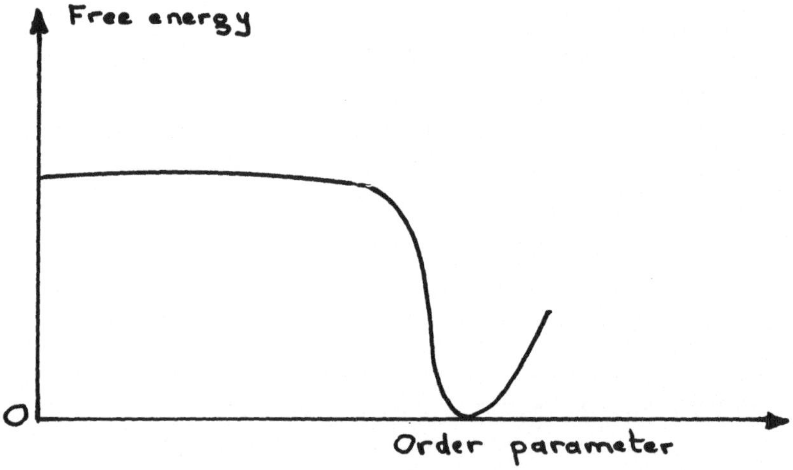

(b)

Fig. 4 : Original inflation/new inflation. In the original inflation (a) the Universe becomes trapped in the metastable state and is eventually reheated once the transition is completed. In the new scenario (b) a fluctuation of the order parameter is slowly growing and "rolls down" the free energy hill giving rise to a reheating of the Universe (the fluctuation region itself).

the transition (essentially, the energy stored in the false vacuum since the energy of the true vacuum (∿ cosmological constant) is negligibly small) is released through bubbles collisions. Thence the Universe is reheated to nearly the critical temperature allowing thereby the eventual baryosynthesis in a baryon-antibaryon symmetric universe and its subsequent evolution is that of the standard hot big bang model.

This is a summary of the original Guth[1] scenario. Unfortunately, it is hardly tenable[1,39] in the context of Grand Unified Theories : indeed, with acceptable values of the parameters[39,40] the phase transition can never be completed and the Universe would appear as formed of islands of large bubbles (of true vacuum) surrounded by smaller ones, in a sea of false vacuum and such a picture has nothing to do with what is actually observed furthermore, these bubbles - whose walls contain most of their energy - could never collide[39] and hence the universe could never be reheated. This is the "fatal flaw" of the original inflation : i.e. the difficulty of a "graceful exit" to the standard big bang model.

In spite of these difficulties though, this general inflationary scenario is very interesting and might perhaps be used in other contexts than GUT's, in other models of the early universe.

... <u>and the new one</u>[13]

In fact, it was soon realized by Linde[13] and Albrecht and Steinhardt[13] that a different scenario was devoided of this "fatal flaw" whilst retaining its main virtues, the "new inflation".

This scenario is not based on the trapping of the universe in a metastable vacuum state but rather on a (slow) loss of stability of the (symmetric) vacuum (see fig. (4.b)). Let us assume, indeed, that the vacuum free energy has the form indicated on fig. (4.b) : below a critical temperature T_c (or below a critical macroscopic parameter) this free energy is assumed to be nearly flat in the neighbourhood of the symmetric state (apart from a small bump occuring because of thermal effects ; a bump of width $\sim T$ and of height $\sim T^4$, in order of magnitude). Hence, it can take a long time before the order parameter (the Higgs' field value in the framework of GUT's) depart significantly from its zero value and thus before the universe falls in its non-symmetric state. During this long period of time the scale factor is essentially growing exponentially. In this scenario, the universe starts as a single fluctuation (of the order parameter) which slowly rolls down the free energy hill driving the order parameter to its true vacuum value. While the universe inflates the corresponding fluctuation region (who e size is presumably of the order of H_o^{-1}) expands more and more so as to encompass our own visible universe : in this scenario the Universe is itself a fluctuation region ! After this period of inflation - where the free energy is approximately constant - the order parameter oscillates violently around its true vacuum equilibrium value. These oscillations are then strongly damped by the emission of various radiations (gravitational, electromagnetic, ultra-relativistic particles, etc) which re-heat the universe slightly below T_c (see fig. 5).

In the new inflation, the usual conceptual problems (horizon, flatness, etc) of the early universe are solved as in the "old" scenario whereas the monopole problem is solved via the fact that monopoles are produced at the frontier of the fluctuation region that constitutes our universe. It should however be noticed that the new scenario is effective insofar as the fluctuation region (i) is homogeneous and (ii) is not sufficiently massive as to modify (at least locally) the global evolution of the universe. When these conditions are assumed to be satisfied, it remains to implement the physical scenario itself by providing a suitable model. In the framework of GUT's, one actually finds a vacuum free energy which is flat enough near the origin (as to make the new scenario effective) via the use of a Coleman-Weinberg potential[41] for the (average) Higgs' field (which is the order parameter of the theory). There are however a number of difficulties in this model (see the reviews by Guth and by Linde in Ref. 10). One of them is the necessary fine tuning of the parameters of the theory (the inflationary universe was invented just for avoiding the fine tuning of the initial conditions of the standard big bang theory !) : this problem is solved in the context of sypersymmetries but...

Another important problem deals with the spectrum of homogeneities arising from such a phase transition : while a detailed study of the perturbations in the new inflation leads to a

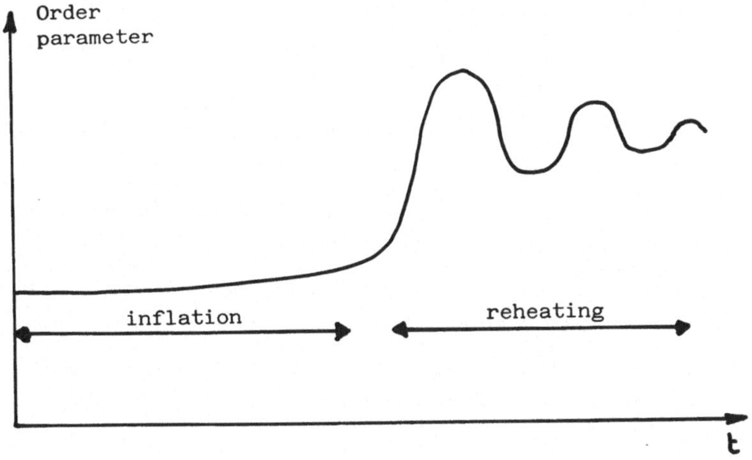

Fig. 5 : General shape of the order parameter as a function of time.

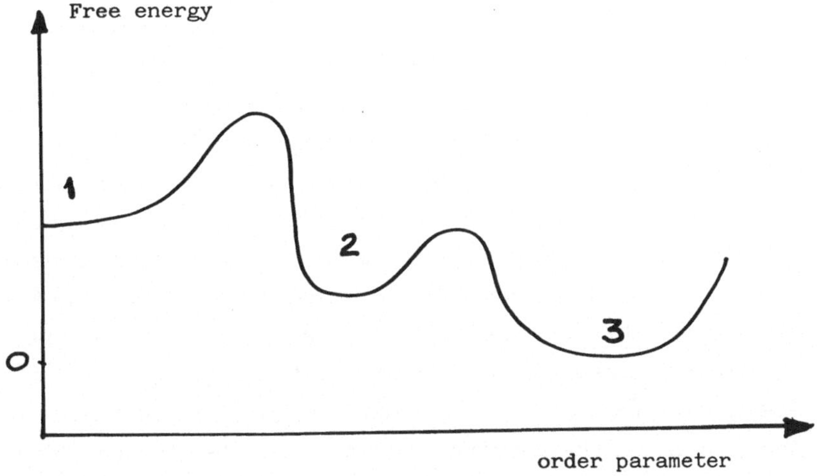

Fig. 6 : Another possible inflationary scenario : the universe is trapped in the false vacuum 1 ; then a bubble universe is nucleated in the false vacuum 2 where it is trapped and thus inflates ; finally it is reheated when it reaches the true vacuum 3.

quasi-scale free spectrum (a property required[42] by some models for galaxy formation[43]), the relative energy density fluctuation $\delta\rho/\rho$ is found to have a much too high value ($\sim 10^2$ instead of 10^{-2}) ; this drawback does not appear to be fatal and this problem is presently under current investigation.

Besides this new inflationary universe there are many other possibilities and one of them is depicted on fig. 6 : if we assume that the free energy curve possesses two relative minima (or, two metastable vacua) corresponding e.g. to two successive phase transitions then one can imagine that a bubble can be nucleated in the second false vacuum, then be trapped there (while most of the universe is still trapped in the first false vacuum (inflation)) and inflates more and more until the true vacuum is reached...

The role of scalar fields

In the above models mention was made that (average) Higgs' fields (Lorentz scalar fields with possibly internal symmetries and thus several components) do constitute in actual practice the order parameter that characterizes the transition although it may be conceivable that other types of transitions (i.e. other than Gut's) and thus other types of order parameters) could take place in the early universe. As a matter of fact, the basic ingredient in the (old or new) inflationary scenarios is the existence of scalar fields. Whether they are necessitated by a spontaneous symmetry breaking or corresponds to scalar modes of other fields or do exist in their own right is immaterial : provided they lead to the required shapes of the vacuum free energy with the correct values of the parameters, one gets an inflationary behavior of the universe. Therefore, it seems highly desirable to study their properties per se (an interesting step in this direction can be found in Ref. 44), starting e.g. with a typical lagrangian

$$L = \frac{1}{2}(\partial\phi)^2 - V(\phi) \tag{4.1}$$

to which phenomenological terms (supposed to account of the coupling with other (assumed more or less important) fields or of quantum fluctuations) should be added. Of course, such studies have also to be extended from Minkowski to de Sitter spacetime.

REFERENCES

1. A.H. Guth, Phys. Rev. D 23, 347, 1981
2. P. Ramond, Ann. Rev. Nucl. Part. Sci. 33, 31, 1983
3. See e.g. E.S. Abers, B.W. Lee, Phys. Report 9, 1, 1973
4. J.C. Taylor, Gauge Theories of Weak Interactions. (Cambridge Univ. Press, Cambridge, 1979)
   Selected papers on "Gauge Theory of Weak and Electromagnetic Interactions" (C.H. Lai, ed.) (World Scientif. Singapore, 1981).
5. An elementary account can be found in D.B. Lichtenberg, Contemp. Phys. 22, 311, 1981
6. See e.g. an elementary account in : N.K. Nielsen Am. J. Phys. 49, 1171, 1981, or, H.D. Politzer, Phys. Reports 14, 129,1974
7. P. Langacker, Phys. Reports 72, 185, 1981
8. See e.g. J. Preskill "Monopoles in the Very Early Universe" (in Ref. 10) or S. Coleman "The Magnetic Monopole Fifty Years Later" in : The Unity of Fundamental Interactions (A. Zichichi ed. ; Plenum Press, New-York, 1983), or, R.A. Carrigan Jr.,W.P Trower, Nature, 305, 673, 1983.
9. T.W.B. Kibble, Phys. Reports 67, 183, 1980 ; J. Phys. A9, 1387 1976. See also the connected review by N.D. Mermin, Rev. Mod. Phys. 51, 591, 1979
10. The Very Early Universe (G.W. Gibbons, S.W., Hawking, S.T.C. Siklos ed. Cambridge Univ. Press, Cambridge, 1983).
11. T.D. Lee, Particle Physics and Introduction to Field Theory. (Harwood Acad. Pub. New-York, 1981)
12. Ya B. Zeldovich, Soviet Phys. Usp. 24, 216, 1981
13. A.D. Linde, Phys. Letts. 108B, 389, 1982
    A. Albrecht, P.J. Steinhardt, Phys. Rev. Letts. 48, 1220,1982
14. S. Weinberg, Gravitation and Cosmology (Wiley ; New-York,1972)
15. A.D. Linde, Rep. Progr. Phys. 42, 389, 1979
16. See e.g. E.W. Kolb, S. Wolfram, Ap. J. 239, 428, 1980
    D. Kazanas, Ap. J. 241, L59, 1980
    A.A. Starobinsky, Phys. Letts. 91B, 99, 1980
17. S.W. Hawking, I.G. Moss, Phys. Letts. 110B, 35, 1982
18. P. Frampton, G. Lipton, preprint (1983).
19. S.M. Faber, J.S. Gallagher, Ann. Rev. Astron. Astrophys. 17, 135,1979

20.See e.g. J.P.J. Lafon, Astron. Astrophys. 46, 461, 1976 ; idem. in the 1984 Moriond Meeting.

21.P.J.E. Peebles, Ap. J. (in press).

22.G. Steigman, Ann. Rev. Astron. Astrophys. 14, 339, 1976

23.E.W. Kolb, M.S. Turner, Ann. Rev. Nucl. Part. Sci., 33, 645, 1983

24.See e.g. P. Weiss. Ann. Rev. Astron. Astrophys. 18, 489, 1980
R.A. Sunyaev, Ya B. Zeldovich ; idem 18, 537, 1980

25.S.W. Hawking, Nucl. Phys. B144, 349, 1978

26.R.H. Dicke, P.J.E. Peebles "The Big Bang Cosmology, Enigmas and Nostrums" in : General Relativity, an Einstein Centenary Survey (S.W. Hawking, W. Israel ed. ; Cambridge Univ. Press, Cambridge 1979).

27.J. Ellis, K.A. Olive, Nature 303, 679, 1983

28.J.D. Barrow, M.S. Turner, Nature 292, 35, 1981
G. Steigman, M.S. Turner, Phys. Letts. 128B, 295, 1983

29.G.F.R. Ellis, Relativistic Cosmology in : General Relativity and Cosmology (1969 Varenna Summer School, edited by R.K.Sachs Acad. Press, New-York 1971.

30.A. Raychaudhuri, Phys. Rev. 98, 1123, 1955

31.M. Demianski, Nature 307, 140, 1984

32.B. Cabrera, Phys. Rev. letts. 48, 1378, 1982

33.See A. Vilenkin in Ref. 10 and papers quoted therein.

34.Ya B. Zeldovich, I. Yu Kobzarev, L.B. Okun, Soviet Phys.JETP 40, 1, 1974

35.See e.g. J.R. Ficenec, V.L. Teplitz "Magnetic Charges" in: "Electromagnetism" (D. Teplitz ed.; Plenum Press, New York 1983)

36.A. Aurilia, G. Denardo, F. Legovini, E. Spalluci, ICTP preprint 1983.

37.F. Englert, "Quantum Field Theory and Cosmology" in : Cargèse Summer School of Fundamental Interactions (1981)

38.See e.g. E.V. Shuryak, Phys. Reports 61, 71, 1980, and references quoted therein

39.A.H. Guth, E.J. Weinberg, Nucl. Phys. B212, 321, 1982

40.S.W. Hawking, I.G. Moss, Phys. Letts. 110B, 35, 1982
S.W. Hawing, I.G. Moss, J.M. Stewart, Phys. Rev. D26, 2681, 1982.

41.S. Coleman, E.J. Weinberg, Phys. Rev. D7, 1888, 1973.

42.E.R. Harrison, Phys. Rev. D1, 2726, 1970
Ya B. Zeldovich, M.N.R.A.S. 160, 1P, 1972

43. G. Efstathiou, J. Silk, Fund. Cosmic Phys., 9, 1, 1983.
44. P.J. Steinhart, M.S. Turner, Phys. Rev.? D29, 2162, 1984.

APPENDIX : bibliography

Abbott L.F., Farhi E., Wise M.B. (1982) Particle production in the new inflationary cosmology (Phys. Letters, 117B, 29, 1982).

Abbott L.F., Burges C.J.C. (1983) Homogeneous transitions in an inflating universe (Phys. Letters, 131B, 49, 1983).

Abbott L.F., Wise M.B. (1984) Anisotropy of the microwave background in inflationary cosmology (Phys. Letters, 135B, 279, 1984).

Albrecht A., Steinhardt P.J. (1982) Cosmology for grand unified theories with radiatively induced symmetry breaking (Phys. Rev. Lett., 48, 1220, 1982).

Albrecht A., Steinhardt P.J., Turner M.S. and Wilczek F. (1982) Reheating an inflationary universe, (Phys. Rev. Lett., 48, 1437, 1982).

Albrecht A., Steinhardt P.J. (1983) Inflation and supersymmetry (Phys. Lett., 131B, 45, 1983).

Albrecht A., Dimopoulos S., Fischler W., Kolb W., Raby S., Steinhardt P.J. (1983) New Inflation in supersymmetric theories (Nucl. Phys., B229, 528, 1983).

Albrecht A., Steinhardt P.J. (1983) Inflation and supersymmetry (Phys. Lett., 131B, 45, 1983).

Albrecht A., Jensen L.G., Steinhardt P.J. (1984) Inflation in SU(5) GUT models coupled to gravity (Nucl. Phys., B239, 290, 1984).

Allen B. (1983) The effective potential in de Sitter space (in Ref. 10 above).

Aurilia A., Denardo G. Legovini F., Spallucci E. (1983) An effective action functional for the inflationary cosmology (ICTP preprint IC/83/238, 1983).

Bardeen J.M., Steinhardt P.J., Turner M.S. (1983) Spontaneous creation of almost scale-free density perturbations in an inflationary universe (Phys. Rev., D28, 679, 1983).

Barrow J.D., Turner M.S. (1981) Inflation in the universe (Nature, 292, 35, 1981).

Barrow J.D., Turner M.S. (1982) The inflationary universe - birth, death and transfiguration (Nature, 298, 801, 1982).

Barrow J.D. (1983) Cosmology and elementary particle physics (Fund. Cosmic Phys. 8, 83, 1983).

Barrow J.D. (1983) Perturbations of a de Sitter universe (in Ref. 10 above).

Boucher W., Gibbons G.W. (1983) Cosmic baldness (in Ref. 10 above).

Branderberger R.H. (1983) An alternate derivation of radiation in an inflationary universe (Phys. Lett., 129B, 397, 1983).

Brandenberger R., Kahn R. (1982) Hawking radiation in an inflationary universe (Phys. Lett., 119B, 75, 1982).

Brandenberger R., Kahn R., Press W.H. (1983) Cosmological perturbation in the early Universe (Phys. Rev., D28, 1809, 1983).

Brandenberger R., Kahn R. (1984) Cosmological perturbations in inflationary-universe models (Phys. Rev., D29, 2172, 1984).

Breit J.D., Gupta S., Zaks A. (1983) Problems with the new inflationary universe (Phys. Rev. Lett., 51, 1007, 1983).

Collins W., Turner M.S. (1984) Thermal production of superheavy monopoles in the new inflationary-universe scenario (Phys. Rev., D29, 2158, 1984).

Davies P.C.W. (1983) Inflation and time asymmetry in the Universe (Nature, 301, 398, 1983).

Demianski M. (1984) Large anisotropy in the Universe does not prevent inflation (Nature, 307, 140, 1984).
Dolgov A.D., Linde A.D. (1982) Baryon asymmetry in inflationary universe (Phys. Lett., 116B, 239-334, 1982).
Ellis J., Linde A.D., Nanopoulos D.V. (1982) Inflation can save the gravitation (Phys. Lett., 118B, 59, 1982).
Ellis J., Nanopoulos D.V., Olive K.A., Tamvakis K. (1982) Cosmological inflation cries out for supersymmetry (Phys. Lett., 118B, 335-339, 1982).
Ellis J., Nanopoulos D.V., Olive K.A., Tamvakis K. (1983) Fluctuations in a supersymmetric inflationary Universe (Phys. Lett., 120B, 331, 1983).
Ellis J., Olive K.A. (1983) Inflation can solve the rotation problem (Nature, 303, 679, 1983).
Ford L.H. (1983) Quantum fluctuations and phase transitions in cosmology (in ref. 10 above).
Ford L.H., Vilenkin A. (1982) Gravitational effects upon cosmological phase transitions (Phys. Rev., D26, 1231, 1982).
Frieman J.A., Will C.M. (1982) Evolution of perturbations in an inflationary universe (Ap. J., 259, 437, 1982).
Gelmini G.B., Nanopoulos D.V., Olive K.A. (1983) Finite temperature effects in primordial inflation (Phys. Lett., 131B, 53, 1983).
Gott III J.R., Statler T.S. (1984) Constraints on the formation of bubble universes (Phys. Lett., 136B, 157, 1984).
Gunzig E., Nardone P. (1984) From unstable Minkowski space to the inflationary universe (Gen. Rel. Grav., 16, 305, 1984).
Guth A.H. (1981) The inflationary universe : a possible solution to the horizon and flatness problems (Phys. Rev., D32, 347, 1981).
Guth A.H., Pi S.Y. (1982) Fluctuations in the new inflationary universe (Phys. Rev. Lett., 49, 1110, 1982).
Guth A.H., Weinberg E. (1982) Could the universe have recovered from a slow first-order phase transition ? (Nucl. Phys., 212, 321, 1982).
Guth A.H. (1982) 10^{-35} seconds after the big bang (in : The birth of the universe, Moriond meeting,1982).
Guth A.H. (1982) Phase transitions in the embryo Universe (Philos. Trans. R. Soc. London, 307, 141, 1982).
Guth A.H. (1983) Phase transitions in the very early universe (in ref. 10 above).
Guth A.H., Steinhardt P.J. (1984) The inflationary universe (Scientif. Amer., may 1984).
Goncharov A.S., Linde A.D. (1984) Chaotic inflation in supergravity (Phys. Lett., 139B, 27, 1984).
Hawking S.W., Moss I.G., Stewart J.M. (1981) Bubble collisions in the very early universe (Phys. Rev., D26, 268, 1981).
Hawking S.W. (1982) The development of irregularities in a single bubble inflationary universe (Phys. Lett., 115B, 295-297, 1982).
Hawking S.W. (1983) Euclidean approach to the inflationary universe (in ref. 10 above).
Hawking S.W., Moss I.G. (1982) Supercooled phase transition in the very early universe (Phys. Lett., 110B, 35-38, 1982).
Hawking S.W., Moss I.G. (1983) Fluctuations in the inflationary universe (Nucl. Phys., B224, 180, 1983).
Holman R., Ramond P., Ross G.C. (1984) Supersymmetric inflationary cosmology (Phys. Lett., 137B, 343, 1984).
Krauss L.M. (1983) Baryosynthesis and primordial inflation reexamined (Phys. Lett., 133B, 169, 1983).
Levi B.G. (1983) New Inflationary Universe : an alternative to Big Bang ? (Phys. Today, 36, 17, 1983).
Linde A.D. (1982) A new inflationary universe scenario : a possible solution of the horizon, flatness, homogeneity, isotropy and primordial monopole problems (Phys. Lett., 108B, 389-393, 1982).

Linde A.D. (1982) Coleman-Weinberg theory and the new inflationary universe scenario (Phys. Lett., 114B, 431-435, 1982).
Linde A.D. (1982) Scalar field fluctuations in expanding universe and the new inflationary universe scenario (Phys. Lett., 116B, 335, 1982).
Linde A.D. (1982) Temperature dependence of coupling constants and the phase transition in the Coleman-Weinberg theory (Phys. Lett., 116B, 340, 1982).
Linde A.D. (1983) The new inflationary scenario (in ref. 10 above).
Linde A.D. (1983) Supergravitation and the inflationary universe (Pis'ma v Zhehtf, 37, 606, 1983 ; JETP Lett., 37, 724, 1983).
Linde A.D. (1983) Chaotic inflationary universe (Pis'ma v. Zhehtf, 38 149, 1983 ; JETP Lett., 38, 176, 1983).
Linde A.D. (1983) Primordial inflation without primordial monopoles (Phys. Lett., 132B, 317, 1983).
Linde A.D. (1983) Chaotic inflation (Phys. Lett., 129B, 177, 1983).
Linde A.D. (1983) Inflation can break symmetry in SUSY (Phys. Lett., 131B, 330, 1983).
Lindley D. (1984) The appearance of bubbles in de Sitter space (Nucl. Phys., B236, 522, 1984).
Lukash V.N., Novikov I.D., Inflationary universe, primordial sound waves and galaxy formation (in : Early evolution of the universe and its present structure ; IAU symposium n° 104).
Mathiazhagan C., Johri V.B. (1984) An inflationary universe in Brans-Dicke theory : a hopeful sign of theoretical estimation of the gravitational constant (Class. Quant. Grav., 1, L29, 1984).
Moss I.G. (1983) Monopoles in the inflationary universe (Phys. Lett., 128B, 385, 1983).
Moss I.G. (1984) More fluctuations in the inflationary universe (Nucl. Phys., B238, 385, 1984).
Moss I.G., Wright W.A. (1984) Wave function of the inflationary universe (Phys. Rev., D29, 1067, 1984).
Mottola E. (1983) Fluctuations in the homogeneous inflationary universe (Phys. Rev., D27, 2294, 1983).
Mottola E., Lapedes A. (1983) Inflationary universe with gravity (Phys. Rev., D27, 2285, 1983).
Nair V.P. (1983) Second order phase transitions, inflationary universe and formation of galaxies (Phys. Rev., D27, 2856, 1983).
Nanopoulos D.V., Olive K.A., Sredniki M. (1983) After primordial inflation (Phys. Lett., 127B, 30, 1983).
Nanopoulos D.V., Olive K.A., Sredniki M., Tamvakis K. (1983) Primordial inflation in simple supergravity (Phys. Lett., 124B, 171, 1983).
Nanopoulos D.V., Sredniki M. (1983) Before primordial inflation (CERN preprint TH.3673, july 1983).
Ovrut B.A. (1983) Supersymmetry and inflation : a new approach (Phys. Lett., 133B, 161, 1983).
Page D.N. (1983) Is inflation needed to suppress monopoles ? (in ref. 10 above).
Page D.N. (1983) Inflation does not explain time asymmetry (Nature, 304, 39, 1983).
Parke S. (1983) Gravity, the decay of the false vacuum and the new inflationary universe scenario (Phys. Lett., 121B, 313, 1983).
Peebles P.J.E. (1984) Tests of cosmological models constrained by inflation (Ap. J., in press).
Rubakov V.A., Sazhin M.V., Veryaskin A.V. (1982) Graviton creation in the inflationary universe and the grand unification scale (Phys. Lett., 115B, 189, 1982).
Starobinski A.A. (1982) Dynamics of phase transitions in the new inflationary universe scenario and generation of perturbations (Phys. Lett. 117B, 175, 1982).
Sasaki M. (1983) Generation and growth of perturbations in the new inflationary scenario (in : Grand Unified Theories and the early Universe ; KEK-13, M. Fukugita and M. Yoshimura eds. ; Ibaraki-ken 1983).

Sasaki M. (1983) The inflationary cosmology (in : Grand Unified Theories and the early Universe ; KEK-13, M. Fukugita and M. Yoshimura eds. ; Ibaraki-ken 1983).
Sasaki M. (1983) Gauge-invariant scalar perturbations in the new inflationary Universe (Progr. Theor. Phys., 70, 394, 1983).
Shafi Q. (1984) Inflation with SU(5), (Phys. Rev. Lett., 52, 691, 1984).
Sikklos S.T.C., Wu Z.C. (1983) Bubble collisions in General Relativity (in ref. 10 above).
Smith D.H. (1983) The inflationary universe lives ? (Sky and Telescope, march 1983, p. 207).
Steigman G., Turner M.S. (1983) Inflation in a shear or curvature dominated universe (Phys. Lett., 128B, 295, 1983).
Steinhardt P.J. (1982) A new cosmology (in : The birth of the Universe, Moriond meeting 1982 ; J. Audouze and J. Tran Than Van ed. ; Frontières, Gif, 1982).
Steinhardt P.J. (1983) Natural inflation (in ref. 10 above).
Steinhardt P.J. (1983) Progress and prospects for the inflationary universe (preprint).
Steinhardt P.J., Turner M.S. (1984) Prescription for successful new inflation (Phys. Rev., D29, 2162, 1984).
Tomita J. (1983) Gravitational instability in an inflationary Universe (in : Grand Unified Theories and the early Universe ; KEK-13, M. Fukugita and M. Yoshimura eds. ; Ibaraki-ken 1983).
Turner M.S. (1983) Particle physics and cosmology : the inner space/outer space connection (talk at the APS division of Particles and Fields Meeting, sept. 1983).
Turner M.S. (1982) A graceful end to inflation (in : The birth of the Universe, Moriond Meeting 1982).
Turner M.S. (1983) The origin of density fluctuations in the "new inflationary Universe" (in ref. 10 above).
Vagonakis C.E. (1983) Natural values of coupling constants and cosmological inflation in a supersymmetric model (Phys. Lett., 123B, 396, 1983).
Veryaskin A.V., Rubakov V.A., Sazhin M.V. (1983) Primordial gravitational waves in the inflationary model, and the grand unification scale (Astron. Zh., 60, 26, 1983 ; Sov. Astron., 27, 16, 1983).
Vilenkin A. (1982) Gravitational effects in Guth cosmology (Phys. Lett., 115B, 91, 1982).
Vilenkin A. (1982) Creation of universes from nothing (Phys. Lett., 117B, 25n 1982).
Vilenkin A. (1983) Birth of inflationary universes (Phys. Rev., D27, 2848, 1983).
Vilenkin A. (1983) Phase transitions in de Sitter space (Nucl. Phys., B226, 504, 1983).
Vilenkin A. (1983) Quantum fluctuations in the new inflationary Universe (Nucl. Phys., B226, 527, 1983).
Waldrop M.M. (1983) The new inflationary universe (Science, 219, 375, 1983).
Wu Z.C. (1983) Gravitational effects in bubble collisions (Phys. Rev., D28, 1898, 1983).

Journées Relativistes 1984 held at Aussois

List of participants

J.M. AGUIRREGABIRIA
Laboratoire de Physique Théorique
Institut Henri Poincaré
11, rue Pierre et Marie Curie
75231 Paris Cedex 05, France

B. BARBERIS
Istituto di Fisica Matematica J.L.Lagrange
Università di Torino
Via Carlo Alberto 10
10123 Torino, Italie

C. BARRABES
Département de Physique
Université de Tours
37200 Tours, France

L. BEL
Laboratoire de Physique Théorique
Institut Henri Poincaré
11, rue Pierre et Marie Curie
75231 Paris Cedex 05, France

H. BENAMOR
19, rue des Clématites, Apt. 169
21300 Chenove, France

G. et N. BESSIS
Laboratoire de Spectroscopie Théorique
Université Claude Bernard - Lyon I
43, boulevard du 11 novembre 1918
69622 Villeurbanne Cedex, France

A. BLANCHARD
Groupe d'Astrophysique Relativiste
Observatoire de Meudon
92195 Meudon Principal Cedex, France

L. BLANCHET
Groupe d'Astrophysique Relativiste
Observatoire de Meudon
92195 Meudon Principal Cedex, France

E. BLANCHETON
Département de Mathématiques et Mécanique
Université de Caen
14032 Caen Cedex, France

B. BOISSEAU
Département de Physique
Université de Tours
37200 Tours, France

C. BOUCHER
Groupe de recherche de Géodésie Spatiale
I.G.N./S.G.N.M./ D.T.I.G.
2, avenue Pasteur
94160 Saint Mandé, France

J.P. BRIAND
Lab. de Physique Théorique et Corpusculaire
Institut du Radium - Laboratoire Curie
11, rue Pierre et Marie Curie
75231 Paris Cedex 05, France

A. BRILLET
Laboratoire de l'Horloge Atomique
Université Paris XI - Bât. 221
91405 Orsay Cedex, France

D. CANARUTTO
Istituto di Matematica Applicata
Università di Firenze
Via S. Marta 3
50139 Firenze, Italie

M. CARFORA
Dipartimento di Fisica Nucleare e Teorica
Università di Pavia
Via Bassi 6
27100 Pavia, Italie

R. CATENACCI
Dipartimento di Matematica
Strada Nuova 165
27100 Pavia, Italie

M. CHEVALIER
Département de Mathématiques
Université de Caen
14032 Caen Cedex, France

Y. CHOQUET-BRUHAT
Département de Mécanique - Tour 65/66
Université Pierre et Marie Curie
4, place Jussieu
75230 Paris Cedex 05, France

P. CHRUSCIEL
Département de Mécanique - Tour 65/66
Université Pierre et Marie Curie
4, place Jussieu
75230 Paris Cedex 05, France

B. COLL
Département de Mécanique - Tour 65/66
Université Pierre et Marie Curie
4, place Jussieu
75230 Paris Cedex 05, France

T. DAMOUR
Groupe d'Astrophysique Relativiste
Observatoire de Meudon
92195 Meudon Principal Cedex, France

E. DA SILVA MAIA
Departamento de Matematica
Universidade de Coimbra
3000 Coimbra, Portugal

N. DERUELLE
Laboratoire de Physique Théorique
Institut Henri Poincaré
11, rue Pierre et Marie Curie
75231 Paris Cedex 05, France

D.M. DOS SANTOS
Département de Mécanique - Tour 65/66
Université Pierre et Marie Curie
4, place Jussieu
75230 Paris Cedex 05, France

J.P. DURUISSEAU
Département de Mécanique - Tour 65/66
Université Pierre et Marie Curie
4, place Jussieu
75230 Paris Cedex 05, France

J. EISENSTAEDT
Laboratoire de Physique Théorique
Institut Henri Poincaré
11, rue Pierre et Marie Curie
75231 Paris Cedex 05, France

R. FABBRI
Istituto di Fisica Superiore
Via S. Marta 3
50139 Firenze, Italie

F. FAYOS
Departamento de Fisica Teorica
Facultad de Fisica
Universidad de Barcelona, Diagonal 645
Barcelona 28, Espagne

M. FERRARIS
Istituto di Fisica Matematica J.L.Lagrange
Università di Torino
Via Carlo Alberto 10
10123 Torino, Italie

D. GALLETTO
Istituto di Fisica Matematica J.L. Lagrange
Università di Torino
Via Carlo Alberto 10
10123 Torino, Italie

J. GOLDBERG
Laboratoire de Physique Théorique
Institut Henri Poincaré
11, rue Pierre et Marie Curie
75231 Paris Cedex 05, France

J. HAJJ BOUTROS
Université Libanaise, Faculté des Sciences
Mansourieh El Metn, B.P. 72, Liban

R. HAKIM
LAM, Observatoire de Meudon
92195 Meudon Principal Cedex, France

F. HAMMER
L.A. 173, Observatoire de Meudon
92195 Meudon Principal Cedex, France

A. HEIDMANN
Laboratoire de Spectroscopie Hertzienne
Ecole Normale Supérieure
24, rue Lhomond
75231 Paris Cedex 05, France

J. HOUGH
Natural Philosophy Department
University of Glasgow
Glasgow G12 8QQ, Royaume-Uni

HU HESHENG
Department of Mathematics
Fudan University
Shanghaï, République Populaire de Chine

T. JACOBSON
Department of Physics
University of California
Santa Barbara, CA 93106, U.S.A.

J.B. KAMMERER
Ecole Centrale des Arts et Manufactures
92290 Chatenay Malabry, France

R. KERNER
Département de Mécanique - Tour 65/66
Université Pierre et Marie Curie
4, place Jussieu
75230 Paris Cedex 05, France

J. KIJOWSKI
Institute for Theoretical Physics
Polish Academy of Sciences
al. Lotnikow 32/46
02-668 WARSAW, Pologne

J. KLEIN
Laboratoire de Mathématiques Pures
Institut Fourier
Université de Grenoble I, B.P. 116
38402 Saint Martin d'Heres Cedex, France

C. KOLASSIS
Laboratoire de Physique Théorique
Institut Henri Poincaré
11, rue Pierre et Marie Curie
75231 Paris Cedex 05, France

J. LACAZE
Lotissement du Pin, Rebigue
31320 Castanet Tolosan, France

G. LE DENMAT
Laboratoire de Physique Théorique
Institut Henri Poincaré
11, rue Pierre et Marie Curie
75231 Paris Cedex 05, France

A. LICHNEROWICZ
6, avenue Paul Appell
75014 Paris, France

B. LINET
Laboratoire de Physique Théorique
Institut Henri Poincaré
11, rue Pierre et Marie Curie
75231 Paris Cedex 05, France

J. LLOSA
Departamento de Fisica Teorica
Universidad de Barcelona, Diagonal 645
Barcelona 28, Espagne

J.C. LUCQUIAUD
Laboratoire de Physique Mathématique
Collège de France
11, place Marcelin Berthelot
75231 Paris Cedex 05, France

R. Mc LENAGHAN
Department of Applied Mathematics
University of Waterloo
Waterloo, Ontario N2L 3G1, Canada

J. MADORE
Laboratoire de Physique Théorique
Institut Henri Poincaré
11, rue Pierre et Marie Curie
75231 Paris Cedex 05, France

A. MAGNON-ASHTEKAR
Laboratoire de Physique Théorique
Institut Henri Poincaré
11, rue Pierre et Marie Curie
75231 Paris Cedex 05, France

N. MAN
Laboratoire de l'Horloge Atomique
Université Paris XI - Bât. 221
91405 Orsay Cedex 05, France

J. MARTIN
Departamento de Fisica Teorica
Facultad de Ciencias
Universidad de Salamanca
Salamanca, Espagne

J.M. MARTIN SENOVILLA
Departamento de Fisica Teorica
Facultad de Ciencias
Universidad de Salamanca
Salamanca, Espagne

A.L. MARZUOLI
Dipartimento di Fisica Nucleare e Teorica
Università di Pavia
Via Bassi 6
27100 Pavia, Italie

M. MODUGNO
Istituto di Matematica Applicata
Via S. Marta 3
50139 Firenze, Italie

I. MORET-BAILLY
Service de Mathématiques
Université d'Angers
49035 Belle Beille Cedex, France

B. MOSCA
Laboratoire de Physique Théorique
Institut Henri Poincaré
11, rue Pierre et Marie Curie
75231 Paris Cedex 05, France

A. PAPADOPOULOS
B.P. 829
Libreville, Gabon

D. PAVON
Departamento de Termologia
Facultad de Ciencias
Universidad Autonoma de Barcelona
Bellaterra, Barcelona, Espagne

F. PIAZZESE
Politecnico di Torino
Corso Duca degli Abruzzi 24
10129 Torino, Italie

P. PIGEAUD
Département de Physique Mathématique
Université de Dijon
Faculté des Sciences Mirande, B.P. 138
21004 Dijon Cedex, France

G. PLATANIA
Istituto di Fisica Teorica
Università di Napoli
Mostra d'Oltremare, Pad. 19
80125 Napoli, Italie

G. RIZZI
Politecnico di Torino
Corso Duca degli Abruzzi 24
10129 Torino, Italie

C. ROCHE
U.E.R. de Sciences
Université de Caen
14032 Caen Cedex, France

A. RUDIGER
Max-Planck Institut für Quantenoptik
Karl-Schwarzschild Strasse 1
8046 Garching bei München, R.F.A.

E. RUIZ
Departamento de Fisica Teorica
Facultad de Ciencias
Universidad de Salamanca
Salamanca, Espagne

F. SALMISTRARO
Dipartimento di Fisica Nucleare e Teorica
Università di Pavia
Via Bassi 6
27100 Pavia, Italie

M.J. SENOSIAIN
Departamento de Fisica Teorica
Facultad de Ciencias
Universidad de Salamanca
Salamanca, Espagne

C. SGARRA
Istituto di Fisica Matematica J.L. Lagrange
Via Carlo Alberto 10
10123 Torino, Italie

H. SIROUSSE ZIA
Laboratoire de Physique Théorique
Institut Henri Poincaré
11, rue Pierre et Marie Curie
75231 Paris Cedex 05, France

P. TEYSSANDIER
Laboratoire de Physique Théorique
Institut Henri Poincaré
11, rue Pierre et Marie Curie
75231 Paris Cedex 05, France

Y. THIRY
Département de Mécanique Céleste
Tour 56/66, 5e étage
Université Pierre et Marie Curie
4, place Jussieu
75230 Paris Cedex 05, France

P. TOURRENC
Laboratoire de Physique Théorique
Institut Henri Poincaré
11, rue Pierre et Marie Curie
75231 Paris Cedex 05, France

W. UNRUH
Department of Physics
University of British Columbia
Vancouver, B.C. VGT 2A6, Canada

C. VANDERRIEST
Groupe d'Astrophysique Relativiste
Observatoire de Meudon
92195 Meudon Principal Cedex, France

C. VILAIN
Groupe d'Astrophysique Relativiste
Observatoire de Meudon
92195 Meudon Principal Cedex, France

Vol. 173: Stochastic Processes in Quantum Theory and Statistical Physics. Proceedings, 1981. Edited by S. Albeverio, Ph. Combe, and M. Sirugue-Collin. VIII, 337 pages. 1982.

Vol. 174: A. Kadić, D.G.B. Edelen, A Gauge Theory of Dislocations and Disclinations. VII, 290 pages. 1983.

Vol. 175: Defect Complexes in Semiconductor Structures. Proceedings, 1982. Edited by J. Giber, F. Beleznay, J. C. Szép, and J. László. VI, 308 pages. 1983.

Vol. 176: Gauge Theory and Gravitation. Proceedings, 1982. Edited by K. Kikkawa, N. Nakanishi, and H. Nariai. X, 316 pages. 1983.

Vol. 177: Application of High Magnetic Fields in Semiconductor Physics. Proceedings, 1982. Edited by G. Landwehr. XII, 552 pages. 1983.

Vol. 178: Detectors in Heavy-Ion Reactions. Proceedings, 1982. Edited by W. von Oertzen. VIII, 258 pages. 1983.

Vol. 179: Dynamical Systems and Chaos. Proceedings, 1982. Edited by L. Garrido. XIV, 298 pages. 1983.

Vol. 180: Group Theoretical Methods in Physics. Proceedings, 1982. Edited by M. Serdaroğlu and E. İnönü. XI, 569 pages. 1983.

Vol. 181: Gauge Theories of the Eighties. Proceedings, 1982. Edited by R. Raitio and J. Lindfors. V, 644 pages. 1983.

Vol. 182: Laser Physics. Proceedings, 1983. Edited by J. D. Harvey and D. F. Walls. V, 263 pages. 1983.

Vol. 183: J. D. Gunton, M. Droz, Introduction to the Theory of Metastable and Unstable States. VI, 140 pages. 1983.

Vol. 184: Stochastic Processes – Formalism and Applications. Proceedings, 1982. Edited by G. S. Agarwal and S. Dattagupta. VI, 324 pages. 1983.

Vol. 185: H. N. Shirer, R. Wells, Mathematical Structure of the Singularities at the Transitions between Steady States in Hydrodynamic Systems. XI, 276 pages. 1983.

Vol. 186: Critical Phenomena. Proceedings, 1982. Edited by F.J.W. Hahne. VII, 353 pages. 1983.

Vol. 187: Density Functional Theory. Edited by J. Keller and J.L. Gázquez. V, 301 pages. 1983.

Vol. 188: A. P. Balachandran, G. Marmo, B.-S. Skagerstam, A. Stern, Gauge Symmetries and Fibre Bundles. IV, 140 pages. 1983.

Vol. 189: Nonlinear Phenomena. Proceedings, 1982. Edited by K. B. Wolf. XII, 453 pages. 1983.

Vol. 190: K. Kraus, States, Effects, and Operations. Edited by A. Böhm, J. W. Dollard and W. H. Wootters. IX, 151 pages. 1983.

Vol. 191: Photon Photon Collisions. Proceedings, 1983. Edited by Ch. Berger. V, 417 pages. 1983.

Vol. 192: Heidelberg Colloquium on Spin Glasses. Proceedings, 1983. Edited by J. L. van Hemmen and I. Morgenstern. VII, 356 pages. 1983.

Vol. 193: Cool Stars, Stellar Systems, and the Sun. Proceedings, 1983. Edited by S. L. Balliunas and L. Hartmann. VII, 364 pages. 1984.

Vol. 194: P. Pascual, R. Tarrach, QCD: Renormalization for the Practitioner. V, 277 pages. 1984.

Vol. 195: Trends and Applications of Pure Mathematics to Mechanics. Proceedings, 1983. Edited by P. G. Ciarlet and M. Roseau. V, 422 pages. 1984.

Vol. 196: WOPPLOT 83. Parallel Processing: Logic, Organization and Technology. Proceedings, 1983. Edited by J. Becker and I. Eisele. V, 189 pages. 1984.

Vol. 197: Quarks and Nuclear Structure. Proceedings, 1983. Edited by K. Bleuler. VIII, 414 pages. 1984.

Vol. 198: Recent Progress in Many-Body Theories. Proceedings, 1983. Edited by H. Kümmel and M. L. Ristig. IX, 422 pages. 1984.

Vol. 199: Recent Developments in Nonequilibrium Thermodynamics. Proceedings, 1983. Edited by J. Casas-Vázquez, D. Jou and G. Lebon. XIII, 485 pages. 1984.

Vol. 200: H. D. Zeh, Die Physik der Zeitrichtung. V, 86 Seiten. 1984.

Vol. 201: Group Theoretical Methods in Physics. Proceedings, 1983. Edited by G. Denardo, G. Ghirardi and T. Weber. XXXVII, 518 pages. 1984.

Vol. 202: Asymptotic Behavior of Mass and Spacetime Geometry. Proceedings, 1983. Edited by F. J. Flaherty. VI, 213 pages. 1984.

Vol. 203: C. Marchioro, M. Pulvirenti, Vortex Methods in Two-Dimensional Fluid Dynamics. III, 137 pages. 1984.

Vol. 204: Y. Waseda, Novel Application of Anomalous (Resonance) X-Ray Scattering for Structural Characterization of Disordered Materials. VI, 183 pages. 1984.

Vol. 205: Solutions of Einstein's Equations: Techniques and Results. Proceedings, 1983. Edited by C. Hoenselaers and W. Dietz. VI, 439 pages. 1984.

Vol. 206: Static Critical Phenomena in Inhomogeneous Systems. Edited by A. Pękalski and J. Sznajd. Proceedings, 1984. VIII, 358 pages. 1984.

Vol. 207: S. W. Koch, Dynamics of First-Order Phase Transitions in Equilibrium and Nonequilibrium Systems. III, 148 pages. 1984.

Vol. 208: Supersymmetry and Supergravity/Nonperturbative QCD. Proceedings, 1984. Edited by P. Roy and V. Singh. V, 389 pages. 1984.

Vol. 209: Mathematical and Computational Methods in Nuclear Physics. Proceedings, 1983. Edited by J. S. Dehesa, J. M. G. Gomez and A. Polls. V, 276 pages. 1984.

Vol. 210: Cellular Structures in Instabilities. Proceedings, 1983. Edited by J. E. Wesfreid and S. Zaleski. VI, 389 pages. 1984.

Vol. 211: Resonances – Models and Phenomena. Proceedings, 1984. Edited by S. Albeverio, L. S. Ferreira and L. Streit. VI, 369 pages. 1984.

Vol. 212: Gravitation, Geometry and Relativistic Physics. Proceedings, 1984. Edited by Laboratoire "Gravitation et Cosmologie Relativistes", Université Pierre et Marie Curie et C.N.R.S., Institut Henri Poincaré, Paris. VI, 336 pages. 1984.

Selected Issues from

Lecture Notes in Mathematics

Vol. 909: Numerical Analysis. Proceedings, 1981. Edited by J.P. Hennart. VII, 247 pages. 1982.

Vol. 912: Numerical Analysis. Proceedings, 1981. Edited by G. A. Watson. XIII, 245 pages. 1982.

Vol. 920: Séminaire de Probabilités XVI, 1980/81. Proceedings. Edité par J. Azéma et M. Yor. V, 622 pages. 1982.

Vol. 921: Séminaire de Probabilités XVI, 1980–81 Supplément: Géométrie Différentielle Stochastique. Proceedings. Edité par J. Azéma et M. Yor. III, 285 pages. 1982.

Vol. 922: B. Dacorogna, Weak Continuity and Weak Lower Semicontinuity of Non-Linear Functionals. V, 120 pages. 1982.

Vol. 923: Functional Analysis in Markov Processes. Proceedings, 1981. Edited by M. Fukushima. V, 307 pages. 1982.

Vol. 926: Geometric Techniques in Gauge Theories. Proceedings, 1981. Edited by R. Martini and E.M.de Jager. IX 219 pages. 1982.

Vol. 927: Y. Z. Flicker, The Trace Formula and Base Change for GL (3). XII, 204 pages. 1982.

Vol. 928: Probability Measures on Groups. Proceedings 1981. Edited by H. Heyer. X, 477 pages. 1982.

Vol. 929: Ecole d'Eté de Probabilités de Saint-Flour X – 1980. Proceedings, 1980. Edited by P.L. Hennequin. X, 313 pages. 1982.

Vol. 930: P. Berthelot, L. Breen, et W. Messing, Théorie de Dieudonné Cristalline II. XI, 261 pages. 1982.

Vol. 931: D.M. Arnold, Finite Rank Torsion Free Abelian Groups and Rings. VII, 191 pages. 1982.

Vol. 932: Analytic Theory of Continued Fractions. Proceedings, 1981. Edited by W.B. Jones, W.J. Thron, and H. Waadeland. VI, 240 pages. 1982.

Vol. 934: M. Sakai, Quadrature Domains. IV, 133 pages. 1982.

Vol. 935: R. Sot, Simple Morphisms in Algebraic Geometry. IV, 146 pages. 1982.

Vol. 936: S.M. Khaleelulla, Counterexamples in Topological Vector Spaces. XXI, 179 pages. 1982.

Vol. 937: E. Combet, Intégrales Exponentielles. VIII, 114 pages. 1982.

Vol. 938: Number Theory. Proceedings, 1981. Edited by K. Alladi. IX, 177 pages. 1982.

Vol. 942: Theory and Applications of Singular Perturbations. Proceedings, 1981. Edited by W. Eckhaus and E.M. de Jager. V, 363 pages. 1982.

Vol. 953: Iterative Solution of Nonlinear Systems of Equations. Proceedings, 1982. Edited by R. Ansorge, Th. Meis, and W. Törnig. VII, 202 pages. 1982.

Vol. 956: Group Actions and Vector Fields. Proceedings, 1981. Edited by J.B. Carrell. V, 144 pages. 1982.

Vol. 957: Differential Equations. Proceedings, 1981. Edited by D.G. de Figueiredo. VIII, 301 pages. 1982.

Vol. 963: R. Nottrot, Optimal Processes on Manifolds. VI, 124 pages. 1982.

Vol. 964: Ordinary and Partial Differential Equations. Proceedings, 1982. Edited by W.N. Everitt and B.D. Sleeman. XVIII, 726 pages. 1982.

Vol. 968: Numerical Integration of Differential Equations and Large Linear Systems. Proceedings, 1980. Edited by J. Hinze. VI, 412 pages. 1982.

Vol. 970: Twistor Geometry and Non-Linear Systems. Proceedings, 1980. Edited by H.-D. Doebner and T.D. Palev. V, 216 pages. 1982.

Vol. 972: Nonlinear Filtering and Stochastic Control. Proceedings, 1981. Edited by S.K. Mitter and A. Moro. VIII, 297 pages. 1983.

Vol. 978: J. Ławrynowicz, J. Krzyż, Quasiconformal Mappings in the Plane. VI, 177 pages. 1983.

Vol. 979: Mathematical Theories of Optimization. Proceedings, 1981. Edited by J.P. Cecconi and T. Zolezzi. V, 268 pages. 1983.

Vol. 982: Stability Problems for Stochastic Models. Proceedings, 1982. Edited by V. V. Kalashnikov and V. M. Zolotarev. XVII, 295 pages. 1983.

Vol. 989: A.B. Mingarelli, Volterra-Stieltjes Integral Equations and Generalized Ordinary Differential Expressions. XIV, 318 pages. 1983.

Vol. 994: J.-L. Journé, Calderón-Zygmund Operators, Pseudo-Differential Operators and the Cauchy Integral of Calderón. VI, 129 pages. 1983.

Vol. 999: C. Preston, Iterates of Maps on an Interval. VII, 205 pages. 1983.

Vol. 1000: H. Hopf, Differential Geometry in the Large, VII, 184 pages. 1983.

Vol. 1003: J. Schmets, Spaces of Vector-Valued Continuous Functions. VI, 117 pages. 1983.

Vol. 1005: Numerical Methods. Proceedings, 1982. Edited by V. Pereyra and A. Reinoza. V, 296 pages. 1983.

Vol. 1007: Geometric Dynamics. Proceedings, 1981. Edited by J. Palis Jr. IX, 827 pages. 1983.

Vol. 1015: Equations différentielles et systèmes de Pfaff dans le champ complexe – II. Seminar. Edited by R. Gérard et J.P. Ramis. V, 411 pages. 1983.

Vol. 1021: Probability Theory and Mathematical Statistics. Proceedings, 1982. Edited by K. Itô and J.V. Prokhorov. VIII, 747 pages. 1983.

Vol. 1031: Dynamics and Processes. Proceedings, 1981. Edited by Ph. Blanchard and L. Streit. IX, 213 pages. 1983.

Vol. 1032: Ordinary Differential Equations and Operators. Proceedings, 1982. Edited by W. N. Everitt and R. T. Lewis. XV, 521 pages. 1983.

Vol. 1035: The Mathematics and Physics of Disordered Media. Proceedings, 1983. Edited by B.D. Hughes and B.W. Ninham. VII, 432 pages. 1983.

Vol. 1037: Non-linear Partial Differential Operators and Quantization Procedures. Proceedings, 1981. Edited by S.I. Andersson and H.-D. Doebner. VII, 334 pages. 1983.

Vol. 1041: Lie Group Representations II. Proceedings 1982–1983. Edited by R. Herb, S. Kudla, R. Lipsman and J. Rosenberg. IX, 340 pages. 1984.

Vol. 1045: Differential Geometry. Proceedings, 1982. Edited by A.M. Naveira. VIII, 194 pages. 1984.

Vol. 1047: Fluid Dynamics. Seminar, 1982. Edited by H. Beirão da Veiga. VII, 193 pages. 1984.

Vol. 1048: Kinetic Theories and the Boltzmann Equation. Seminar, 1981. Edited by C. Cercignani. VII, 248 pages. 1984.

Vol. 1049: B. Iochum, Cônes autopolaires et algèbres de Jordan. VI, 247 pages. 1984.

Vol. 1054: V. Thomée, Galerkin Finite Element Methods for Parabolic Problems. VII, 237 pages. 1984.

Vol. 1055: Quantum Probability and Applications to the Quantum Theory of Irreversible Processes. Proceedings, 1982. Edited by L. Accardi, A. Frigerio and V. Gorini. VI, 411 pages. 1984.